Springer Series in
Electronics and Photonics 24

Edited by D. H. Auston

Springer Series in Electronics and Photonics

Editors: D.H. Auston W. Engl T. Sugano

Volume 21 **Picosecond Electronics and Optoelectronics**
Editors: G.A. Mourou, D.M. Bloom, C.-H. Lee

Volume 22 **High-Speed Electronics**
Basic Physical Phenomena and Device Principles
Editors: B. Källbäck and H. Beneking

Volume 23 **Gyided-Wave Acoustooptic**
Interactions, Divices, and Applications
Editor: C.S. Tsai

Volume 24 **Picosecond Electronics and Optoelectronics II**
Editors: F.J. Leonberger, C.H. Lee, F. Capasso, and H. Morkoc

This series was originally published under the title *Springer Series in Electrophysics* and has been renamed starting with Volume 22.

Volumes 1–20 are listed on the back inside cover

Picosecond Electronics and Optoelectronics II

Proceedings of the Second OSA-IEEE (LEOS)
Incline Village, Nevada, January 14–16, 1987

Editors: F. J. Leonberger, C. H. Lee,
F. Capasso, H. Morkoc

With 235 Figures

Springer-Verlag Berlin Heidelberg New York
London Paris Tokyo

Dr. Frederick John Leonberger

United Technologies Research Center, East Hartford, CT 06108, USA

Chi H. Lee

University of Maryland, Dept. of Electrical Engineering,
College Park, MD 20742, USA

Federico Capasso

AT&T Bell Laboratories, Room 7A-209, 600 Mountain Avenue
Murray Hill, NJ 07974, USA

Hadis Morkoc

University of Illinois, Dept. of Electrical Engineering,
Champaign Urbana, IL 61801, USA

Series Editors:

Dr. David H. Auston

Columbia University, Dept. of Electrical Engineering, New York, NY 10027, USA

Professor Dr. Walter Engl

Institut für Theoretische Elektrotechnik, Rhein.-Westf. Technische Hochschule,
Templergraben 55, D-5100 Aachen, Fed. Rep. of Germany

Professor Takuo Sugano

Department of Electronic Engineering, The Faculty of Engineering,
The University of Tokyo, 7-3-1, Hongo, Bunkyo-ku, Tokyo, 113, Japan

ISBN 3-540-18329-9 Springer-Verlag Berlin Heidelberg New York
ISBN 0-387-18329-9 Springer-Verlag New York Berlin Heidelberg

Printing: Druckhaus Beltz, 6944 Hemsbach/Bergstr.
Binding: J. Schäffer GmbH & Co. KG., 6718 Grünstadt
2153/3150-543210

Preface

Over the past five years, there has been an enormous increase in the interest in and understanding of electronic and optoelectronic devices operating in the picosecond (multigigahertz) range. This has been fueled in a significant way by the spectacular advances in picosecond laser technology, electrooptic sampling, III–V devices, and wideband fiber optic systems. Partly to address these advances, a new conference jointly sponsored by the IEEE Lasers and Electrooptics Society (IEEE (LEOS)) and the Optical Society of America (OSA) was founded and its first meeting held in March 1985. The purpose of this meeting was to bring together workers in the areas of electronics and optoelectronics who share a common interest in the physics and technology of picosecond solid-state electronic and optoelectronic devices, their multigigahertz applications, and ultrafast measurement techniques. Emphasis was placed on the interdisciplinary aspects of these areas, since each area is covered by its own topical meeting. This meeting was quite successful and led to a second meeting, of which this volume forms the proceedings.

The Second Picosecond Electronics and Optoelectronics Meeting was held at Incline Village, Nevada, January 14–16, 1987. Subjects covered at the meeting included picosecond and femtosecond optical probing of electronic devices and integrated circuits, multigigahertz digital and microwave electronic devices and circuits, cryogenic operation of electronic circuits and striplines, high-speed lasers, detectors and integrated optic modulators, picosecond carrier transport, resonant tunneling and quantum-well devices, photoconductive switches, and optical microwave devices and applications. The meeting was attended by over 215 scientists and 65 papers were presented. The conference was co-sponsored by the IEEE Lasers and Electrooptics Society and the Optical Society of America in cooperation with the IEEE Electron Device and Microwave Theory and Techniques Societies.

We gratefully acknowledge the generous grants from the Air Force Office of Scientific Research (AFOSR), the National Science Foundation (NSF) and the Office of Naval Research (ONR) which provided funding for the meeting. Finally, we would specially like to thank Joan Carlisle of the Optical Society of America for her personal and thoroughly professional management of this meeting.

East Hartford, June 1987 *The Editors*

Contents

Part VI **Fast Lasers and Detectors**

Part VII Photoconductive Switching Devices

Part VIII Optical Microwave Techniques

Introduction

F.J. Leonberger, C.H. Lee, F. Capasso, and H. Morkoc

This book is composed of papers that were presented at the 1987 Picosecond Electronics and Optoelectronics Meeting. This introduction is intended to serve as a brief summary of the meeting and as a guide to this volume.

Part I of the volume addresses high-speed probing techniques. Here picosecond and subpicosecond laser pulse trains are used to probe or sample both digital and microwave circuits and to study basic nonstationary transport phenomena. In most cases propagating electric signal fields on metal lines are used to electrooptically alter the polarization of the probe signal with a spatial resolution of $<10\,\mu$m. Results for both GaAs and InP-based circuits are presented and a contactless electrooptic microprobe is described. In addition, a photoconductive switch is used to generate short electrical pulses for characterizing Si circuits. In a related area, optical probing techniques are reviewed for characterizing GaAs etalon arrays for envisioned high-speed computation applications.

Part II of the book addresses transient transport in III–V semiconductors. Topics covered include the use of subpicosecond pulses to probe the carrier mobility of hot electrons, simulations of transient transport in MODFET devices, and femtosecond heat transport in thin gold films.

The third part is concerned with ballistic transport and resonant tunneling. Here emphasis is on hot electron current-gain devices, and resonant tunneling structures, both as devices, such as oscillators and mixers, and as a means of performing hot electron spectroscopy.

In Part IV, quantum-well (QW) physics and devices are described. Here ultrashort pulses are used to study electronic transient effects in GaAs quantum wells. Processes associated with optical high-speed modulation due to the Stark effect in InGaAs/InP QWs and the dynamics of below-gap optical absorption in GaAs QW structures are covered as well.

Part V contains reports of a variety of high-speed electronic devices. Reviews are given of MBE technology for high-speed devices, millimeter-wave integrated circuits and high performance MODFETS (including a device with an $In_{0.15}Ga_{0.85}As$ channel layer having a noise figure of 2.3 dB at 60 GHz). Progress in cryogenic HEMT LSI technology is described including development of a 16K static RAM and a 1.5K gate array. Other papers consider broadband distributed amplifiers, multigigahertz InP logic gates and 18-GHz GaAs static frequency dividers. Also described are the millimeter-wave characteristics of microstrip lines on GaAs/Si structures.

High-speed diode lasers and detectors are discussed in Part VI. Characteristics of a variety of laser and detector structures are reviewed, including the achievement of 30-GHz bandwidth for cooled lasers and 100-GHz bandwidth for GaAs Schottky detector diodes. The reduction of frequency chirping in quantum-well lasers under high-speed modulation and a front-side-illuminated InGaAs PIN detctor with a response time below 18 ps are reported. An interdigitated metal-semiconductor-metal detector which is compatible with GaAs integrated circuit fabrication is described. A new method for measuring the picosecond response time of photodetectors using intrinsic nonlinearities in the response is presented. An AlGaAs/GaAs quantum well photodetector based on intersubband absorption and operating at 10 μm with a projected response time of 45 ps is also reported.

Photoconductive switching devices are described in Part VII. Detailed studies are presented on both optimizing the performance of silicon-on-sapphire (SOS) switches and on bias effects on the speed of InGaAs and GaAs/Si switches. The SOS switches have shown response times as short as 0.6 ps. In addition, several papers describe fast high-voltage switches. A switch capable of generating 10-kV pulses with no long-term degradation is described, as is a 3-kV pulse sharpening diode with a 60 ps rise time.

Finally, in Part VIII, optical microwave techniques are described. The use of optoelectronic components and optical fibers for signal processing applications is reviewed. Integrated optic modulators are discussed with emphasis on their >10-GHz bandwidth applications. In a related talk, the achievement of >40-GHz bandwidth with a $LiNbO_3$ traveling wave modulator is described. An optically controlled millimeter wave phase shifter is proposed and its associated pulse dispersion is calculated. In addition, a cw microwave source time synchronized by a pulsed laser source is demonstrated with 3-ps phase noise, and a novel optoelectronic millimeter wave transceiver and opto-isolator are reported.

Part I

High-Speed Probing Techniques

A Non-contact Electro-optic Prober for High Speed Integrated Circuits

J.A. Valdmanis and S.S. Pei

AT&T Bell Laboratories, 600 Mountain Avenue,
Murray Hill, NJ 07974, USA

1. Introduction

Electro-optic sampling has been used extensively for the characterization of picosecond electronic devices such as photodetectors, transistors, and diodes [1,2]. For the large part, these studies have been with discrete devices coupled to specialized electro-optic modulators in a hybrid electrode geometry. However, there is great interest in being able to probe internal points of integrated circuits in order to characterize device and circuit operation in situ. Recently, a specialized embodiment of electro-optic sampling [3] was developed to perform sampling directly in the substrate of, specifically, GaAs intégrated circuits. With this technique, almost any internal point of the circuit could be accessed. However, GaAs is the only commonly used semiconducting material that is also electro-optic. The substrate probing technique, or *internal* electro-optic sampling, also requires the surfaces of the integrated circuit to be of optical quality and the sampling laser beam to have a photon energy below the band gap energy of the substrate material.

Ideally one would like a non-perturbative means of probing integrated circuits with high temporal and spatial resolution that is *generally* applicable to circuits fabricated on any type of substrate. We have developed and demonstrated a new, non-contact, picosecond, electro-optic technique for probing internal nodes of high speed integrated circuits fabricated on any substrate material [4]. This technique, referred to as *external* electro-optic sampling, has the potential for single picosecond temporal resolution and micron spatial resolution, and is designed to operate at the wafer level with conventional wafer-probing equipment without any special circuit preparation. Optical pulses from any picosecond, high repetition rate laser throughout the visible and near infrared can be used. A similar system, employing a GaAs injection laser, has also been demonstrated and employed to measure electric fields around integrated circuit packaging connections [5].

2. Principles of External Electro-Optic Probing

External electro-optic probing is based on the use of an extremely small electro-optic crystal as a proximity electric field sensor near the surface of an integrated circuit (see Fig. 1). This technique exploits the open electrode structure of two-dimensional circuits where there exists a fringing field above the surface of the circuit between metalization lines at different potentials. "Dipping" an electro-optic tip into a region of fringing field induces a birefringence change in the tip that can be measured from above by an optical beam directed through the tip. In

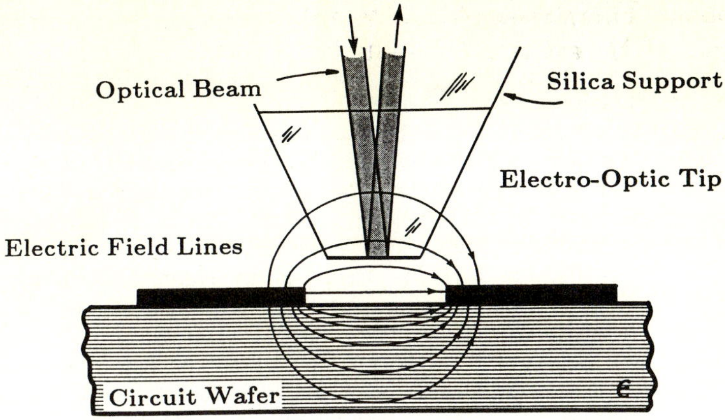

Fig. 1. External electro-optic probing

this way, the electro-optic tip is employed as the modulator in a conventional electro-optic sampling system [1,2]. It is because the electro-optic medium is separated from the integrated circuit that we call this technique *external* electro-optic sampling. By using an external electro-optic medium, the sampling system does not rely on any optical properties of the integrated circuit itself, hence making it generally applicable to a wide variety of circuit embodiments. This technique also allows for the use of electro-optic crystals that are transparent in the visible portion of the spectrum, thus enabling visible laser sources to be used.

The spatial resolution of the system is fundamentally limited by the focussed spot size of the sampling beam on the probe tip face. With the current 10X objective, this results in a spatial resolution of $\sim 5\ \mu m$, however one must also consider the effects of field distortion due to the presence of the tip which could alter the realized spatial resolution. This question will be addressed in a future publication. Temporal resolution is fundamentally dictated by the duration of the optical pulses used for sampling the induced birefringence. For the current measurements we use 100 fs long pulses at 630 nm which exceeds the response of the lithium tantalate crystal sensor. The crystal limit is set by the onset of the first lattice resonance which limits risetimes to $\sim$300 fs. One must also consider the effects of dielectric loading of the circuit due to the presence of the probing tip. Loading can vary greatly with the size of the tip and the sampling location, but preliminary experiments indicate the effects could be on the order of a few picoseconds. A more thorough study of loading effects will also be included in a future publication. Sensitivity is a very strong function of probe tip height above the circuit, h, and also circuit line spacing, since both parameters control the extent to which field penetrates into the electro-optic tip. Initial measurements indicate that at reasonable working distances ($h \simeq 2\ \mu m$) our system has a sensitivity of $\sim 8\ mV/\sqrt{Hz}$ which, for $10\mu m$ line spacings and a 1 Hz measurement bandwidth, implies we can measure a 1 volt potential difference with a signal to noise ratio of $\sim$125.

3. Sampling System Configuration

Figure 2 shows the arrangement of the optical system. Sampling pulses ~100 fs long are generated by a balanced colliding pulse mode-locked dye laser [6] and are directed into the sampling system via a polarization preserving single mode optical fiber and 20X microscope objective. The probe tip is mounted optically between crossed polarizers and below a dichroic beamsplitter in such a way as to facilitate viewing of tip face from above via a microscope system. Light of a wavelength different from that of the laser is injected through the beamsplitter to illuminate the circuit below the tip. In this way, both the circuit and the sampling beam spot can be seen together. Quartz compensating plates are included between the polarizers to operate the modulator at a "zero-order" quarter wave point. A wollaston analyzer is used to separate orthogonal polarizations and direct the output to dual differential detectors. The detector output is then fed to a lock-in amplifier and signal averager. The entire sampling system (minus laser) is mounted vertically on a conventional probing station above the wafer stage. The probe tip can then be brought down to any suitable point on the wafer for sampling.

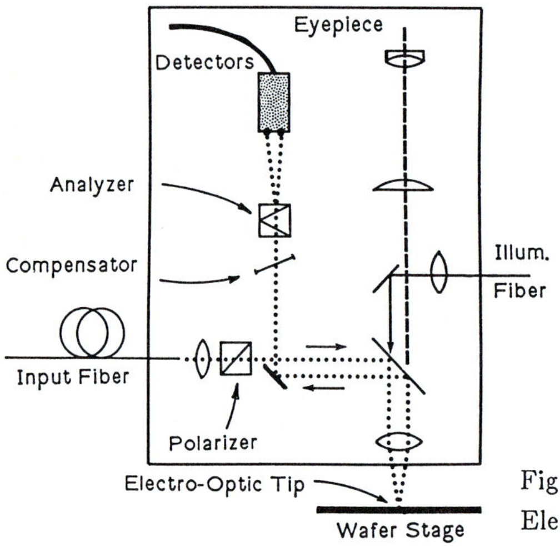

Fig. 2
Electro-optic prober optical system

As depicted in Fig. 1 the probe tip consists of a 100 μm thick piece of lithium tantalate bonded to a short fused silica rod for support. The tip is polished as a four-sided pyramid with a half angle of 30 degrees to a tip size of less than 40 μm. The z-axis of the lithium tantalate is in the plane of the tip face. A high reflectance dichroic coating is evaporated onto the tip face so the sampling beam can be reflected back to the optical system. A 10X microscope objective is used to both focus the sampling beam onto the tip face and recollimate the reflected beam.

The circuit wafer, or chip, is mounted below the probe tip on the standard vacuum chuck of the probe station. The circuit is driven via either a waveform synthesizer synchronized with the laser, or an electrical signal from a photodetector illuminated by the laser pulses. Electrical connections to the circuit are made by conventional probes, a probe card, or wire bonding. Relative delay between trigger and sampling times is introduced by a motor-driven optical delay line, although purely electronic means, in some cases, can also be implemented [3].

4. Application to Digital Circuits

a. Circuit Description

Initial measurements investigated the operation of a GaAs selectively doped heterostructure transistor (SDHT) prescaler. Measurements were made on the input buffer of the circuit which consists of two banks of three inverter stages each, which generate clock and complementary clock output signals. A scale drawing of the inverter stages and output lines is shown in Fig. 3. Linewidths are 3.5 μm with a closest spacing of 3.5 μm. A ground bus separates the two banks of inverters with the upper bank generating the clock signal and the lower bank the comp. clock signal. Each inverter consists of two SDHT's, one operating as the switching stage, and the other as an active load. The reference input enables a limited adjustment of the relative phase between the clock outputs. Vdd is the supply bus at 2 volts, and the switching threshold of each inverter is 0.4 - 0.5 volts.

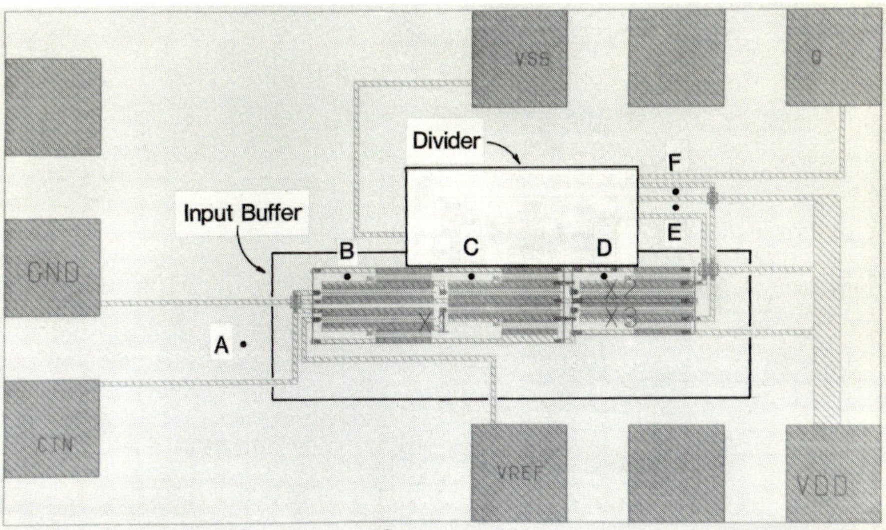

Fig. 3 GaAs SDHT pre-scaler input buffer with sampling points A - F (see text)

b. Waveform Measurements

We used the external prober to monitor the propagation and shaping of an input pulse as it progressed through the input buffer inverter stages. The input signal was a single pulse generated by a fast photodiode brought to the circuit by miniature coaxial cable and terminated at the chip. The input pulse had a risetime of ~1 ns and a peak amplitude of ~0.8 V. Waveforms were measured at points A-F in Fig. 3 by bringing the probe tip to within a few microns of the surface of the circuit at each of these points. In order to make a measurement, the prober tip was positioned between the line that carries the signal of interest and a line that behaves as a DC reference, ie. typically either Vdd or ground. Point A measures the field between the input signal line and the ground bus. Points B, C, and D measure the output signal of inverter stages 1, 2, and 3 respectively, by monitoring the field between the source metalization of the switching transistor and Vdd. The clock and comp. clock outputs were measured with respect to Vdd at points E and F.

Figure 4 shows the waveforms obtained at points A-D labeled in the figure as Input, Inv. 1, Inv. 2, and Inv. 3, respectively. The input waveform was measured with the circuit powered up and exhibits extra structure near the first stage switching points. This structure is believed to arise from the switching transistor's input capacitance (dis)charging at the threshold levels. The next three traces show that for each successive stage, the switching transitions become more clearly defined, the pulse polarity alternates (as it should), and a clear clock pulse evolves with a risetime of 56 ps, a fall time of 66 ps, and a duration of 0.93 ns.

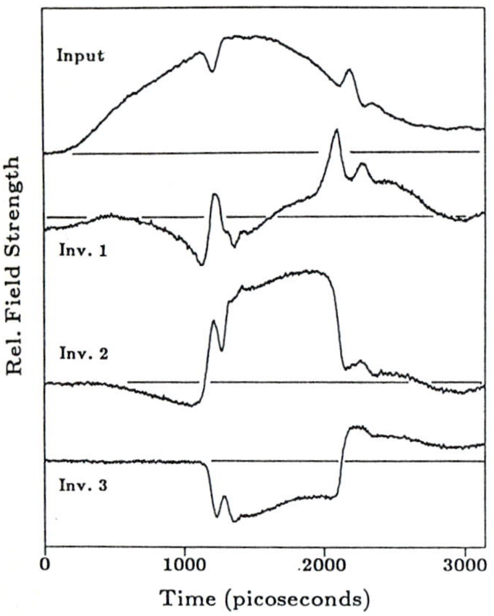

Fig. 4 Input buffer waveforms at points A - D (see text)

Two other effects are observed in Fig. 4. The first effect is an apparent skew of the traces from low to high, especially for stages 1 and 3. We believe this is due in part to loading of the Vdd line (which has a non-negligible resistance) when the inverter stages are drawing current. This is true in particular for Inv. 1 and 3, because they draw current simultaneously. The second effect is the propagation of the structure on the leading edge through the entire input buffer, which does not occur for the trailing edge. This is attributed to the switching characteristics of the inverter chains and is being studied by circuit simulation. The results of this study will be reported in a future publication.

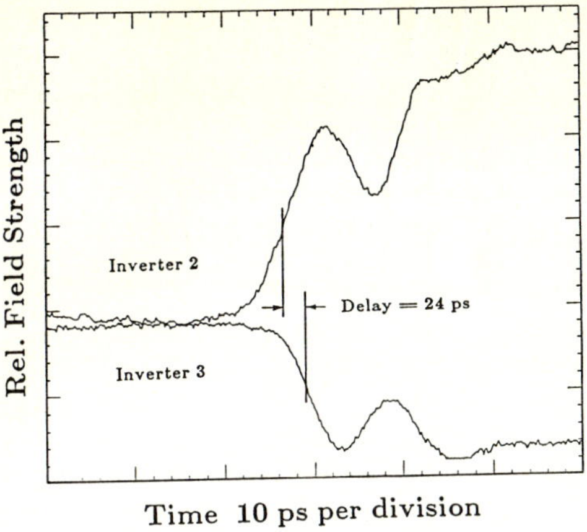

Fig. 5 Propagation delay between inverter stages 2 and 3 (pts. C and D)

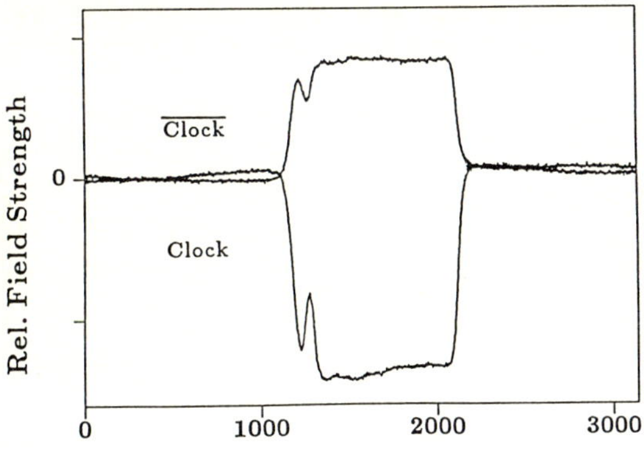

Fig. 6 Input buffer output waveforms at points E and F

Figure 5 shows the propagation delay between stages 2 and 3 to be 24 ps, which agrees well with ring oscillator measurements on the same type of device. Figure 6 shows the clock and comp. clock waveforms measured at points E and F. Here, we see almost no skew in the traces and very good symmetry. The clock line signal is larger because its spacing from the Vdd line is larger thus allowing more field to enter the probing crystal, even though the field due to the increased spacing should be less.

5. Conclusion

We have demonstrated that external electro-optic probing is a viable solution to the problem of non-invasive wafer level testing of integrated circuits. External electro-optic techniques enable internal node testing of integrated circuits fabricated on any type of substrate, in conventional testing configurations, without special circuit preparation.

References

[1] J.A. Valdmanis, G. Mourou and C.W. Gabel, "Picosecond electro-optic sampling system," *Appl. Phys. Lett.*, vol. 41-3, pp. 211-212, Aug. 1982.

[2] J.A. Valdmanis, G.A. Mourou and C.W. Gabel, "Subpicosecond electrical sampling," *IEEE J. Quan. Elec.*, vol. QE-19-4, pp. 664-667, Apr. 1983.

[3] K.J. Weingarten, M.J.W. Rodwell, H.K. Heinrich, B.H. Kolner and D.M. Bloom, "Direct electro-optic sampling of GaAs integrated circuits," *Elect. Lett.*, vol. 21-17, pp. 765-766, Aug. 1985.

[4] J.A. Valdmanis, "Progress in picosecond electro-optic measurement techniques," Conf. on Lasers and Electro-Optics, San Francisco, CA, June 1986.

[5] J. Nees and G. Mourou, "Noncontact electro-optic sampling with a GaAs injection laser," *Elect. Lett.*, vol. 22-17, pp. 918-919, Aug. 1986.

[6] J.A. Valdmanis, R.L. Fork and J.P. Gordon, "Design considerations for a femtosecond pulse laser balancing self-phase modulation, group velocity dispersion, saturable absorption, and saturable gain," *IEEE J. Quan. Elec.*, vol. QE-22-1, pp. 112-118, Jan. 1986.

Photoconductive Generation of Subpicosecond Electrical Pulses and Their Measurement Applications*

*D. Grischkowsky, C.-C. Chi, I.N. Duling III, W.J. Gallagher, N.H. Halas**, J.-M. Halbout, and M.B. Ketchen*

IBM Watson Research Center, Yorktown Heights, NY 10598, USA

The generation of short electrical pulses via optical methods has for some time been performed by driving Auston switches (photoconductive gaps) with short laser pulses.[1] The shape of the electrical pulse depends on the laser pulseshape, the material properties of the semiconductor, the nature of the charge source, and the characteristics of the associated electrical transmission line. The same techniques can also measure the generated electrical pulses by sampling methods. An alternate measurement approach has been to use the electro-optic effect in a crystal.[2] In this case, the field of the electrical pulse is sampled through the rotation of the polarization of the optical sampling pulse. Because it has demonstrated 460 fsec time resolution,[3] the electro-optic method is presently considered to be the fastest sampling technique. However, recent work utilizing photoconductive switches has generated and measured subpicosecond electrical pulses.[4] This large reduction in the generated pulsewidth demonstrates the ultrafast capability of the Auston switches and challenges the ultrafast time resolution of the electro-optic methods. We will now discuss subpicosecond pulse generation using photoconductive gaps,[4] together with the initial interpretation of these experimental results. In addition, we will present some recent measurements, which will allow us to conclude with a relatively complete description of the pulse generation process.

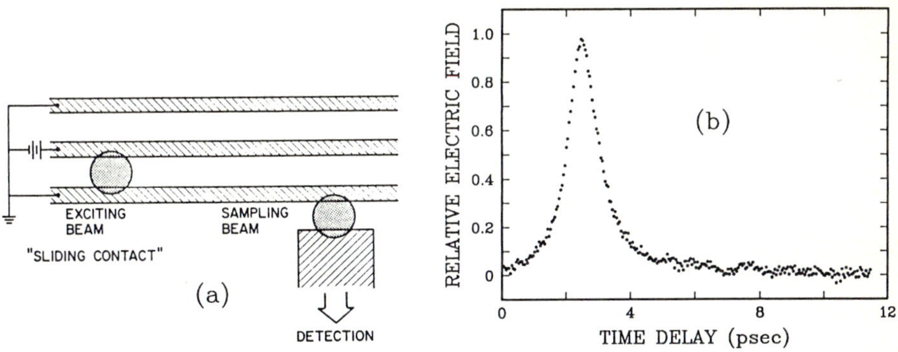

Figure 1. (a) Experimental geometry (b) Measured electrical pulse

* This research was partially supported by the U.S. Office of Naval Research.

**Present address: The Center for Atomic and Molecular Physics at Surfaces, Department of Physics and Astronomy, Vanderbilt University, Nashville, TN, 37235.

The geometry of the experimental arrangement is illustrated in Fig. 1a.[4] The 20-mm-long transmission line with a design impedance of 100 ohms consists of three parallel 5 micron-wide aluminum lines separated from each other by 10 microns. The dc resistance of a single 5 micron line is 200 ohms, and the laser spot diameters are 10 microns. The transmission line, composed of a 0.5 micron thick Al film, was fabricated on an undoped silicon on sapphire (SOS) wafer, which was then heavily implanted with O^+ ions to ensure the required short carrier lifetime.

The laser source is a compensated, colliding-pulse, passively mode-locked dye laser producing 80 fsec pulses at a 100 MHz repetition rate. The measurements were made with the standard excite and sample arrangement for the beams of optical pulses. The time delay between the exciting and sampling beams was mechanically scanned by moving an air-spaced retro-reflector with a computer controlled stepping motor synchronized with a multichannel analyzer. All the measured pulses shown are displayed with respect to the delay of the sampling beam. The actual electrical pulses are significantly narrower than the directly measured values due to the finite resolution of the optical sampling. The measured electrical pulse with an excellent signal-to-noise-ratio is shown in Fig. 1b. Here, the measured pulsewidth of 1.1 psec corresponds to a generated electrical pulse of approximately 0.6 psec. For this result the spatial separation between the exciting and sampling beams was approximately 50 microns.

We will now describe our initial theoretical procedure for determining that the actual pulsewidth was approximately 0.6 psec. We follow the theory presented by Auston and use his notation throughout.[1] An important experimental observation illustrated in Fig. 2 was that when, instead of using the sliding contact as in Fig. 2a, we generated the electrical pulse with a photoconductive gap of the same type as shown in Fig. 1, the pulsewidth broadened by 1.5 times to 1.7 psec as shown in Fig. 2b. This measurement (Fig. 2b.) had the characteristic asymmetric shape (Fig. 10, Ref.1) of a gap generated and detected pulse for which the carrier lifetimes and the laser pulses were short compared to the measured electrical pulse. In order to

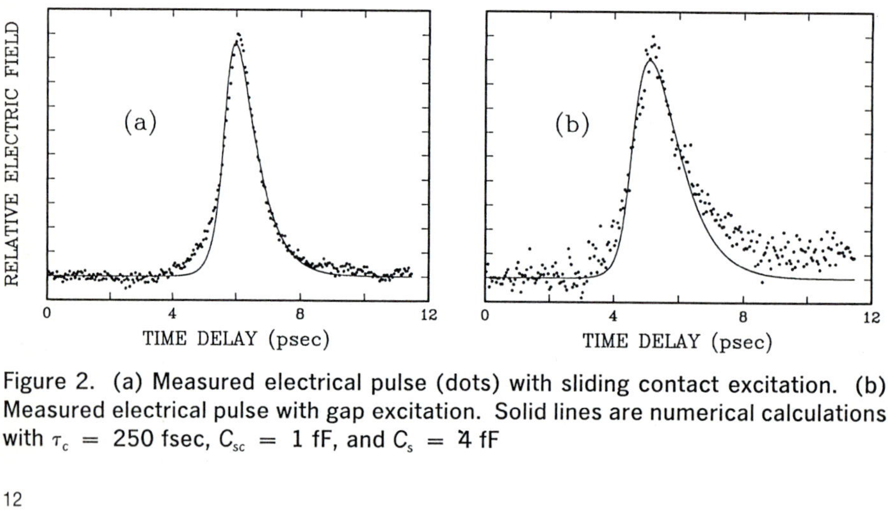

Figure 2. (a) Measured electrical pulse (dots) with sliding contact excitation. (b) Measured electrical pulse with gap excitation. Solid lines are numerical calculations with τ_c = 250 fsec, C_{sc} = 1 fF, and C_s = 4 fF

explain this result, the sampling gap capacitance C_s and the sliding contact capacitance C_{sc} were estimated in Ref. [4], using the short pulse limit of Ref. [1]. For this, the photoconductances g_1 and g_2 for the excitation and sampling gaps, respectively, were assumed to be delta functions, thereby implying infinitely short laser pulses and carrier lifetimes. With similar gaps we obtained $C_s = 4.6$ fF from the relationship $\Delta\tau = 3.67\ ZC_s$, where $\Delta\tau = 1.7$ psec was the measured pulsewidth, and $Z = 100$ ohms is the transmission line impedance. Now the fact that with the sliding contact excitation we measured the shorter pulsewidth $\Delta\tau = 1.1$ psec implies (from Eq. 32 of Ref [1]) that the effective capacitance C_{sc} of the sliding contact is less than $C_s/2.4$. It should be clear that these capacitances obtained in the delta function limit, are only upper limits for the more accurate values obtained from a numerical fit to the measured electrical pulses using realistic $g_1(t)$ and $g_2(t)$, i.e., the inclusion of other broadening mechanisms only broadens the observed pulsewidth and thereby less capacitance is required for a fit.

We then made the more reasonable assumption that $g_1(t)$ and $g_2(t)$ were given by the convolution of the laser pulse (allowing for its spatial extent) with an exponential response function describing the carrier lifetime τ_c. We numerically calculated (via Eq. (32) of Ref.[1]) the pulseshapes, shown as the solid lines in Fig. 2, using the parameters $\tau_c = 250$ fsec, $C_{sc} = 1$ fF, and $C_s = 4$ fF. These theoretical results agree reasonably well with the measurements and correspond to an electrical pulse duration of less than 0.6 psec. An additional point is that the value of C_{sc} compares well with that obtained from the calculated capacity per unit length of the transmission line (0.7 pF per cm) multiplied by the laser spot size.

However, the numerical results shown in Fig. 2 have a much sharper leading edge for the pulses than we observed, indicating that the carrier lifetime is probably longer than our somewhat arbitrary value of 250 fsec. Because of this situation, we set up an experiment to determine τ_c directly by time-resolved reflectivity measurements.[5] With the newly measured value of $\tau_c = 600$ fsec, we again analyzed the results of Fig. 2 and obtained the significantly better fits as illustrated in Fig. 3. The agreement between theory and experiment is particularly impressive for

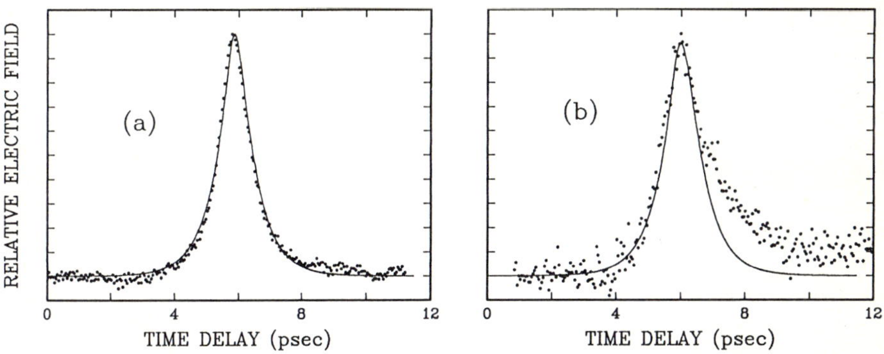

Figure 3. (a) Measured electrical pulse (dots) with sliding contact excitation. (b) Measured electrical pulse with gap excitation. Solid lines are numerical calculations with $\tau_c = 600$ fsec, $C_{sc} = 0$ fF, and $C_s = 1$ fF

Fig. 3a. The deviation from the trailing edge of the predicted curve of Fig. 3b is now considered to result from the multi-mode excitation of the line caused by the gap excitation, rather than the gap capacitance. This point has important implications for the spectroscopic applications to be discussed later. For these results the gap capacitance was lowered significantly to $C_s = 1$ fF, and the sliding contact capacitance was set equal to zero. Directly exciting a long ideal transmission line does not have any parasitic capacitance which must be charged to generate (or alternatively measure) the propagating electrical pulses. For this situation we calculated the actual generated electrical pulse shown in Fig. 4, which has a pulsewidth of only 0.52 psec. The sharp rising edge is due to the short laser pulse excitation, while the slower falling edge of the pulse is due to the 600 fsec carrier lifetime. This pulse is directly proportional to the time-dependent current flow $g_1(t)$.

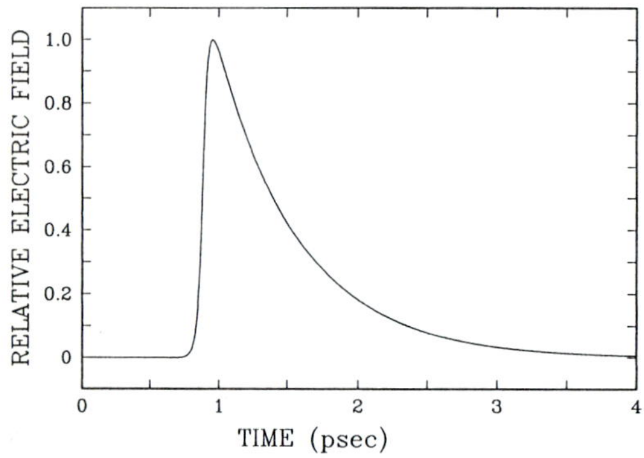

Figure 4. Calculated actual electrical pulse with $\tau_c = 600$ fsec and $C_{sc} = 0$ fF

A more symmetric experimental arrangement with no observable capacitance is illustrated in Fig. 5a, which is called the double sliding contact.[6] The electrical pulse is generated as before, but here the sampling measurement is also made by shorting the transmission line with the sampling laser pulse. For this case the two laser beams are mechanically chopped at incommensurate frequencies and the signal is measured as a modulation of the photocurrent on the transmission line at the sum of the two chopping frequencies. This arrangement appears to have no capacitance at both the generation and sampling sites. Consequently the measured electrical pulse has a pulsewidth of only 0.85 psec, compared to previous measurement of 1.1 psec (Fig.1b) using a sampling gap. The double sliding contact measurement was made on a smaller geometry two-line co-planar line with line dimensions of 1.2 microns separated by 2.4 microns, due to signal-to-noise problems on the larger line of Fig. 1a. The bias was 1.5 volts. A fit to this data, using no adjustable parameters, is shown as the solid line in Fig. 5b. The theoretical prediction is equivalent to the autocorrelation of the electrical pulse shown in Fig. 4. The agreement between theory and experiment showing the sharpness of the peak of the pulse is quite good. This extremely sharp peak is due to the sharp leading edge of the generated pulse shown in Fig. 4. The disagreement on the wings is not presently understood.

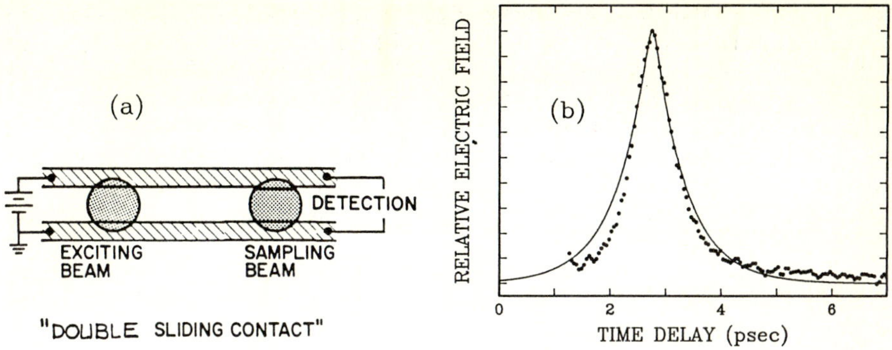

Figure 5. (a) Double sliding contact experimental geometry. (b) Measured electrical pulse. Solid line is theory with $\tau_c = 600$ fsec, and $C_{sc} = 0$ fF

The fact that we can now generate and propagate subpsec electrical pulses with frequency bandwidths extending to beyond 1 THz allows us to consider spectroscopic applications in this frequency range for which there are no convenient sources. Another advantage is that, because the electrical field is measured, it is possible to obtain both the frequency-dependent absorption and dispersion of the transmission line. Weak absorption features can be observed by simply propagating a longer distance. However, in order to make these measurements, the pulse reshaping should be determined by the dielectric properties of the transmission line materials, and not by spatial dispersion or multi-mode effects. These complications can be eliminated in the following ways. Firstly, by reducing the dimensions of the co-planar transmission line to 5 micron lines separated by 10 microns, the spatial dispersion effects were eliminated. This occurred because now the maximum transverse dimension (20 microns) is small compared to the spatial length (approximately 100 microns) of the propagating electrical pulse. Secondly, by using the sliding contact method of pulse generation only the differential mode was excited. Ordinary side gap generation equally excites both the differential mode and the common mode.

Consequently, our initial propagation studies [4] with the Al coplanar transmission line showed less dispersion than had been previously measured for any transmission line.[7,8] As shown in Fig. 6a after 8 mm of propagation on this line the pulse broadened to only 2.6 psec. The broadening is mainly due to the frequency dependent absorption of the Al film which more strongly attenuates the higher frequency components. The 65.4 psec delay indicates a group velocity of c/2.45, faster than that for sapphire (c/3.15). However, for our case the field is partially in air so that the pulse should travel faster. The energy of this pulse is half that of the initial pulse (Fig. 1b), indicating an overall absorption length of 12 mm for our line. Via Fourier analysis of the input (Fig. 1b) and propagated (Fig. 6a) pulses, we were able to characterize the dielectric response of the transmission line to beyond 1 THz,[9] as shown in Figure 7. The general trend of absorption increasing with frequency is correct, although the irregular structure on the absorption curve is due to noise in the experimental data.

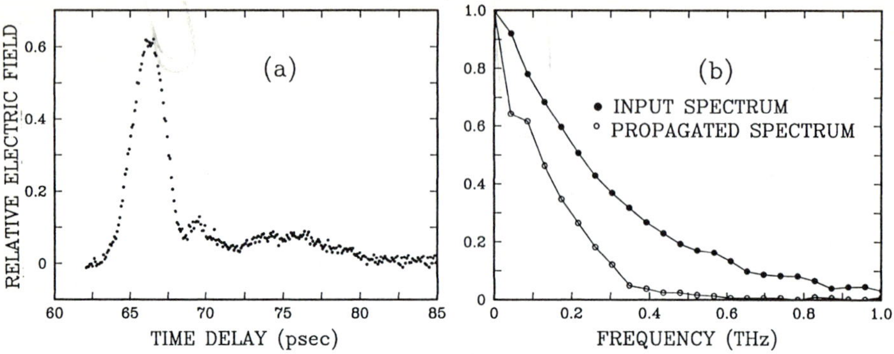

Figure 6. (a) Measured electrical pulse after 8 mm propagation on the transmission line shown in Fig.1a. (b) Amplitude spectrum of the initial pulse (Fig.1b) and the propagated pulse (Fig.6a)

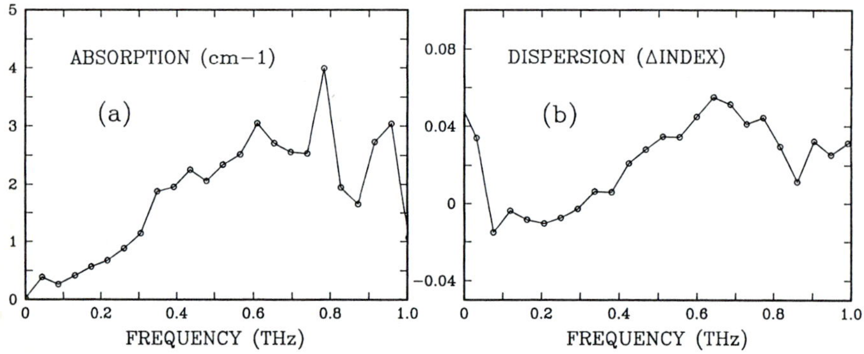

Figure 7. (a) Measured absorption vs frequency for the transmission line of Fig.1a. (b) Dispersion vs frequency for the same line

Our next spectroscopic application was to study superconducting niobium lines.[10] Following the same methods, we were able to demonstrate that reshaping of the propagating pulse was determined by the superconductor. Fourier analysis of the propagated pulses allowed us to obtain the line absorption as a function of frequency,which showed an abrupt absorption increase at a frequency corresponding to that of the superconducting gap. Furthermore, we were able to measure the temperature dependence of the gap frequency.

1. D.H. Auston, in 'Picosecond Optoelectronic Devices, edited by C.H. Lee (Academic, London, 1984), pp. 73-116.

2. J.A. Valdmanis, G.A. Mourou, and C.W. Gabel, IEEE J. Quantum Electron. QE-19, 664 (1983).

3. G.A. Mourou and R.E. Meyer, Appl. Phys. Lett. 45, 492 (1984).

4. M.B. Ketchen, D. Grischkowsky, T.C. Chen, C-C. Chi, I.N. Duling III, N.J. Halas, J-M Halbout, J.A. Kash, and G.P. Li, Appl. Phys. Lett. 48, 751 (1986).

5. F.E. Doany, D. Grischkowsky, and C-C. Chi, 'Proceedings of the 2nd Topical Meeting on Picosecond Electronics and Optoelectronics' Incline Village, Nevada, January 14-16, 1987; also to be published in Appl. Phys. Lett.

6. P.G. May, G.P. Li, J-M. Halbout, M.B. Ketchen, C-C. Chi, M. Scheuermann, I.N. Duling III, D. Grischkowsky and M. Smyth, 'Proceedings of the 5th International Conference on Ultrafast Phenomena', Snowmass, Colorado, June 16-19, 1986.

7. D.E. Cooper, Appl. Phys. Lett. 47, 33 (1985).

8. C.J. Kryzak, K.E. Meyer and G.A. Mourou, 'Picosecond Electronics and Optoelectronics', Proceedings of the Topical Meeting, Lake Tahoe, Nevada, March 13-15, 1985, Editors, G.A. Mourou, D.M. Bloom, and C.H. Lee, (Springer-Verlag, New York 1985).

9. N.J. Halas, I.N. Duling III, M.B. Ketchen, and D. Grischkowsky, Digest of Technical Papers, Conference on Lasers and Electro-Optics, San Francisco, June 9-13, 1986.

10. I.N. Duling III, C-C. Chi, W.J. Gallagher, D. Grischkowsky, N.J. Halas, M.B. Ketchen and A.W. Kleinsasser, 'Proceedings of the 5th International Conference on Ultrafast Phenomena' Snowmass, Colorado, June 16-19, 1986; also to be published in Appl. Phys. Lett.

Picosecond Sampling of GaAs Integrated Circuits

K.J. Weingarten, M.J.W. Rodwell, and D.M. Bloom

Eward L. Ginzton Laboratory, Stanford University,
Stanford, CA 94305, USA

1. Introduction

Gallium Arsenide (GaAs) microwave integrated circuits (IC's) are now being developed for operation in the millimeter-wave range, while GaAs digital IC's have demonstrated ring-oscillator propagation delays of 5-10 ps, with gate delays of 50-100 ps for larger scale circuits. Digital IC's are currently tested only by indirect techniques (multi-stage propagation delay or cycle times), while microwave circuits are tested by external scattering parameter measurement; if the circuit does not perform to expectations, the cause is not easily identified. Electrooptic sampling was initially developed to measure the response of photoconductors and photodetectors faster than the time resolution of sampling oscilloscopes and used an external electrooptic modulator connected to the device under test [1,2]. The system developed at Stanford uses the GaAs IC substrate as the electrooptic material, permitting detailed internal-node circuit evaluation with picosecond time resolution.

2. The Sampling System

GaAs is electrooptic; the electric fields associated with conductor voltages induce optical birefringence, causing a polarization change to a probe beam passing through these fields. Figure 1 shows the standard probing arrangements. For standard [100] cut GaAs substrates the probe beam is effected by the longitudinal field components. Passing the reflected light through a polarizer results in a change in intensity proportional to the voltage across the substrate at the probe point [3]. For frontside probing, the fields extend from the conductor roughly the distance of the substrate thickness and the probe spot size is set to approximately one-tenth of the substrate thickness to accurately measure the conductor potential. For backside probing, the characteristic extent of the fields into the substrate is the distance between the signal conductor and the ground; the substrate thickness must be much greater than the signal-ground spacing and the probe spot size should be less than or equal to the conductor width.

The sampling system is shown schematically in Fig. 2. A Nd:YAG laser produces 1.06 μm, 90 ps pulses at an 82 MHz rate. The laser has free-running pulse-to-pulse timing fluctuations of 4 ps rms, reduced to less than 0.5 ps rms by a feedback system [4]. A fiber-grating pulse compressor shortens the pulses to 1.5 ps FWHM. The beam passes through a polarizing beamsplitter and two waveplates, then is focused through the IC substrate with a microscope objective to a 3 μm spot on the probed conductor (backside probing) or a 10 μm spot on the ground plane adjacent to the probed conductor (frontside probing). The reflected light is analyzed by the polarizing beamsplitter; the change in intensity is proportional to the voltage across the GaAs substrate [2] and is detected by a photodiode connected to a vector receiver.

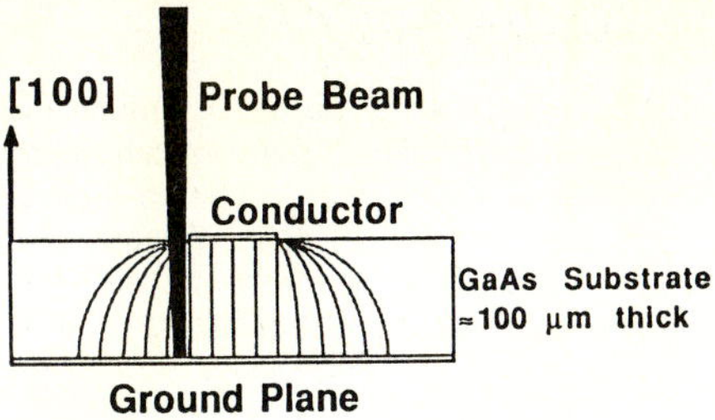

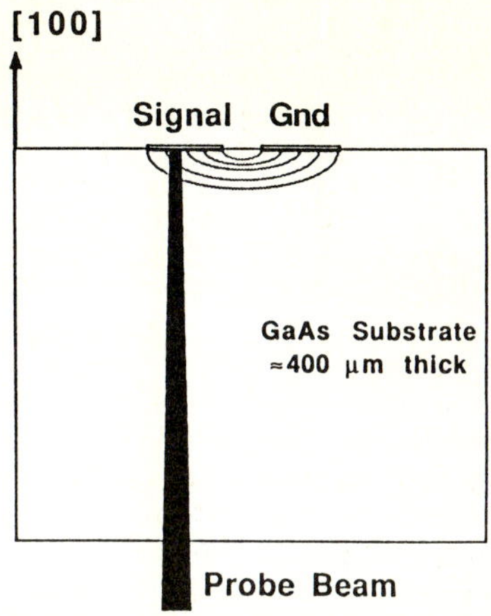

Fig. 1 Probing geometries for longitudinal electrooptic effect in [100] cut GaAs substrates. (a) Frontside probing for microstrip transmission lines (b) Backside probing for coplanar transmission lines or digital interconnects

To drive the IC a microwave synthesizer generates either sinusoidal excitation for microwave circuits or the clock or data signals for digital circuits. For wafer-level testing of IC's a microwave probe station (Cascade Microtech Model 42) was modified to allow for backside electrooptic probing. The transmission line probes used with this test station allow for repeatable, low-reflection connections in a 50 Ω environment to 26 GHz. The synthesizer is set to an exact multiple of the laser repetition rate (82 MHz), plus a small frequency offset Δf; the resulting intensity modulation at the photodiode receiver varies at this offset rate Δf. To enhance the system sensitivity, the synthesizer is also modulated at 10 MHz to allow synchronous detection in a spectral region where low-frequency laser amplitude noise is below the shot noise limit.

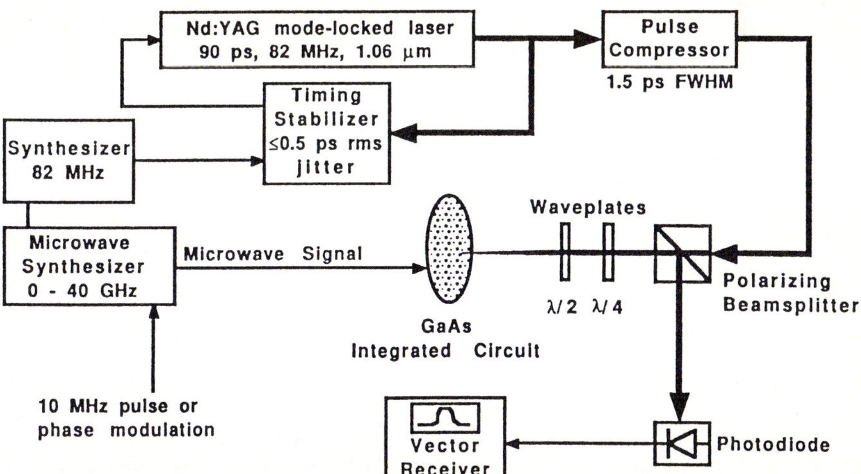

Fig. 2 Schematic of the electrooptic sampler

3. Bandwidth and Noise Considerations

The system's bandwidth is determined by the optical transit time of the pulse through the GaAs substrate where fields are present, the optical pulsewidth, and pulse-to-pulse timing jitter of the laser with respect to the microwave synthesizer driving the circuit. In general the optical transit time of the pulse in the GaAs substrate can be neglected. Because the optical and microwave dielectric constant in GaAs are nearly equal, microwave transmission lines have a cutoff frequency for higher-order modes roughly equal the inverse of the optical transit time. Well-designed microwave circuits operate at frequencies well below the multimode cutoff frequency. Only when measuring interconnects near or above the cutoff frequency (where dispersive characteristics are of interest) must the optical transit time be considered.

To accurately measure the shape of circuit waveforms, the impulse response of the system should be of short duration and free of "wings" (long-duration substructure). Using a 1 km fiber in the pulse compressor introduces significant group velocity dispersion, producing a more linear frequency chirp [5], reducing the wings on the compressed pulse. To accurately determine the extent of wings on the compressed pulse, the autocorrelation is numerically Fourier transformed to obtain the power spectral density of the pulse.

Timing jitter influences both bandwidth and sensitivity; the impulse response of the sampling system is the convolution of the optical pulse with the probability distribution of its arrival time (neglecting optical transit time), while those Fourier components of the jitter lying within the detection bandwidth of the receiver introduce noise proportional to the time derivative of the measured waveform as shown in Fig. 3. Stabilization of the laser timing is thus imperative for low-noise measurements of microwave or picosecond signals. To address this issue a timing stabilization feedback system was implemented to reduce the jitter to a level less than the optical pulsewidth. Figure 4 shows the reduction of the timing jitter of our laser, measured as phase noise with an RF spectrum analyzer. With the addition of a low-phase noise synthesizer (HP8662) and an improved modelocker the laser exhibits jitter of ≤300 fs rms.

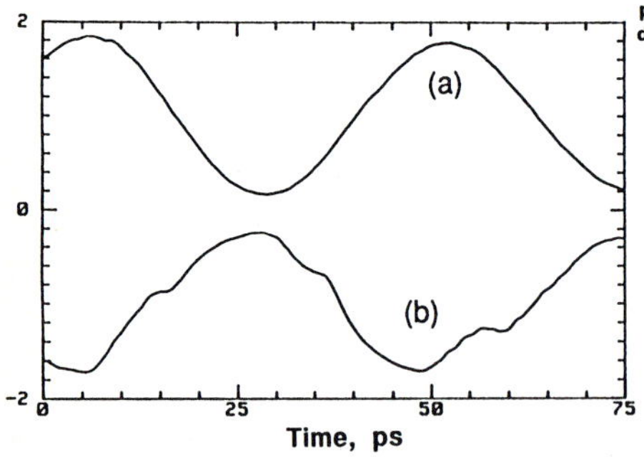

Fig. 3 A 20 GHz signal, measured with the electrooptic sampler. (a) Timing stabilizer on. (b) Timing stabilizer off. Excess timing jitter introduces excess amplitude noise, timing error, and reduces measurement bandwidth.

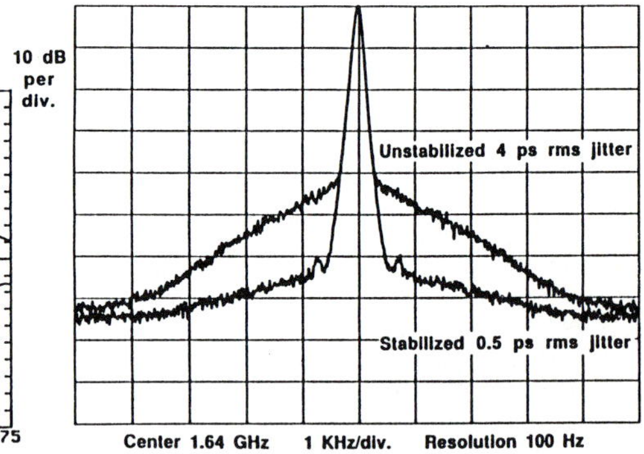

Fig. 4 Timing jitter measured as phase noise measuring the laser output with a photodiode connected to an RF spectrum analyzer. Integrating the phase noise over frequency gives timing jitter.

To achieve accuracy, repeatability, and fast acquisition of circuit measurements requires shot-noise-limited sensitivity in the detection system [2]. Low-frequency noise on the laser contributes excess amplitude noise above the shot noise limit to frequencies of several hundred kilohertz. To translate the signal detection to a frequency where this 1/f laser noise is below the shot-noise limit, the microwave signal to the IC is modulated at 10 MHz. With a narrowband 10 MHz vector receiver, the resulting shot-noise-limited sensitivity is sufficient to acquire measurements at scan rate of 10-100 Hz. For microwave circuits or simple digital circuits, the input signal is pulse modulated at 10 MHz. For sequential digital circuits which do not operate correctly with chopped excitation, a small-deviation 10 MHz phase modulation is used. In this case the received signal, proportional to the derivative of the sampled waveform, is integrated in software [6].

The pulse compressor is also a source of several types of excess amplitude noise. Stimulated Raman scattering (SRS) with gain proportional to the pump intensity and interaction length, sets an upper limit to the intensity in the fiber and thus an upper limit to its compression. However, as the pump intensity approaches the SRS threshold, we observe a substantial increase in amplitude noise above the shot noise limit on the compressor output (Fig. 5). The period of the repetitive noise spectrum in this figure corresponds to the free spectral range of the 1 km fiber; the high gain of the SRS and 4% Fresnel reflection at each end facet form a parasitic synchronously-pumped fiber-Raman laser [7]. Self-phase modulation (SPM) occurs over the enter length of the fiber while the interaction length for SRS is set by dispersion-induced walkoff between the input and Stokes-shifted wavelengths, approximately 60 meters. This dispersion also causes the 1.06 μm pump and the 1.12 μm Raman pulse to separate by 1.8 ns over the length of the 1 km fiber. The weakly reflected Raman pulse is further amplified if after its first round trip through the fiber it is synchronized to within 1.8 ns of a pump pulse. With this condition the Raman threshold with the 1 km fiber is 400 mW average power from the fiber output. Trimming the fiber length a few inches defeats this synchronism and increases the Raman threshold to 700 mW. We routinely obtain 50X compression ratios at power levels (350 mW average power from the fiber output) well below the Raman threshold.

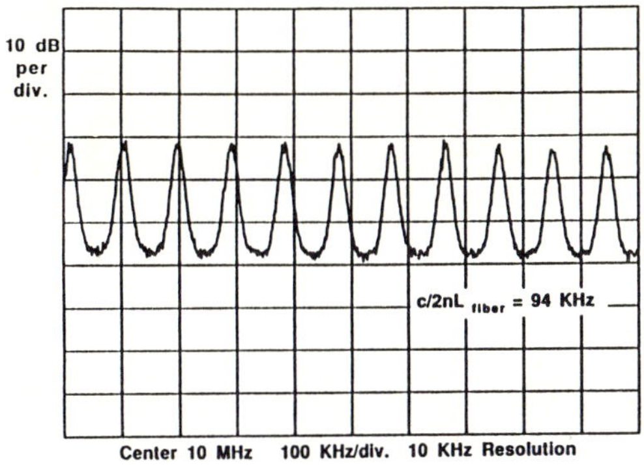

Fig. 5 Noise spectrum of pulse compressor output near the SRS threshold. The free spectral range of the 1 km fiber is 94 kHz, indicating the fiber forms a parasitic fiber-Raman laser.

In addition to SRS the fiber generates broadband polarization noise, possibly arising from guided acoustic wave Brillioun scattering [8]. The polarization noise is converted to amplitude noise after passing through the grating pair in the compressor. Adjusting the polarization from the fiber to maximize transmission through the grating path results in second-order intensity variations due to polarization fluctuations reducing this excess noise to a level below the shot noise limit. To suppress polarization drift, the non-polarization-preserving fiber is placed in a temperature-stabilized environment.

4. Circuit Measurements

Propagation delays of switching devices in simple test circuits, measured by either electrooptic sampling or by conventional methods, are used to project the maximum clock frequency of these devices used in digital systems; unless the test circuit provides representative switching voltages, interface impedances, and fan-in/fan-outs, the measured delays will not correlate well with the maximum clock frequency of circuits such as shift registers, binary multipliers, and memory. In particular, the risetime of a single device driven by a low-impedance photoconductor and loaded by a low-impedance transmission line will be much shorter than the propagation delay of a logic gate driven by the moderate output impedance of a driving gate and loaded by the input capacitance of a cascaded gate. The simplest representative test structures, ring oscillators and inverter strings, load the gates with unity fan-out and hence give optimistically small delay measurements. Ring oscillators often operate small-signal (without full logic-level swings), however, while inverters strings allow large-signal operation to be verified. Using such a string of 1 μm gate buffered-FET-logic (BFL) inverters, we have measured large-signal gate propagation delays of 75 ps [6].

Master-slave flip-flops, connected as binary frequency dividers, operate with realistic signal levels and loading conditions, and serve as better performance indicators. We have tested 23 GHz dynamic and 18 GHz static frequency dividers implemented in BFL and capacitively enhanced (CEL) MESFET logic from Hughes Research Laboratories [9] using the modified Cascade probe station for electrooptic sampling. These circuits used 0.2 μm e-beam written gates, molecular-beam epitaxy-grown channels, air-bridge interconnects, and optimized feedback to achieve such clock rates. Correct divide-by-two operation was verified by direct waveform measurement (Fig. 6), gate propagation delays of 30-40 ps were measured and correlated to maximum clock frequencies (Fig. 7), and the internal delays through the inverting and source-follower stages of individual BFL gates were identified.

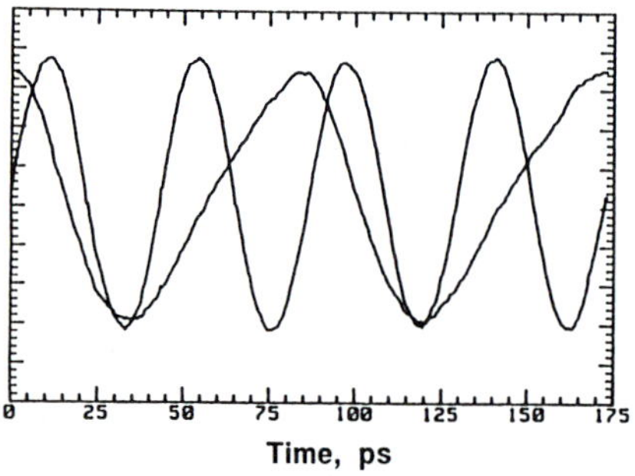

Fig. 6 BFL dynamic frequency divider, clock and output waveforms, clocked at 23 GHz.

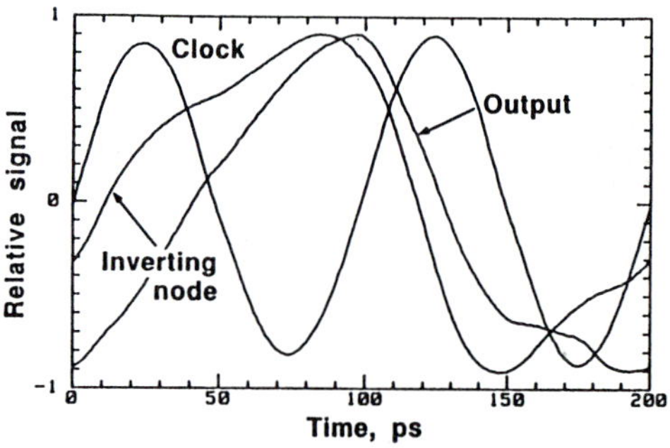

Fig. 7 BFL static frequency divider clocked at 10 GHz. Twice the delay from the clock to the output sets the circuit's maximum clock frequency.

Configured as a vector voltmeter, the sampler has been used to measure voltage transfer functions of microwave structures and circuits, including slotted-line measurements of standing wave ratio (Fig. 8), phase velocity, and lateral potential distribution [10] of microstrip and coplanar waveguide transmission lines. On GaAs microwave amplifiers, propagation of microwave signals internal to microstrip and coplanar-waveguide MESFET distributed amplifiers [11,12] have been measured (Fig. 9), and the gain saturation behavior of these circuits investigated (Fig. 10).

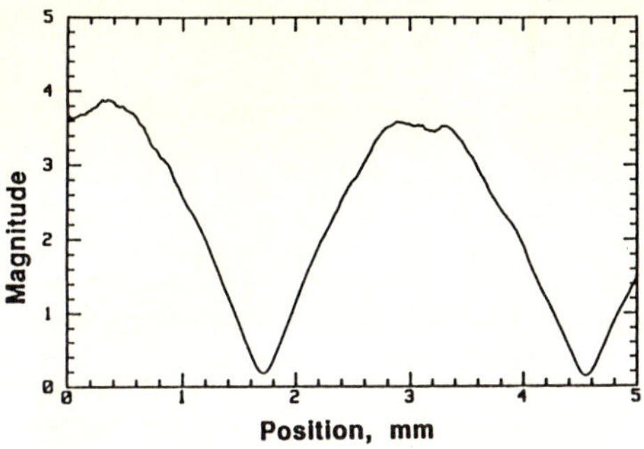

Fig. 8 Standing wave on a CPW transmission line at 20 GHz with an open termination at one end.

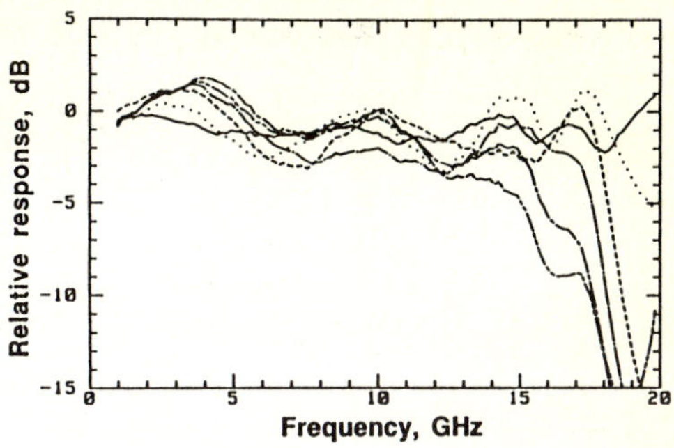

Fig. 9 Small-signal gate voltages vs. frequency of a 5-stage 2-18 GHz CPW distributed amplifier.

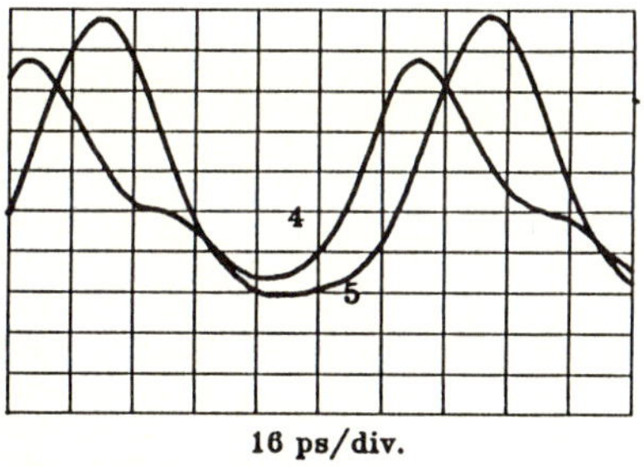

Fig. 10 Saturation waveform at drains 4 and 5 of a 2-18 GHz distributed amplifier at 10 GHz and 1 dB gain compression

5. Conclusion

With a pulsewidth of 1.5 ps FWHM and a demonstrated timing jitter ≤300 fs, the electrooptic sampler has a measurement bandwidth exceeding 100 GHz. A microwave probe station has been modified to allow for electrooptic probing of GaAs IC's at the wafer level. With a spatial resolution of 3 μm and a noise floor of 300 μV(1 Hz), the system is suitable for characterization of ultrafast digital and analog GaAs IC's, such as inverter chains, multiplexers, frequency dividers, and microwave amplifiers.

6. Acknowledgments

We thank M. Riaziat of Varian Associates and J. Jensen of Hughes Research Laboratories for their invaluable assistance with their GaAs IC's and the generous equipment donations of Cascade Microtech, Inc., Tektronix, Inc, and the Hewlett-Packard Co. M. Rodwell acknowledges an IBM Pre-doctoral Fellowship and K. Weingarten acknowledges a Newport Research Award. This work was supported in part by the Air Force Office of Scientific Research contract F49620-85K-0016 and by Wright-Patterson Air Force Base Avionics Laboratory contract F33615-86-C-1126.

7. References

1. J.A. Valdmanis, G.A. Mourou., and C.W. Gabel: IEEE J. Quant. Elec., **19**, 664 (1983)

2. B.H. Kolner and D.M. Bloom: IEEE J. Quant. Elec., **22**, 69, (1986), J.A. Valdmanis and G.A. Mourou: ibid., **22**,79 (1986)

3. J.L. Freeman, S.K. Diamond, H. Fong, and D.M. Bloom: Appl. Phys. Lett., **47**, 1083 (1985)

4. M.J.W. Rodwell, K.J. Weingarten, D.M.Bloom, T. Baer, and B.H. Kolner: Optics Letters, **11**, 638, (1986)

5. D. Grischowski and C.A. Balant: Appl. Phys. Lett., **41**, 1, (1982)

6. M.J.W. Rodwell, K.J. Weingarten, J.L. Freeman, and D.M.Bloom: Electron. Lett., **22**, 499 (1986)

7. M.N. Islam, L.F. Mollenauer, and R.H. Stolen: Ultrafast Phenomena V, ed. by G.R. Fleming and A.E. Siegman, Springer Ser. Chem. Phys., Vol. 46 (Springer-Verlag, New York 1986) p. 46

8. M. Levenson, R.M. Shelby, and P.W. Bayer: Phys. Rev. B, **31**, 5244 (1985)

9. J.F. Jensen, L.G. Salmon, D.S. Deakin, and M.J. Delaney: Technical Digest of the 1986 International Electron Device Meeting, p. 476

10. K.J. Weingarten, M.J.W. Rodwell, J.L. Freeman, S.K. Diamond, and D.M. Bloom: Ultrafast Phenomena V, ed. by G.R. Fleming and A.E. Siegman, Springer Ser. Chem. Phys., Vol. 46 (Springer-Verlag, New York 1986) p. 98

11. M.J.W. Rodwell, M. Riaziat, K.J. Weingarten, B.A. Auld, and D.M. Bloom: IEEE Trans. Microwave Theory Tech., **MTT-34**, 1356, (1986)

12. G. Zdasiuk, M. Riaziat, R. LaRue, C. Yuen, and S. Bandy: To appear in Picosecond Electronics and Optoelectronics, (Springer-Verlag, 1987)

Electro-optic Sampling of High-Speed, InP-Based Integrated Circuits

J.M. Wiesenfeld, R.S. Tucker, A. Antreasyan, C.A. Burrus, and A.J. Taylor***

AT&T Bell Laboratories, Crawford Hill Laboratory, Holmdel, NJ 07733, USA

We report the application of electro-optic sampling[1-3] to measurements of waveforms internal to a high-speed InP integrated circuit. This is possible because InP is an electro-optic material with electro-optic coefficient comparable to that of GaAs.[4] The measurement uses an electro-optic sampling system based on a 1.3 μm InGaAsP injection laser that is gain-switched.[5] Waveforms and propagation delays through a two stage InGaAs/InP metal-insulator-semiconductor FET (MISFET) inverter circuit[6] have been measured. Propagation delays as short as 15 ps have been observed in a single inverter stage. This work extends to InP-based circuits previously developed techniques for non-invasive sampling using the electro-optic effect in GaAs[2,5].

The experimental arrangement is shown in Fig. 1. Two microwave frequency synthesizers are phase-locked and offset in frequency by an amount Δf. One synthesizer drives the integrated circuit, and the other is used to gain-switch an InGaAsP double-channel planar buried heterostructure (DCPBH) laser. The 1.3 μm radiation from this laser is below the bandgap of InP. For these experiments, the test frequency f_0 is varied between 1 and 5 GHz, and the laser produces pulses

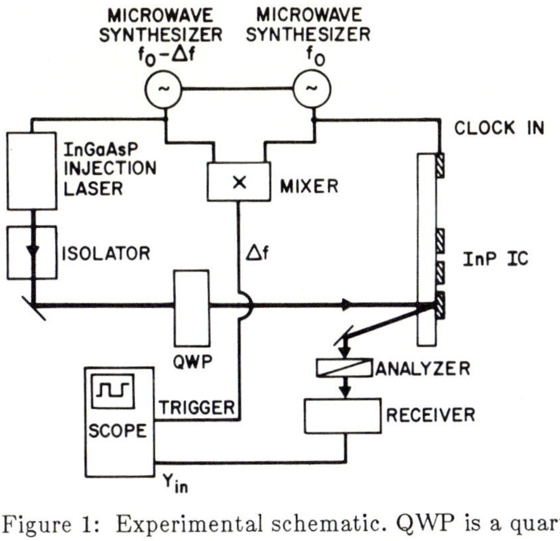

Figure 1: Experimental schematic. QWP is a quarter-wave plate.

* AT&T Bell Laboratories, Murray Hill, N.J. 07974

** Present address: Los Alamos National Laboratory, Los Alamos, N. M. 87545 of duration 18 - 20 ps. The beam from the laser, made circularly polarized to maximize the sensitivity of the experiment[2], impinges from the back side of the InP chip and is reflected from the active side of the chip. The back side of the chip is chemically polished to reduce scattering. The polarization of the optical beam is changed by the longitudinal electro-optic effect in InP, and the change in polarization, which is proportional to the electric field at the sampling point, is measured by a sensitive optical receiver placed after the analyzer. Since the laser and the circuit are driven at frequencies offset by Δf, the laser pulses sample the waveform on the circuit at this rate. The offset frequency was chosen to be 4 kHz. A "real time" reconstruction of the sampled waveform is displayed on an oscilloscope triggered at the offset frequency Δf, using a signal generated by a microwave mixer. This trigger arrangement permits measurement of absolute propagation delays between various points in the circuit. Timing jitter between the laser pulse and the frequency synthesizer driving the circuit is less than 1 ps,[7] so the timing resolution of the sampling system is limited by the 18 - 20 ps optical pulsewidth from the gain-switched laser. However, it is possible to measure peak positions of pulses, and hence pulse propagation delays, to within 5 ps. The ohmic contacts in the present circuit are made by alloying, and scatter the reflected light. Therefore, it is necessary to align the probe beam to pass adjacent to the probed contact, so that it reflects off the front surface of the InP substrate. The reflected beam intensity is too small to permit shot-noise limited operation of the receiver. Consequently, the voltage sensitivity of the present experiment, $20\ \mathrm{mV}/\sqrt{Hz}$, is poorer than the observed sensitivity of the system, $1.5\ \mathrm{mV}/\sqrt{Hz}$, when the receiver is operating in a shot-noise limited regime.[5]

The circuit used in the present experiments is an InGaAs/InP enhancement mode MISFET two-stage inverter[6], with the drain of the first FET connected to the gate of the second FET. The MISFETs have integrated 500 Ω load resistors. The supply voltage is 1.5 V. We sample at the input (gate) of the first FET, the output (drain) of the first FET, and the output of the second FET.

The first measurements we report used a 5 GHz sinusoidal input waveform of amplitude 1.5 V p-p with a dc offset of 1 V. The input and output waveforms of the first stage are shown in Fig. 2. The waveform at the output is inverted and

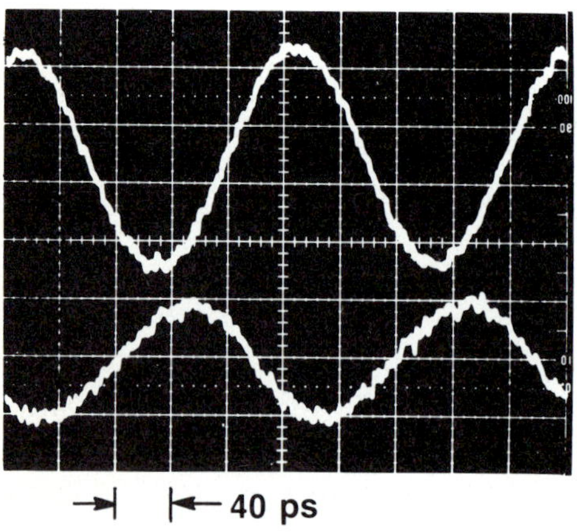

Figure 2: Measured waveforms for a 5 GHz sinusoid. Top is sampled at the input of the first FET, bottom is sampled at the output of the first FET. Vertical scale is 0.4 V/div.

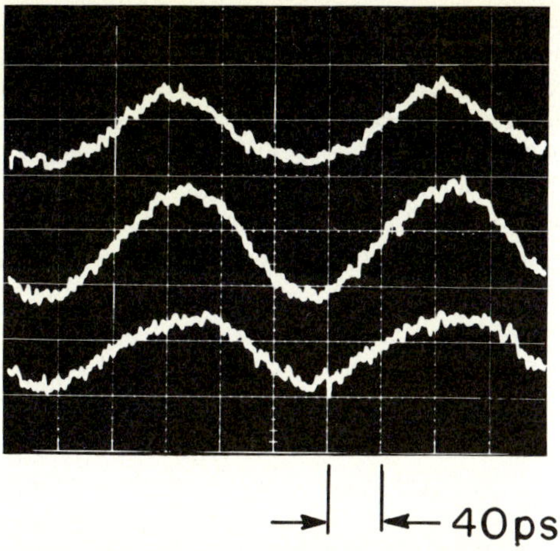

Figure 3: Waveform at output of first FET for 5 GHz sinusoid. Value of the dc offset at the input is 1.25 V, 1.0 V, and 0.75 V from top to bottom. Vertical scale is 0.4 V/div.

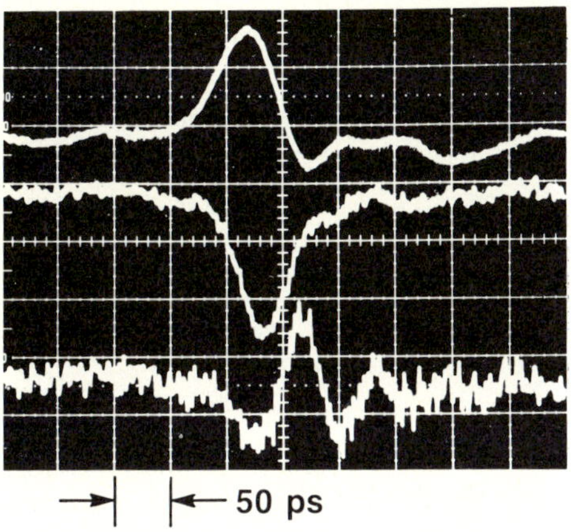

Figure 4: Measured waveforms for input pulse from the comb generator. Traces are input to the first FET (top), output of the first FET (middle), and output of the second FET (bottom). Vertical scale is 1.0V/div (top), 0.2 V/div (middle) and 0.1 V/div (bottom).

shows approximately 20 ps delay with respect to the input. The voltage swing at the output (0.8 V) is close to the maximum possible swing for this device with a 1.5 V bias voltage. The waveform at the output of the first stage is shown in Fig. 3 for several values of the dc offset to the input. The effect of the offset is evident.

For measurement of pulse propagation delays, the circuit is driven by a comb generator which produces pulses of 2 V amplitude and 43 ps (FWHM) duration[3]. Fig. 4 shows the resulting waveforms. The upper curve is the input to the first FET, the middle curve is the output of the first FET, and the lower curve is the output of the second FET. No significant broadening of the pulse occurs in the MISFETs, but there is an observable propagation delay in each stage. The propagation delay of the first inverter is shorter than that of the second inverter. From this measurement and other measurements of propagation delays for different offsets and polarity, the average propagation delays are determined to be 15$\pm$5 ps and 33$\pm$5 ps for the first and second inverter stages, respectively. The negative voltage swings at the output of the second inverter, seen in Fig. 4 just before and after the main pulse, may be due to capacitive or inductive coupling from the input of the circuit.

In summary, we have extended the application of non-invasive characterization of integrated circuits by electro-optic sampling to InP-based circuits. We have measured waveforms internal to an InGaAs/InP MISFET inverter circuit, and have measured propagation delays through *individual* MISFETs. Thus, the propagation delay for a particular type of transistor can be measured without the need for fabrication of a ring oscillator circuit. By virtue of the propagation delay of 15 ps measured for one InGaAs/InP MISFET, this FET ranks with the highest speed transistors reported to date.

References

1. J. A. Valdmanis and G. Mourou: IEEE J. Quant. Electron. *QE-22,* 69 (1986).
2. B. H. Kolner and D. M. Bloom: IEEE J. Quant. Electron. *QE-22,* 79 (1986).
3. A. J. Taylor, J. M. Wiesenfeld, G. Eisenstein, R. S. Tucker, J. R. Talman, and U. Koren: Electron. Lett. *22,* 61 (1986).
4. N. Suzuki and K. Tada: Jpn. J. Appl. Phys. *23*, 291 (1984).
5. A. J. Taylor, R. S. Tucker, J. M. Wiesenfeld, C. A. Burrus, G. Eisenstein, J. R. Talman, and S. S. Pei: Electron. Lett. *22*, 1068 (1986).
6. A. Antreasyan, P. A. Garbinski, V. D. Mattera, N. J. Shah, and H. Temkin: Electron. Lett. *22*, 1014 (1986).
7. A. J. Taylor, J. M. Wiesenfeld, G. Eisenstein, and R. S. Tucker: Appl. Phys. Lett. *49*, 681 (1986).

Electro-optic Sampling Analysis of Timing Patterns at Critical Internal Nodes in Gigabit GaAs Multiplexers/Demultiplexers

X.-C. Zhang[1], *R.K. Jain*[1], *and R. Hickling*[2]

[1]Amoco Research Center, P.O. Box 400, MS F-4, Naperville, IL 60566, USA

[2]GigaBit Logic, Inc., 1908 Oak Terrace Lane, Newbury Park, CA 91320, USA

Subsequent to its initial demonstration for the measurement of picosecond electrical transients in travelling-wave electro-optic modulators,[1-3] the technique of electro-optic sampling (EOS) has been shown by several researchers[4-7] to represent a unique capability for direct non-invasive probing of GaAs integrated circuits with high temporal resolution. In this article, we describe the use of this capability for a detailed analysis of on-chip waveforms at critical internal nodes in high speed sequential logic ICs, exemplified by gigabit rate 8:1 multiplexers/demultiplexers. Besides providing vital timing information between such internal nodes (and thereby providing the first clear measurements of timing parameters such as set-up and hold timing margins in flip-flops, and gate propagation delays in normally-loaded and normally-exercised gates), such EOS measurements enable direct on-chip observation of non-ideal logic circuit operation, including effects such as clock feedthrough at internal nodes.

Figure 1 shows the experimental arrangement, which is similar to that reported earlier[7], with ∿ 3.5 picosecond pulses from a mode-locked and compressed YAG laser being used for the EOS measurements described here. All the measurements reported here were performed directly on wafers (obtained from GigaBit Logic, Inc.) with conventional probe cards used for biasing and feeding of the gigabit rate clock inputs. A clock frequency of 1.31 Gigahertz, synchronized to within a few Hertz of a precise integer multiple of the repetition rate of the mode-locked YAG laser, was used for all the data reported here.

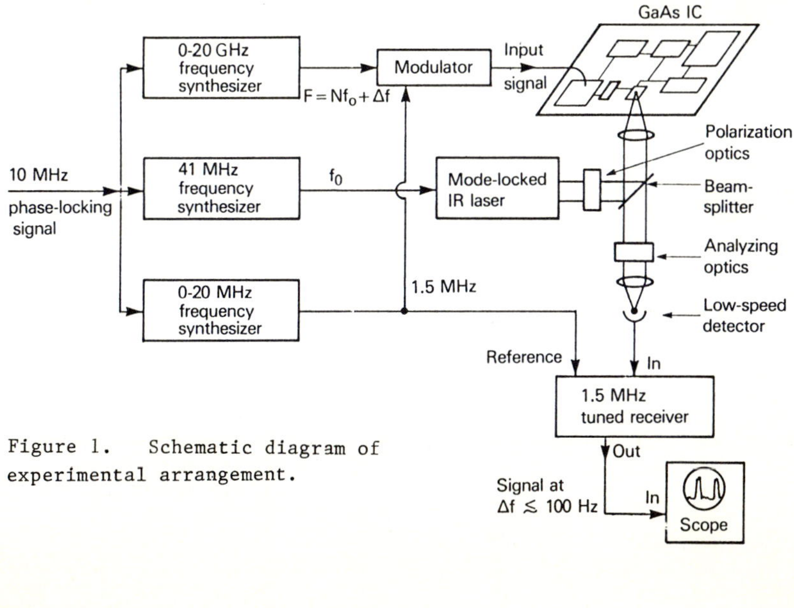

Figure 1. Schematic diagram of experimental arrangement.

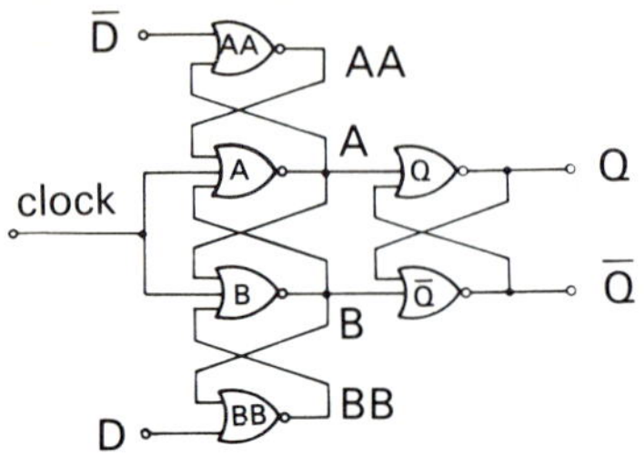

Figure 2. Circuit diagram of D-flip-flop used in the Johnson counter of the MUX/DEMUX circuits, depicting the names (AA,A, etc) of the internal nodes probed. (These correspond to outputs of NOR gates with the same name designations).

Figure 2 shows the circuit diagram of one of the four falling-edge-triggered D-flip-flops (D-FF) that comprise the 4-stage Johnson counter used to establish the timing characteristics of the multiplexer/demultiplexer studied. Figure 3 shows the waveforms measured at the various internal nodes in this flip-flop, along with the measured phase of the clock input to this flip-flop. From the timing relationships between such waveforms, measured on highly-expanded traces similar to those in Figure 4, the timing margins for set-up and hold of this flip-flop and the propagation delays of each of the six NOR gates comprising the D-FF, can be inferred very accurately. For instance, the propagation delays for gates Q (fan-out = 2) and A (fan-out = 3) were measured to be 90 and 97 ps, respectively. The maximum frequency of operation (f_{max}) of each of the flip-flops can be inferred from the timing margins measured with any test clock frequency $f < f_{max}$. (The set-up timing margin is given by the amount of time by which the falling edges in the outputs of AA and BB precede the corresponding clock falling edge). From the smallest timing margin (190ps)

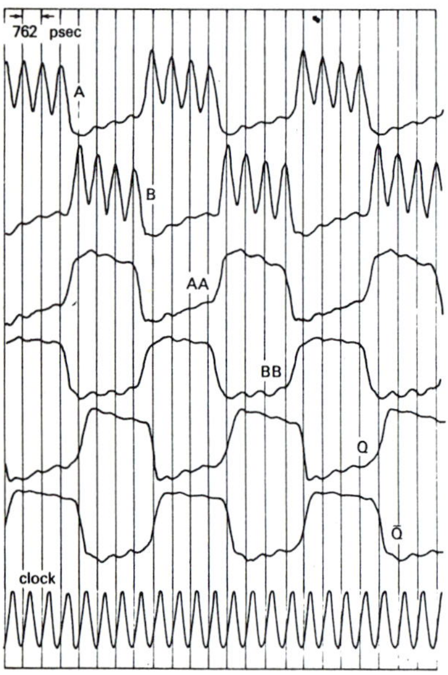

Figure 3. Waveforms measured by electro-optical sampling at the various nodes within a single D-flip-flop in a 4-stage Johnson counter.

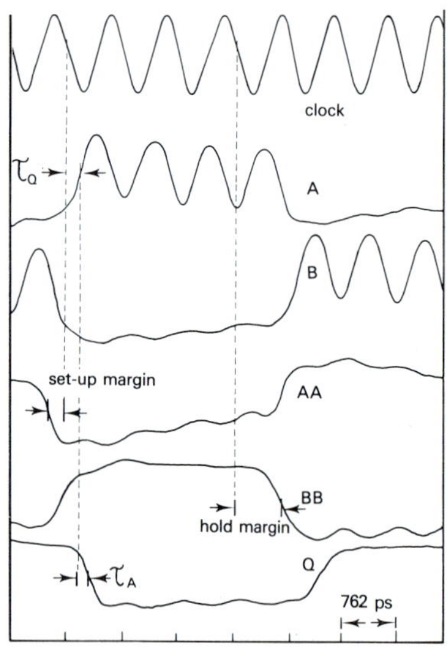

Figure 4. Expanded versions of some of the waveforms of Fig. 3, indicating measurement of timing parameters such as the set-up and hold timing margins and gate propagation delays τ_A, τ_Q.

measured for set-up of this D-FF when operated at f=1.31 GHz, we infer its maximum operating frequency as approximately f_{max}=1.74 GHz.

Timing patterns corresponding to the 8 phases of a 4-stage Johnson counter (constructed with D-FF's similar to that described above) are shown in Figure 5(a) and the 8-phase strobe pulses that control data combining and separation, are depicted in Figure 5(b). (The strobe pulses are decoded from Johnson counter outputs as $S_i=\overline{Q_i+Q_{i+1}}$ by a simple combinational logic circuit). Note that the relatively large ripples and apparent glitches observed in the "zero" state on several of the waveforms depicted in Figs. 3-5 are not random noise, but are a clear manifestation of intrinsic operational characteristics of the logic circuits used. Two "1" inputs in any of the NOR gates used (Fig. 6) result in a "hard zero" (0') that is lower than the zero (0) obtained with a single "1" input, due to the extra current drawn by the second logic FET. The difference between the levels of the single -"1"- input zero (0) and the dual -"1"- input zero (0') is the source of several of the apparent glitches in the observed data. For instance, in the output latch of edge-triggered flip-flops, simple (timing diagram and truth table) analysis shows that a pair of "1"s will appear alternating with a single "1" at the inputs of one of the NOR gates (AA or BB) for exactly half of a clock period, manifesting itself at the flip-flop output waveforms (Q, $\overline{Q}$) as a ripple at precisely the clock period. (This effect of clock feedthrough manifests itself on the zero state of each of the traces of Figs. 3 to 5(a); see $\overline{Q_3}$ in Fig. 5(a) for the clearest illustration). Likewise, when Q_i is represented by the data string (00001111), and $\overline{Q_{i+1}}$ is represented by the data string (01111000), then the strobe pulse $S_i=\overline{Q_i + Q_{i+1}}$ is not simply (10000000), but is more accurately given by (10000'000). The hard zero (0') manifests itself as an apparent negative glitch of one clock period width, exactly midway between the "1" pulses in the strobe waveforms (see S7 in Figure 5(b) for the clearest illustration of this effect).

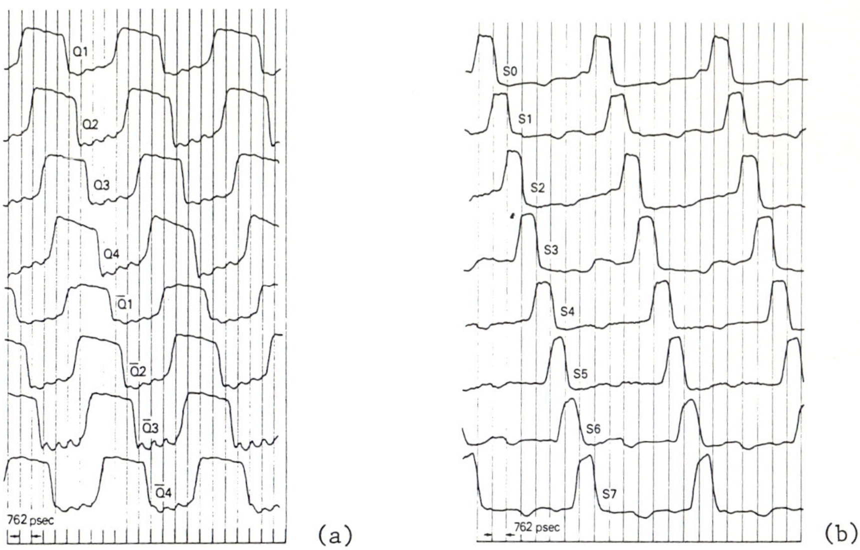

Figure 5. EOS waveforms observed at the outputs of (a) the 4-stage (8-phase) Johnson counter, and (b) the 8-phase strobe decoder.

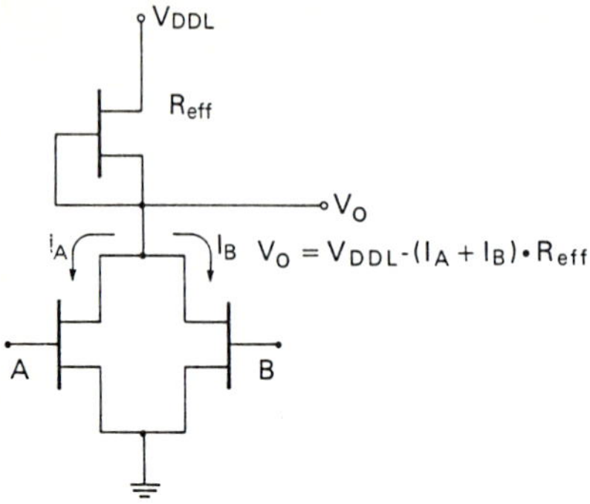

A	B	V_o
0	0	1
1	0	0
0	1	0
1	1	0'

Figure 6. Circuit diagram of the NOR gates used, along with the modified truth tables depicting the presence of a hard zero (0') when both inputs are at logic "1".

The slow risetimes in the "0" state, and the slow falltimes in the "1" state are apparently caused by an RC time constant due to a dynamic capacitance created by a change in state. Such an effect has not been observed in our SPICE simulation of the circuit, and is being presently investigated in further detail.

In conclusion, we have presented the first results of detailed timing analysis of a high speed sequential logic circuit by electro-optic probing of its internal nodes, and have demonstrated several of the key features that can be elucidated with this technique. It is clear that such EOS analysis will represent an important analytical tool to circuit designers, and that it may ultimately facilitate rapid characterization of the essential features of an IC for even the end user.

References

1. D. H. Auston and A. M. Glass, Phys. Rev. Lett. 20, 398 (1972); P. LeFur and D. H. Auston, Appl. Phys. Lett. 28 21 (1976).
2. R. C. Alferness, N. P. Economou, and L. L. Buhl, Appl. Phys. Lett. 37, 597 (1980).
3. J. A. Valdmanis, G. Mourou, and C. W. Gabel, Appl. Phys. Lett. 41, 211 (1982).
4. J.L. Freeman, S.K. Diamond, H. Fong, and D.M. Bloom, Appl. Phys. Lett., 47, 1083 (1985).
5. X.-C. Zhang and R.K. Jain, Electron. Lett., 22, 264 (1986).
6. M.J.W. Rodwell, K.J. Weingarten, J.L. Freeman, and D.M. Bloom, Electron. Lett., 22, 499 (1986).
7. R.K. Jain and X.-C. Zhang, Technical Digest of the IEEE 1986 GaAs IC Symposium (1986), p. 141.

A Technique for Wing Suppressed IR Sampling

J. Nees and S. Williamson

Laboratory for Laser Energetics, University of Rochester,
250 East River Road, Rochester, NY 14623, USA

The development of hot electron techniques and high mobility materials is one more step bringing electronics to the 100 GHz regime. With these new approaches come the problems of measuring device characteristics and performance. Several approaches have now been developed to observe high-speed electrical transients using the electro-optic technique [1]. In the case of many III-V compound semiconductors such as GaAs and InP, direct probing of the semiconductor substrate is possible. Such probing is potentially very useful as it allows for the noncontact noninvasive probing of devices and circuits. Noncontact measurements of devices and circuits have been made using compressed pulses from a cw mode-locked Nd:YAG laser [2]. This paper deals specifically with the use of compressed pulses for directly probing device and circuit substrates.

A primary consideration for direct substrate probing is that the probe light not be absorbed by the semiconductor. In experiments using 1.06 μm light to probe GaAs photogenerated carriers interfere with measurements. Furthermore, in InP 1.06 μm light is above bandgap making noninvasive sampling impossible. These observations suggest a move to longer wavelength lasers. For instance, compressed pulses from mode-locked YAG or color-center lasers in the 1.3 - 1.8 μm range, could be used. In this range, however, optical fibers do not exhibit sufficient positive group velocity dispersion (GVD) to form a wing-free compressed pulse [3]. Consequently the electro-optic sampling window is broadened and time resolution is degraded.

A solution to the problem of probing with broadened pulses is to use some nonlinearity to preferentially attenuate the low intensity wing energy. This can be done by frequency conversion; a process whose efficiency is proportional to input power. The frequency converted pulse shape is proportional to the square of the input pulse and shows very little energy in the wings. This may appear to defeat the intent to use sub-bandgap light as a probe, but it is not necessary to frequency-double before the device under test. Sub-bandgap light can be used to probe a device substrate and frequency conversion can be done after sampling up to the point of detection.

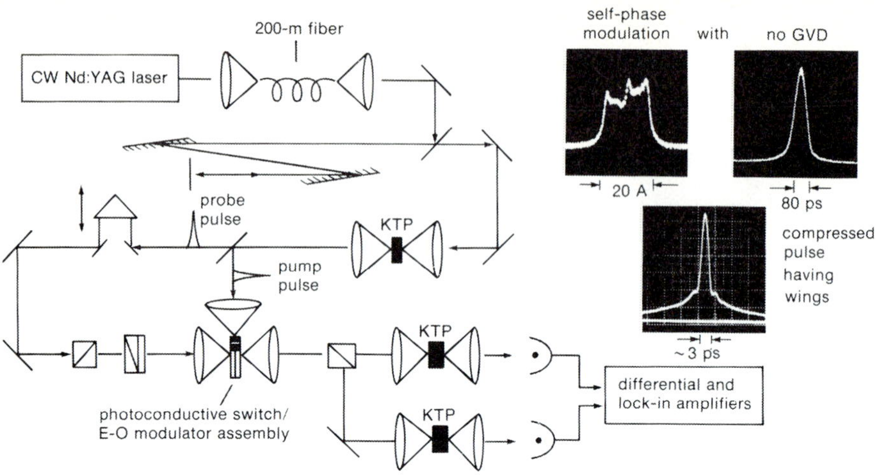

Fig. 1 Schematic of wing suppressed sampling experiment.

The experiment described here was designed to demonstrate the use of wing suppression to improve sampling resolution. Light from a 1.06 μm cw mode-locked Nd:YAG laser was focussed into a 400 m single-mode fiber and the output compressed with a double pass grating (1700 lines/mm) configuration. This yielded 3 ps pulses with a large fraction of energy in low intensity wings. To obtain a fast electrical transient a fraction of this light was doubled and focussed into a photoconductive switch. The remaining 1.06 μm light was used to sample the transient in a lithium tantalate microstripline modulator [4] and finally converted to 532 nm by 5 mm KTP crystals immediately before detection. The main limitation of this technique is the length bandwidth product for the frequency conversion crystal which limits the conversion efficiency of short pulses.

With conversion efficiency of only 0.5 percent before detection corresponding to a factor 10 in the signal-to-noise ratio, the following results were obtained. In Fig. 2 we see that the false leading edge produced by wings and extending more than 50 ps from the main rising edge has been suppressed by the frequency conversion. Expanding this waveform about the rising edge it also (see Fig. 3) becomes clear that there is a significant improvement in the rise time due to wing suppression.

Conclusion

For applications where optical fibers do not exhibit sufficient positive GVD for wing-free compression, we have demonstrated that the effects of pulse wings on electro-optic sampling resolution may be rejected by using a nonlinear process such as frequency conversion after the sampling point and before detection.

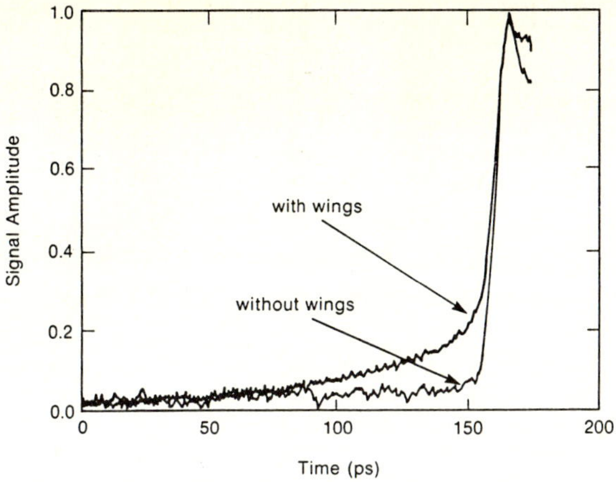

Fig. 2 Sampling with and without wings on a long time scale.

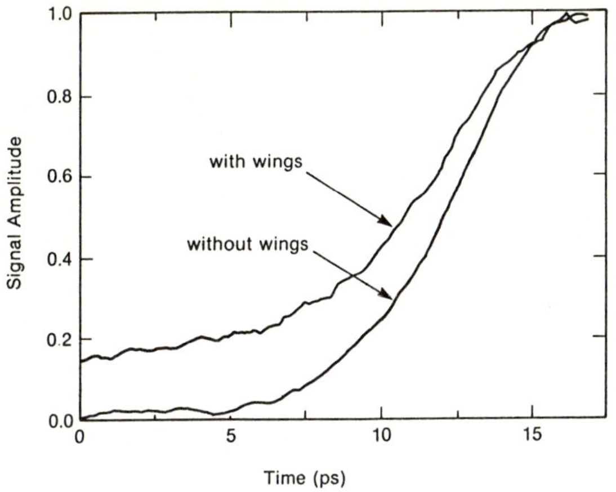

Fig. 3 Sampling with and without wings on a short time scale.

Acknowledgment

This work was supported by the sponsors of the Laser Fusion Feasibility Project at the Laboratory for Laser Energetics.

1. J.A. Valdmanis and G.A. Mourou, Laser Focus, 84-96 (February 1986) and Laser Focus, 96-106 (March 1986).
2. B.H. Kolner and D.M. Bloom, IEEE J. Quant. Elec. QE-22, 79-93 (1986).
3. D. Grischkowsky and A.C. Balout, Appl. Phys. Lett. 41(1), 1-3, (1982).
4. J.A. Valdmanis, G.A. Mourou, and C.W. Gabel, Appl. Phys. Lett. QE-19, 664-667 (1982).

Picosecond Electrical Pulse for VLSI Electronics Characterization

J.-M. Halbout, P.G. May, M.B. Ketchen, H. Jackel, G.P. Li, C.-C. Chi, M. Scheuermann, and M. Smyth*

IBM Thomas J. Watson Research Center, P.O. Box 218,
Yorktown Heights, NY 10598, USA
*IBM Zürich Research Laboratory, CH-8803 Rüschlikon, Switzerland

Picosecond electrical pulses with well resolved frequency components into the terahertz region are ideally suited for the characterization of VLSI electronics, both for the devices themselves and for the interconnect wiring. In this paper, we describe the technology we employ to photoconductively generate and measure these ultrafast electrical signals./1/ A particular emphasis was given to the compatibility with existing VLSI device technology. Although other measurement techniques /2,3/ have shown very good sensitivity and time resolution, the photoconductive sampling technique is particularly attractive for silicon device characterization with a comparable sensitivity and time resolution /4/ and a better amplitude calibration and reproducibility as the sampling geometry is defined by the lithography itself. Impedance matching is facilitated while keeping identical geometries between the device site and the generating and sampling sites. We also present results on the optoelectronic characterization of a $1\mu m$ gate GaAs MESFET. We will not address in this paper the problem of circuit testing which has some stringent additional requirements, in particular that of submicron spatial resolution and of wide field addressing capability. These imperatives may ultimately limit the use of purely optical techniques for the functional testing of high speed circuits at their internal nodes.

The optoelectronic switches we use are fabricated with polysilicon as the photoconductor for compatibility with silicon device processing. The details of the fabrication process follow. A $0.5\mu m$ film of high quality thermal oxide is grown on the underlying silicon substrate for electrical isolation from carriers photogenerated in the substrate. It provides a well defined lower boundary to the photoconductive volume. The photoconductive film itself consist of a $0.5\mu m$ layer of undoped polysilicon grown on top of the oxide. In order to increase the carrier mobility in the polysilicon, a high temperature anneal is performed at $1100°C$ in an argon atmosphere. This anneal causes an enlargement of the polysilicon grain size and a surface roughening with features sizes up to $0.1\mu m$ easily resolved in an electron microscope. The aluminum-copper metallization is deposited on the polysilicon and patterned by a standard lift-off lithographic technique. We have principally studied coplanar transmission lines of $1.2\mu m$ linewidth separated by twice that distance, with a metal thickness of $1.0\mu m$. The metal replicates the roughness of the polysilicon surface. In order to further reduce the carrier lifetime in the photoconductor, it is submitted to two oxygen ion implants at 100 keV and 200 keV at doses between 10^{15} and $10^{16}/cm^3$. This double step insures a uniform implantation throughout the photoconductive film thickness. This entire procedure is compatible with either a MOS or a bipolar device fabrication process. We have fabricated such structure on silicon substrates of widely different dopings, identical to those used in device fabrication, as well as on perfect insulators such as quartz.

The laser source used to drive the photoconductive switches is a Nd:YAG laser, acousto-optically mode-locked at 100 MHz repetition rate. Its 73 ps pulses are compressed by passage through 40 meters of single mode optical fiber followed by a double-pass dispersive optical delay line composed of a pair of high efficiency gratings oriented close to the Littrow angle. The compressed pulses are frequency doubled in a 0.5 cm KTP crystal to yield pulses of autocorrelated width 2 ps and average power 350 mW at 532 nm when 5 W at $1.064\mu m$ is incident on the fiber coupler.

The line geometry as described above provides a characteristic impedance of 100Ω for these coplanar transmission lines. These dimensions are compatible with the most competitive integrated circuit technologies. We routinely generate on these lines electrical pulses of 2 psec duration and 2.5 mA peak current which correspond to an amplitude of 250 mV. Figure 1 shows the propagation of such a pulse on respectively 1 mm and 4 mm of coplanar transmission lines fabricated on an N doped silicon substrate. The experiment shows that, over these lengths which are in keeping with chip dimensions, the risetime of the signal is not degraded catastrophically but that there is a substantial attenuation of the peak amplitude. This strong attenuation as well as the step which appears on the trailing edge of the main peak can be attributed mainly to the resistive nature of the lines. This result is confirmed by modelling. The model also incorporates the dispersion due to the skin effect on these lines which is predominantly responsible for the smearing of the leading edge of the pulse. These results are very encouraging for the propagation of very fast signals in VLSI environment. The small geometry of these coplanar lines is essential in preserving the speed of the signals, when compared to previous studies on microstrip geometries of larger characteristic dimensions./5/

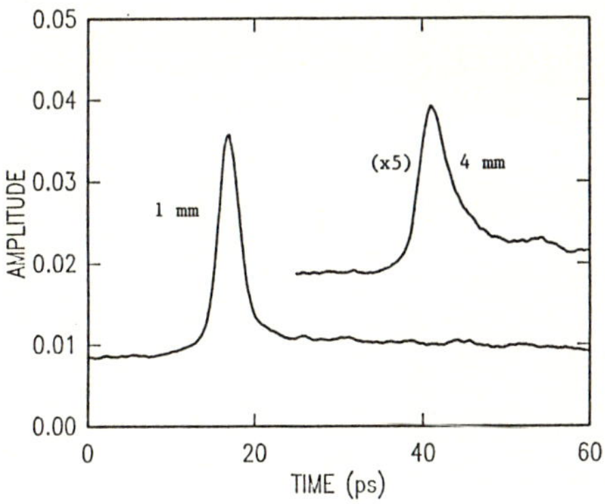

Fig. 1: Propagation of a 2 psec electrical pulse on a coplanar transmission line fabricated on a N doped silicon substrate. The amplitude of the signal propagated 4 mm is multiplied by a factor of 5.

Figure 2 depicts the inverting turn-on response of a $1\mu m$ gate GaAs MESFET to an electrical pulse generated as described above. The pulse is coupled onto the GaAs device chip by very short $300\mu m$ wire bonds and the device output is similarly connected to a sampling chip. We have shown previously that, with care in the lay-out, fast pulses can be easily propagated through such wire bonds./6/ The device is designed in a coplanar geometry with proportional tapering interconnects from the wire bonding sites to the gate and source connections on the input side, and a symmetric lay-out from the drain and source pads to the output bonding pads. This lay-out is shown in figure 3. The outside dimensions of these tapers match those of similar tapers designed in the coplanar lines of the pulse generator and of the sampling chip. Also designed on this device chip are calibration structures (a through and an open) which allow us to fully characterize the very signal which reaches the gate of the device. We have found these constant impedance tapering structures to preserve very well the integrity of high speed signals with rise times down to a few picoseconds. The voltage bias to the device terminals is applied at the end of the coplanar lines on the generating and the sampling chip. The device can be either studied in the turn-on mode by coupling a fast pulse from a biased side gap onto the line, or in the turn-off mode by shorting the biased input line connected to gate and source in a fashion similar to the sliding contact./4/

The device response was measured at a drain-source bias of 2V and a gate-source bias of 0.25V. The drain waveform when the device is kept in its off state has been subtracted from the original data. We have not deconvolved from the data as shown the width of the pulse applied to the gate. We see that the device turns on in less than 15 psec and back off in about 25 psec. The

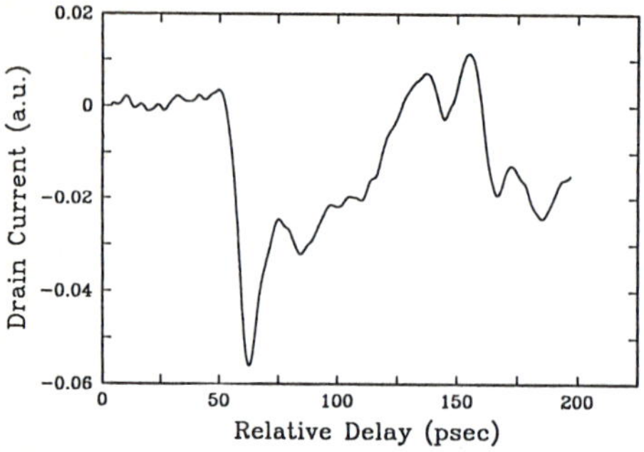

Fig. 2: Response of a $1\mu m$ gate GaAs MESFET

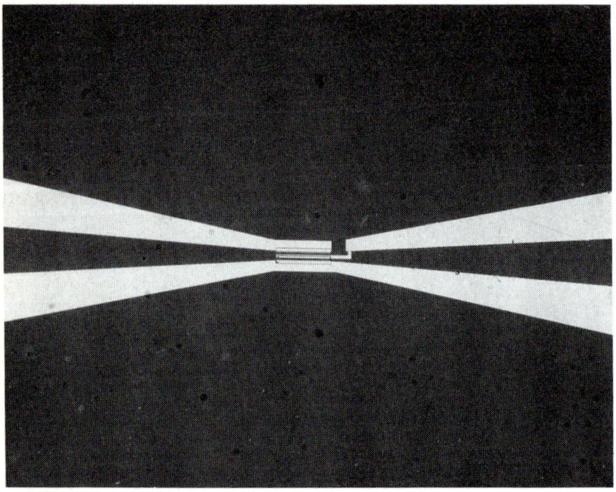

Fig. 3: Device connections lay-out.

fall time is somewhat masked by spurious reflections on the gate side which are amplified and superimposed on the falling edge of the waveform. It is important to note that the device is operating in the large signal regime and therefore these data cannot be compared directly to the frequency domain characterization of the device. It is indeed desirable that devices designed for logic or switching applications be characterized directly in the time domain under operating conditions as close as possible to their working environment, rather than only in the small signal regime, as it is frequently practiced today. Optoelectronics techniques allow us to bridge the two regimes and to compare direct time response with unprecedented resolution to S-parameter characterization./7/

It has been demonstrated that picosecond electrical pulses can be generated and propagated on transmission line structures of dimensions and materials compatible with the most advanced VLSI technology. We also have presented the characterization of an active device using photoconductive sampling. The versatility of this technique makes it very well suited to single device characterization in the time domain.

1. D.H. Auston,in Picosecond Optoelectronic Devices, edited by C.H. Lee (Academic Press, London, 1984), pp.73-116.

2. D.R. Dykaar, R. Sobolewski, J.F. Whitaker, T.Y. Hsiang, G.A. Mourou, M.A. Hollis, B.J. Clifton, K.B. Nichols, C.O. Bozler and R.A. Murphy, in Ultrafast Phenomena V, edited by G.R. Fleming and A.E. Siegman, (Springer-Verlag, 1986), pp. 103-106.

3. H.K. Heinrich, D.M. Bloom and B.R. Hemenway, Appl. Phys. Lett, 48, 1066, (1986).

4. M.B. Ketchen, D. Grischkowsky, T.C. Chen, C.-C. Chi, I.N. Duling III, N.J. Halas, J.-M. Halbout, J.A. Kash and G.P. Li, Appl. Phys. Lett., 48, 751, (1986).

5. K.W. Goossen and R.B. Hammond, in Picosecond Electronics and Optoelectronics, edited by G.A. Mourou, D.M. Bloom and C.H. Lee (Springer-Verlag, 1985), pp. 70-73.

6. P.G. May, G.P. Li, J.-M. Halbout, M.B. Ketchen,C.-C. Chi, M. Scheuermann, I.N. Duling III, D. Grischkowsky and M. Smyth, in Ultrafast Phenomena V, edited by G.R. Fleming and A.E. Siegman (Springer-Verlag, 1986), pp. 120-122.

7. D.E Cooper and S.C. Moss, IEEE J. Quant. Electron., QE-22, 94, (1986).

Ultrafast Optics Applied to Modern Device Research

G. Mourou[1], *K. Meyer*[1], *J. Whitaker*[1], *M. Pessot*[1], *R. Grondin*[2], *and C. Caruso*[2]

[1]University of Rochester, Laboratory for Laser Energetics, Ultrafast Science Center, 250 East River Road, Rochester, NY 14623, USA
[2]Arizona State University, Center for Solid State Research, ERC140, Tempe, AZ 85287, USA

1. Abstract

We show that ultrafast optical techniques have made possible the direct investigation of basic nonstationary transport, i.e. velocity overshoot and resonant tunneling, which governs the operation of modern high-speed devices.

2. Introduction

Heteroepitaxial and lithographic techniques make possible the fabrication of control regions in modern transistors with dimensions comparable to electron ballistic distances. As a consequence, the electron transport is nonstationary, leading to velocities greater than those obtained in the steady state limit.

The unity short-circuit current gain F_T of a transistor is given by:

$$F_T = \frac{\langle v \rangle}{2\pi l_{eff}}$$

where $\langle v \rangle$ is the average electron velocity and l_{eff} is the effective control gate length. Reducing the device dimensions to ballistic dimensions leads to potentially faster devices due to the smaller l_{eff} and possiby larger $\langle v \rangle$. The electrons stay in nonequilibrium for times on the order of 1 ps. Because the performance of all modern devices relies on nonstationary electron transport, it is important to elucidate the physics of hot electron phenomena, which has thus far been studied mainly through numerical modeling. Short optical pulses offer a way to directly monitor these effects. Not only can pulses of a few optical cycles be generated, but they also can be used to generate [1] and characterize waveforms of a few hundred femtoseconds rise time, which corresponds to a spectrum with terahertz bandwidth (see Fig. 1). This ultrafast optical technique has been used to investigate the response of state-of-the-art transistors, such as the permeable base transistor, when subjected to a step function with a fast (subpicosecond) rise time (Fig. 2).

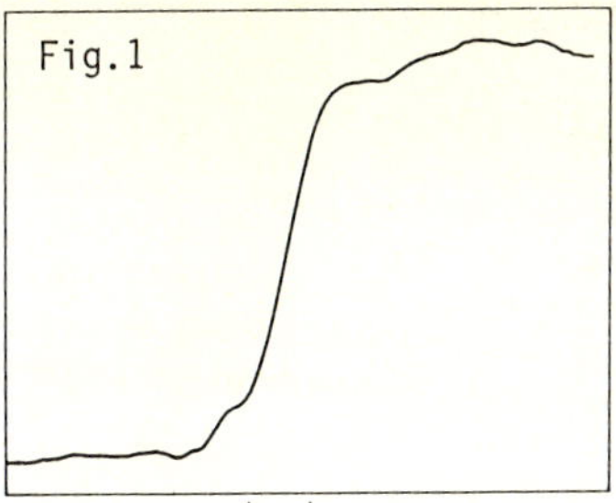

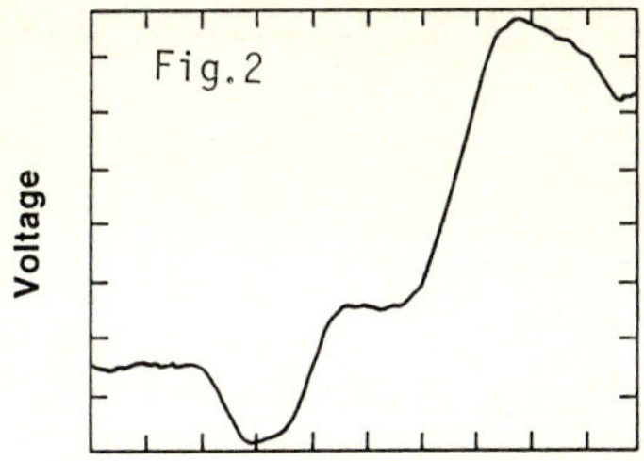

Figure 1 Temporal response of coplanar strip electro-optic sampler.

Figure 2 Low-gain PBT transient response.

The devices, known as MESFET, MODFET, bipolar heterojunction transistors, hot electron transistors, permeable base transistors, resonant tunneling devices, and resonant tunneling hot electron transistors, rely on different types of transport according to the detailed structure of the control region. Device performance may be governed by ballistic, quasi ballistic, velocity overshoot, or resonant tunneling transport.

When carriers are subjected to an electric field E, their average velocity (v) and kinetic energy (ε) are related by the moment equations:

$$\frac{dv}{dt} = \frac{qE}{m} - \frac{v}{\tau_m(\varepsilon)} \tag{1}$$

$$\frac{d\varepsilon}{dt} = qEv - \frac{(\varepsilon - \varepsilon_0)}{\tau_\varepsilon(\varepsilon)} \tag{2}$$

where q is the charge of the electron, m is the effective mass under the assumption of parabolic dispersion and $\tau_m(\varepsilon)$ and $\tau_\varepsilon(\varepsilon)$ are the momentum and energy relaxation times.

In the steady state limit,

$$\text{for } t > \tau_\varepsilon(\varepsilon), \frac{dv}{dt} = 0$$

and the drift velocity v_D given by the classical expression:

$$v_D = \frac{qE}{m}\tau_m(\varepsilon)$$

for the time $t>\tau_m(\varepsilon)$ and $t<\tau_\varepsilon(\varepsilon)$ the carriers will experience different types of nonstationary transport. Transport when $t<\tau_m(\varepsilon)$ is considered to be truly ballistic, as no collision with phonons or impurities will occur. The distance over which this regime exists corresponds to around 500 Å. $t \sim \tau_m(\varepsilon)$ is considered the quasi ballistic regime. The practical length over which this regime dominates can in fact be significant over a few thousand Å. In this regime, large velocities can be achieved through velocity overshoot. This effect is particularly significant in materials with a higher mass valley giving rise to a rapid increase in the momentum relaxation rate when the electron energy exceeds the intervalley energy gap.

It is important to note that when the carrier's energy, ε is less than the optical polar phonon, i.e. 36 meV for gallium arsenide, the carrier will only interact with acoustic phonons with a low scattering rate of $10^{+11}s^{-1}$. As the carrier energy exceeds the optical phonon energy, the scattering rate increases dramatically to $10^{+13}s^{-1}$. Due to the polar phonon, however, this emission is very anisotropic and the electrons will not change significantly in either their direction or momentum. So, even under optical polar phonon scattering, quasi ballistic transport can be sustained over distances of a few thousand Å. As the electron energy increases further and reaches the intervalley energy gap of 0.3 eV for gallium arsenide, the electrons are scattered into the low mass valley. This scattering process is both highly efficient and isotropic and will lead to a sharp reduction of $\tau_m(\varepsilon)$ and large increase in the electron effective mass.

When device dimensions are on the order of the de Broglie wavelength (i.e. 50 Å), such as with the resonant tunneling diode, quantum mechanical transport will dominate. This effect, shown by G. Sollner [2] for devices with alternating layers of potential barriers such as aluminum gallium arsenide, occurs when the injected carrier energy corresponds to the electron energy level in the quantum well. The energy band diagram of a typical resonant tunneling diode is shown in Fig. 3. High frequency experiments up to 2.5 THz have shown that this type of device could respond as fast as 100 fs. A great controversy on resonant tunneling time followed the Sollner group's result leading to calculated response times relying on models based on the WKB [3] approximation, a numerical solution of Schroedinger equation for thin quantum wells [4], or a method simply using the uncertainty [5] principle knowing the resonance level width. According to the model, estimates vary from 50 ps to 0.1 ps. The large difference between the theoretical predictions and the experimental results shows the need for direct experimental investigation.

According to the type of device, ballistic, quasi ballistic, or resonant tunneling transport will be involved. For instance, in MESFETs or MODFETs with gate lengths of a fraction of a micron, the transport

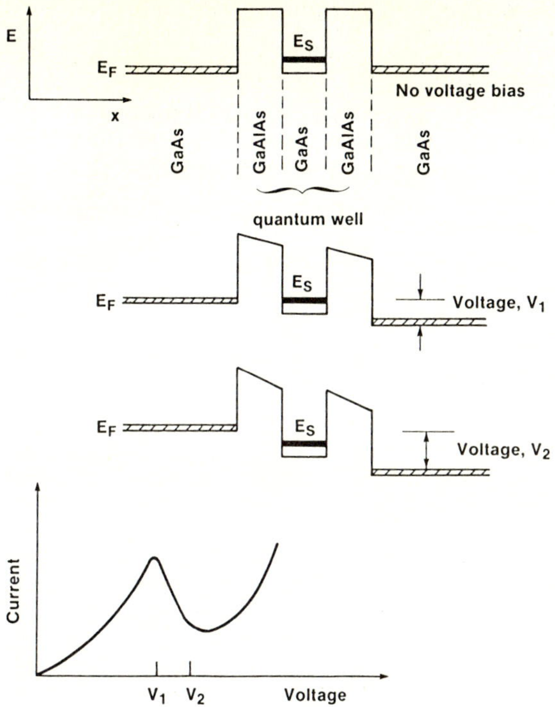

Figure 3 Band diagram and I-V characteristic of resonant tunneling device.

parallel to the interface is quasi ballistic. On the other hand, for heterojunction bipolar transistors where the current flow is perpendicular to the interface, the transport is ballistic due to the ballistic injection produced by the conduction peak at the interface and, of course, the short length of the gate corresponding to a few hundred Å. Also, transistors such as the hot electron transistor, metal base transistor, and permeable base transistor will rely on true or quasi ballistic transport due to their short gate lengths. Ballistic transport recently has been shown in this type of structure by M. Heiblum [6].

The two types of transport, i.e., nonequilibrium and resonant tunneling, can be combined to control high-speed device operation. For instance, Yokoyama and coworkers [7] built a resonant tunneling hot electron transistor where the injected electron energy is precisely selected by means of a resonant tunneling emitter. Current gain of 5 has recently been reported.

Few studies using ultrafast optical techniques have been used to investigate the nonequilibrium or the resonant tunneling transport. Shank et al. [8] have used the Franz-Keldysh effect to monitor the carrier velocity overshoot in gallium arsenide by looking at the field induced by the charge separation when the latter is photogenerated by a femtosecond optical pulse. Looking at the induced absorption change they were able to observe the time integrated velocity where velocity overshoot was detected. Castagne and coworkers [9] studied the photocurrent produced by a cw laser in a submicron gallium arsenide gap. The photocurrent is directly proportional to the carrier density and the velocity. For a submicron gap, the quasi ballistic current dominates, leading to a photocurrent increase as the structure dimension decreases. Finally, R. Hammond made an attempt to time-resolve directly the velocity overshoot of photocarriers produced by a picosecond pulse using a photoconductive gap as a gating element [10]. The instrumental resolution was 6 ps which was not enough to resolve the velocity overshoot conclusively.

3. *Time-Resolved Velocity Overshoot*

We have tried to investigate directly the photocarrier overshoot with a femtosecond optical pulse by using the electro-optic sampling technique [1]. Figure 4 shows the experimental arrangement. The initial photocurrent produced by a femtosecond optical pulse in a gallium arsenide photoconductive gap is given by

$$I_p(t) = Ne\, v(t) \tag{3}$$

where I_p is the transient photocurrent, N is the induced carrier density, and v is the carrier drift velocity. It is assumed that the carrier density does not change over a few picoseconds. The photogenerated current should vary with v(t). In practice, because of the photoconductive gap charging time [11]

$$I_p(t) = e\,N \int_{-\infty}^{t} v(t')\ \exp - \left(\frac{t - t'}{2Z_0 Cg}\right) \cdot\ dt' \tag{4}$$

where Z_0C_g is the charging time, Z_0 is the line impedance, and C_g is the gap capacitance.

The structure used in the experiment is a coplanar strip transmission line built on semi-insulating gallium arsenide. The gap is typically 50 μm and the electrode width equals 50 μm.

The carriers are photogenerated by using 100 fs pulses at 620 nm. The Monte Carlo simulation has recently been performed and shows a sharp velocity overshoot even when carriers are excited at

Fig.4

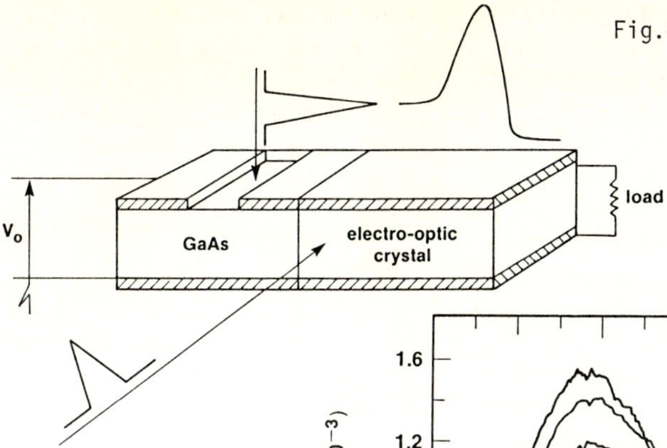

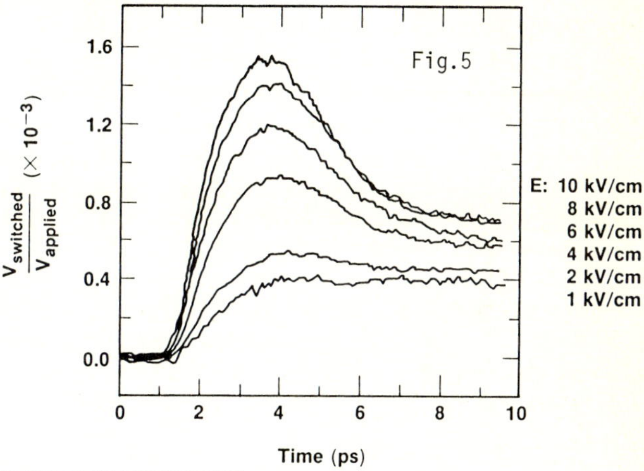

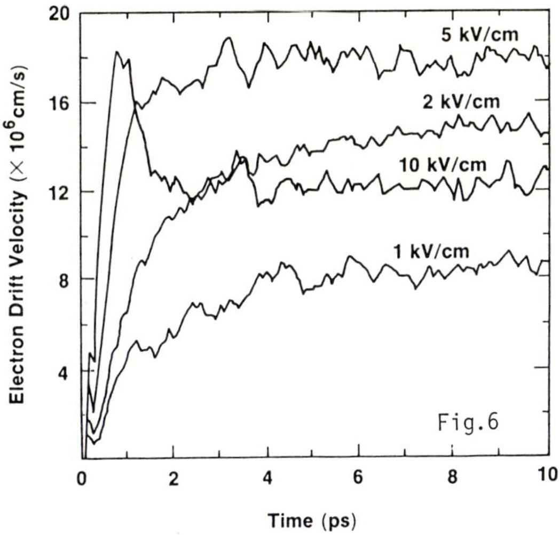

Figure 4 Time-resolved velocity overshoot measurement in the femtosecond time scale.

Figure 5 Room temperature photocurrent transients for λ_{ex}=620 nm.

Figure 6 Monte Carlo transient velocity calculation for λ_{ex}=620 nm, T=300 K.

620 nm (Fig. 5). As predicted by the simulation the velocity overshoot is observed at a field greater than 5 kV (Fig. 6). The experimental results exhibit a rise time limited by the gap charging time of 1.5 ps. Presently experiments are being pursued with excitation wavelengths corresponding to the band gap energy; under these conditions velocity overshoot is expected to be maximal.

4. Time-Resolved Resonant Tunneling

We used the technique of electro-optic sampling to time-resolve the transport of a resonant tunneling device. Several structures given to us by Texas Instruments and MIT Lincoln Laboratories were tested. We are presenting the results obtained with a resonant tunneling diode from MIT Lincoln Laboratories, where the barrier material is aluminum arsenide, barrier thickness is 15 Å, and gallium arsenide well thickness is 45 Å. The device was mounted with a whisker contact. A short step–like electrical pulse is produced by a photoconductive switch with a rise time of a few hundred femtoseconds. This pulse is applied to the double-barrier resonant tunneling diode. The transmitted pulse is monitored as a function of the dc bias voltage using the electro-optic sampling technique, according to Fig. 7. Figure 8 shows the experimental results.

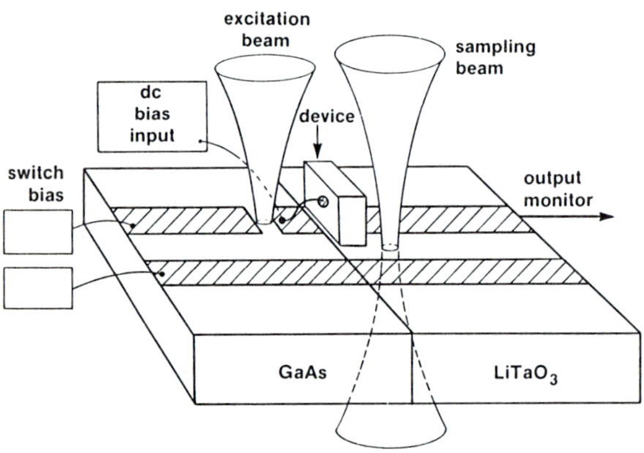

• 12-μm diameter wire with 1-μm contacts 2–8-μm diameter mesa on resonant tunneling diode chip

Figure 7 Test geometry for resonant tunneling diode.

Curve (a) displays the transmitted signal associated with the resonant tunneling current. When the device is biased along the monotonic part of the current-voltage characteristic, no significant changes in the diode response are observed. Curve (b) shows the diode response when the operating point is just below the negative differential resistance (NDR). A marked difference appears in the waveforms. First, the rise time is

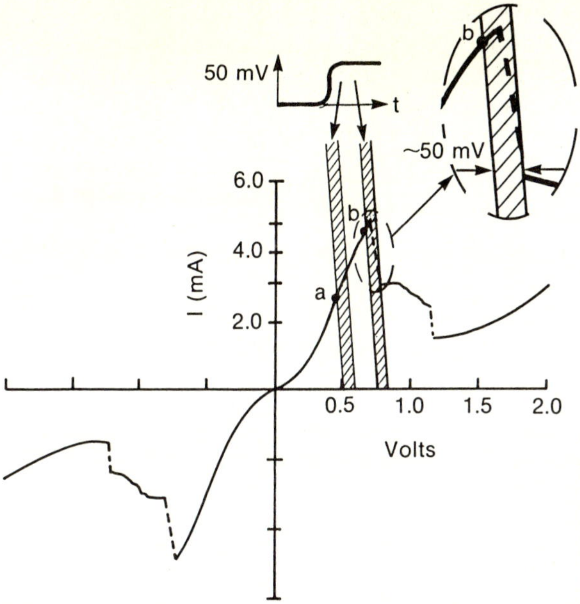

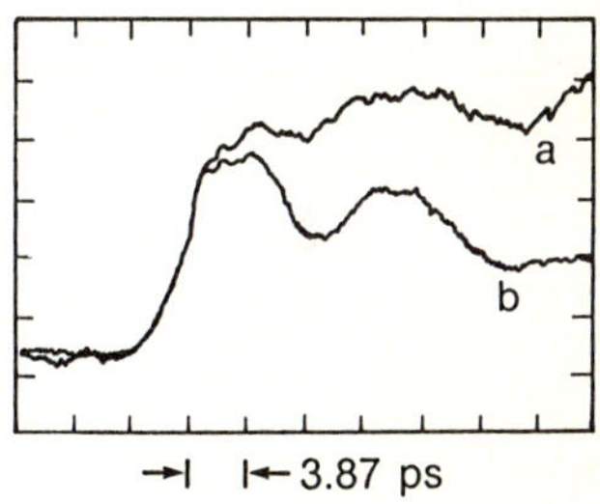

Figure 8 *Time-resolved resonant tunneling current.*

the same as in Curve (a), as expected, because in both cases a limit is imposed by the system response and the circuit parameters. After the initial rise, the voltage in Curve (b) drops quickly as the voltage in Curve (a) continues to rise. This is due to the sharp NDR effect producing a reduction of the resonant tunneling current in (b) as the voltage on Curve (a) rises. The sharp drop observed is characteristic of the change in slope sign and magnitude at operating point (b). It is worth noting that Curves (a) and (b) display the same initial rise (3 ps), because the step function of around 50 mV in amplitude starts from point b, which is slightly below the beginning of the NDR. Unfortunately, because the device was oscillating it could not be biased in the NDR part of the current-voltage characteristics.

5. Conclusion

In conclusion, we have shown that nonstationary transport can be directly time-resolved in the picosecond time scale showing the important role of ultrafast optics in the understanding of the physics of modern devices.

6. Acknowledgment

We wish to express our thanks to MIT/Lincoln Laboratories and Texas Instruments for providing structures necessary to perform our research.

This work was supported by the United States Air Force Office of Scientific Research under contract F49620-87-C-0016 to the Ultrafast Optical Electronics Center at the Laboratory for Laser Energetics of the University of Rochester as well as supported by the Laser Fusion Feasibility Project at the Laboratory for Laser Energetics which has the following sponsors: Empire State Electric Energy Research Corporation, General Electric Company, New York State Energy Research and Development Authority, Ontario Hydro, and the University of Rochester. Such support does not imply endorsement of the content by any of the above parties.

7. *References*

1. *J.A. Valdmanis, G.A. Mourou, and C.W. Gabel, IEEE J. Quantum Electron. QE-17, 664 (1983).*

2. T.C.L.G. Sollner, W.D. Goodhue, P.E. Tannenwald, C.D. Parker, and D.D. Peck, *Appl. Phys. Lett.* 43, 588 (1983).

3. S. Luryi, *Appl. Phys. Lett.* 47, 490 (1985).

4. J. Kundrotas and A. Dargys, *Phys. Status Solidi* B134, 267 (1986).

5. D.D. Coon and H.C. Liu, *Appl. Phys. Lett.* 49, 94 (1986).

6. M. Heiblum, *High Speed Electronics*, (ed.) B. Källbäck and H. Beneking (Springer-Verlag, 1986), p. 11.

7. T. Mori, H. Ohnishi, K. Imamura, S. Muto, and N. Yokoyama, *Appl. Phys. Lett.* 49, 1779 (1986).

8. C.V. Shank, R.L. Fork, B.I. Greene, F.K. Reinhart, and R.A. Logan, *Appl. Phys. Lett.* 38, 104 (1981).

9. C. Bru, D. Pascal, S. Laval, C. Arnodo, and R. Castagne, *The Physics of Submicron Structures*, (ed) H.L. Grubin, K. Hess, G.J. Iafrate, and D.K. Ferry (Plenum Publishing Corporation, 1984), p. 219.

10. R. Hammond, *Physics* 134B, 475 (1985).

11. D.H. Auston, *J. Quantum Electron.* QE-19, 639 (1983).

An Optically Strobed Sample and Hold Circuit *

C. McConaghy

Lawrence Livermore National Laboratory, 7000 East Avenue, Livermore, CA 94550, USA

1.0 Introduction

Radiation damaged photoconductors have been shown to be useful devices for sampling repetitve electrical signals with time resolution to less than a picosecond[1,2]. In previous work, the signal being sampled has generally had to repeat in order to be averaged by a lock-in amplifier for detection. In contrast, the circuit described in this paper is designed to make one sample on a non-repetitive signal with a 10 ps aperture and then hold this sample for a microsecond. The hold time only needs to be long enough for the signal to be A to D converted. This circuit was designed as a prototype with the intent of integrating a number of these circuits in the future for making multiple samples, offset in time, on a single electrical transient signal and thereby capture the transient. This paper will be concerned with the design, fabrication and testing of one of these optically strobed sample and hold circuits.

2.0 Circuit Description

Figure 1 shows a simplified equivalent circuit of the optically strobed sample and hold circuit. The circuit consists of a GaAs photoconductor that has been radiation damaged with protons to speed up the turn off time to 10 ps. Following the photoconductor is a series resistance R_{on} which represents the on resistance of the photoconductor. The photoconductor charges the holding capacitor C which is then read out by a MESFET source follower with an active load. An additional resistance R_l represents the leakage resistance which slowly discharges the holding capacitor. The voltage stored on the capacitor, V_{out}, is related to input voltage, V_{sig}, by

$$V_{out} = V_{sig}[1 - e^{-t_{on}/R_{on}C}].$$

For $R_{on}C$ less than t_{on}, V_{out} will charge up near V_{sig}. However, radiation damaged photoconductors have R_{on} values as high as 100 k Ω and even for integrated capacitance values of tenths of picofarads, the capacitor only gets charged to a small percentage of V_{sig} in its on time

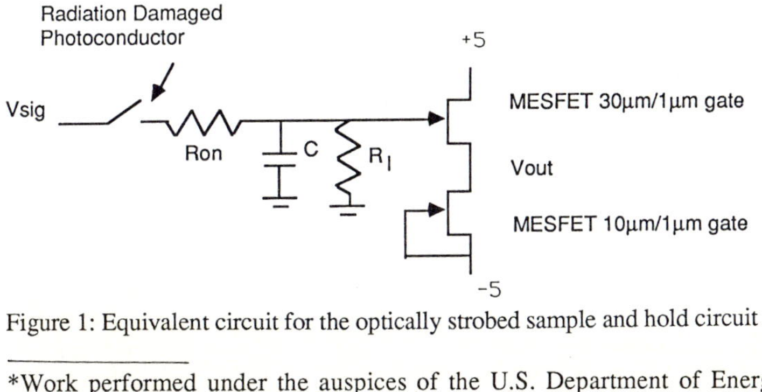

Figure 1: Equivalent circuit for the optically strobed sample and hold circuit

*Work performed under the auspices of the U.S. Department of Energy by the Lawrence Livermore National Laboratory under contract W-7405-Eng-48

of 10 ps. The percentage for the sample and hold circuit described here is 0.2%. This yields millivolts of output signal for input voltages in the one to ten volt range. These millivolt signals, although small, when amplified with low noise amplifiers produce good results.

The circuit shown in Fig.2 was fabricated in LEC GaAs. The circuit consists of a 50 Ω microstrip transmission line which propagates the fast electrical signal onto and back off of the chip. A 25 μm gap photoconductor, which borders the main microstrip transmission line, is connected to the input of a MESFET source follower. The source follower amplifier has an active load. The dimensions of the MESFETs in the amplifier are 1 μm gate lengths with a 30 μm width for the source follower and a 10 μm width for the active load. The input of the source follower has an additional 0.3 pF polyimide storage capacitor which can be bonded into use. It turned out that parasitic capacitance was enough to give a hold time of 1 μs, so this capacitor was not utilized. Additional photoconductors shown in Fig.2 were for electrical crosscorrelation testing of the basic photoconductor response times in optimizing the radiation damaging implant. The radiation damaging was accomplished with 200 keV protons at a dose of $10^{15}/cm.^2$ A thick (3 μm) layer of photoresist was utilized to protect the MESFETs during the proton implant. The radiation damaged photoconductors were independently measured by electrical crosscorrelation to have a 1/e time of 10 ps including the effect of the finite 5 ps wide laser pulse.

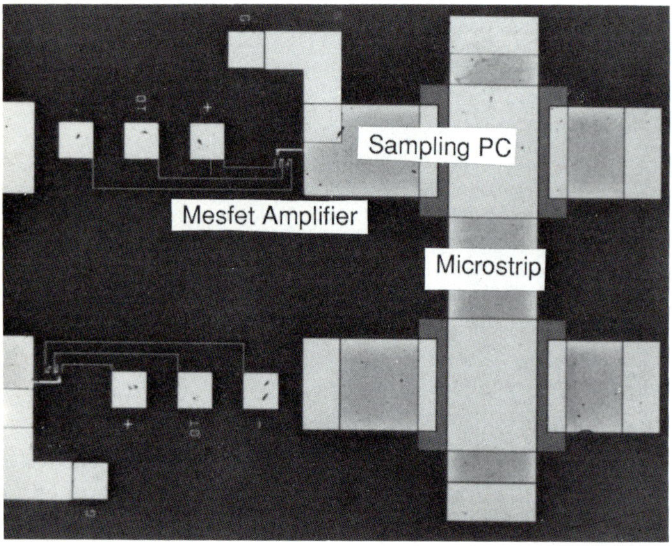

Figure 2: Optically strobed sample and hold circuit showing microstrip, photoconductor and MESFET amplifier

3.0 Experimental Evaluation

The experimental setup is shown in Fig.3. The laser utilized as the strobe generator is a mode locked Nd:Yag laser which has been pulse compressed to provide 5 ps 1.06 μm pulses. One of these pulses is switched out of the continuous train by a Pockels cell at a maximum rate of 10 per second. The pulse is then frequency doubled to 532 nm and focussed onto the 25 μm sample gap with a 5X microscope objective. The energy in the laser pulse at the gap is about 1 nJ. Prior to radiation damaging, the circuit had a sensitivity of 270 mV/volt of input signal. After radiation damaging the circuit had a sensitivity of 2.7 mV/volt of input signal. An electrical test step was generated with an avalanche transistor (RS3500). The top was removed from the avalanche transistor to allow opticaly triggering with the residual 1.06 μm light left after the frequency doubler. This provided a minimum delay, low jitter electrical signal for testing the sample and

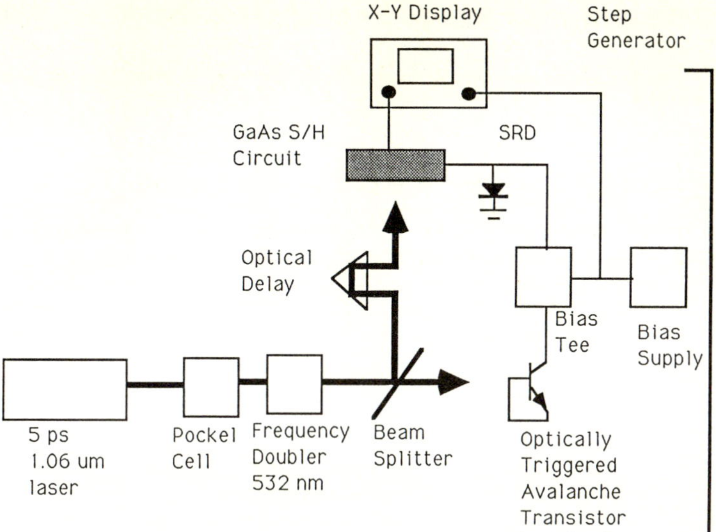

Figure 3: Test setup for evaluating the optically strobed S/H circuit

hold performance. Two step recovery diodes were further utilized to provide a 75 ps fall time in a method similar to Tielert.[3] The timing between the optical strobe pulse and the electrical test step was equalized by careful measurement of optical and electrical path lengths. Fine time adjustment was accomplished by varying the current on the first step recovery diode with additional current resulting in a longer time to snap off. This allowed the single sample and hold circuit to sample an individual point on the 75 ps fall time.

Figure 4a shows 19 samples taken 100 ms apart on the test step. Ten picoseconds of phase delay was added between each sample by biasing the step recovery diode so that it switched later in time for each sample. This phase delay was calibrated by adding a 50 ps optical delay to the strobe pulse which offset the entire waveform by 5 samples. The output of the GaAs circuit was amplified and resampled with a slower (10 ns aperture) sampler and held for 100 ms to produce the display. Each sample is produced with only one laser strobe pulse and is not averaged in any way. The additional bumps in the waveform of Fig.4a, after the step, are believed to be due to

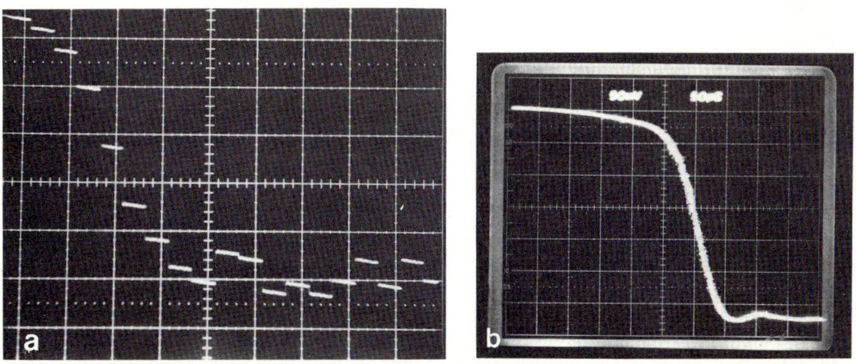

Figure 4: a)Successive samples by the optically strobed S/H circuit made 100ms apart in time on a step with 10 ps of phase delay added per sample. b)Tektronix S4 picture of the same step

the bond wires that interconnect the circuit to off-chip microstrip transmission lines. Figure 4b shows the step as measured by a Tektronix S4 sampling head. However, in order to produce the sampling scope picture, the step recovery diode circuit had to be run at 100 kHz as opposed to the 10 Hz rate used for Fig.4a.

4.0 Summary

An optically strobed sample and hold has been fabricated and utilized to acquire samples on a 75 ps electrical step. Each of these samples is taken only once and not averaged. The ultimate response time of the circuit is believed to be as fast as 10 ps. Present testing has been limited by the available electrical test signal. Future plans call for incorporating multiple numbers of this circuit onto a single chip to capture an electrical transient.

5.0 References

1. P.M. Downey:"Subpicosecond Response Times From Ion Bombarded InP," Picosecond Elect. and Optoelect. Conference, Incline Village, 1985

2.R.B. Hammond, N.G. Paulter, R.S. Wagner: Appl. Phys.Lett.45(3),1 Aug 1984

3. Tielert R. : "Subnanosecond pulse generation employing a two stage pulse step generator," Electronic Letters, 5 Feb. 1976, Vol. 12, No. 3, pp. 84-85

Laser Pulsed E-Beam System for High Speed I.C. Testing

P. May, J.-M. Halbout, and G. Chiu

IBM T.J. Watson Research Center, P.O. Box 218, Yorktown Heights, NY 10598, USA

1. Introduction

There has been much discussion recently on different techniques for non-contact picosecond probing of internal nodes of integrated circuits fabricated on any substrate material/1-3/. All these techniques, however, use an optical probe and as such are limited in spatial resolution to about 1μm which is already large compared to feature sizes in some modern devices and wiring geometries. Electron probe techniques which do not suffer from this disadvantage have been used for many years to test I.C. circuits but have in general been limited by their poor temporal resolution. In this paper we describe a novel type of instrument that combines the advantages of conventional SEM voltage contrast techniques /4/ (including a spatial resolution of order 100A and beam steerability) with an order of magnitude improvement in temporal resolution.

2. Photoelectron Electron Microscope

Figure 1 shows a schematic of the system used. The thermionic gun or field emission gun of standard SEM systems is replaced by a thin film roughened gold photocathode /3/ deposited on

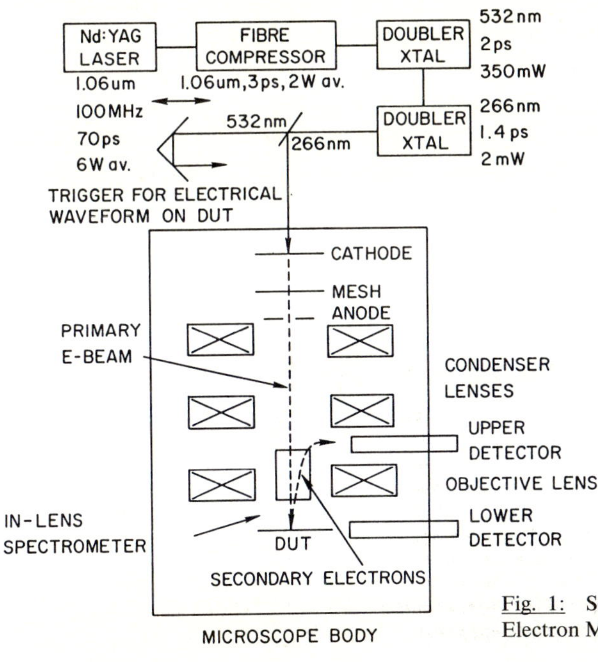

Fig. 1: Schematic of the Photoelectron Electron Microscope

a quartz substrate. The quadrupled output from a mode-locked Nd:YAG laser/fibre compressor arrangement is focussed onto the photocathode (operating in transmission mode) which is mounted at the top of the SEM column. Because the photocathode is operated in transmission mode and the escape depth of the photoelectrons in gold is only a few 10's of angstroms then the film should be as thin as possible consistent with there being electrical continuity. Typically, 1ps pulses at 266nm, 100MHz and 2mW average power impinge on a 200A thick gold film which has an accelerating mesh in close proximity to reduce temporal spread produced by Coulomb interactions. The photoelectron yield is $0.5 10^{-3}$, which gives about 10,000 electrons per pulse or an average photocurrent of 100nA. At these emission levels there seems to be no dependence of photocurrent on extraction field and the measured current is linear with average uv input power. These facts suggest that we are not in a space charge limiting region /5/.

The average brightness of the photocathode is quite sufficient to see real time 2-D images of samples under test, which is necessary so that a stationary spot beam can be easily positioned at the waveform measuring point for voltage measurements (or alternatively for 2-D time resolved voltage mapping). In addition the peak brightness is much larger than in beam blanked SEM systems which results in shorter measurement times. An especially designed in-lens energy spectrometer with an extraction and a retarding grid fits inside the objective lens and analyses the emitted secondary electrons (whose energy depends on the voltage at the emission point) and also allows close working distances for good spatial resolution. Using phase sensitive detection, voltage resolutions of better than $20mV/(Hz)^{1/2}$ have been achieved compared to $10mV/(Hz)^{1/2}$ when the thermionic source is used.

3. Time Resolved Measurements

For time resolved measurements, part of the light beam incident on the photocathode is used as a trigger for the electrical signal whose waveform is to be measured, and the waveform is sampled by variably delaying the cathode incident beam with respect to the trigger. When looking at internal nodes on a clock-driven integrated circuit then the light signal is used to drive the clock in synchronism (less than 1ps jitter is achievable) with the electron pulse.

Figure 2 shows a sampled waveform derived from a step-recovery diode. A pulse generator, triggered by a photodiode signal from part of the light beam, provides an approximately sinusoidal signal at 100MHz to the step recovery diode. The resultant pulse measured with a sampling scope

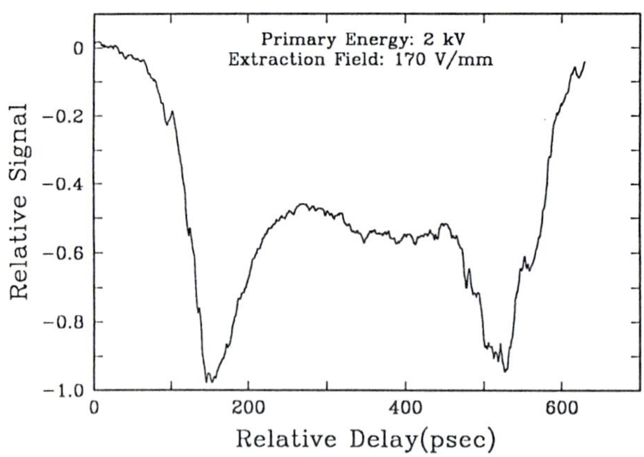

Fig. 2: Measured waveform from a step-recovery diode pulse launched on transmission line. The second peak comes from a reflection at a right angle bend in the transmission line.

has a rise time of 55-60ps and a 5V peak amplitude. The electrical pulse is launched onto a 50 Ohm microstrip transmission line within the chamber where it is probed by the electron beam (primary beam energy of 2keV). The rise time of the resultant waveform, limited by the device, is under 50ps, which implies an instrument resolution better than 20ps. Though the line geometry is large (1mm) a small extraction field of 170V/mm is sufficient to achieve this time resolution; for smaller extraction fields the measured risetime increases due to the transient local field effect on the emitted secondary electrons. Monte Carlo simulations for safe extraction fields of 2kV/cm and for coplanar lines with micron geometries support a predicted instrument resolution of order 5ps.

4. Conclusions

We have developed an electron probe for high speed testing of I.C's which is non-invasive and has the necessary spatial, temporal and voltage resolutions to test the high speed nanostructure devices and wiring geometries currently under development. Its beam steerability should allow for rapid testing of different sites on a particular chip as well as giving the capability for time resolved pictures. The currently demonstrated characteristics of the instrument include a temporal resolution of less than 20ps (presently device limited -actual instrument resolution for reasonable extraction fields is probably closer to 5ps), a voltage resolution of better than $20mV/(Hz)^{1/2}$ and a spatial resolution of 0.1μm. Reduction in laser noise should improve the voltage resolution and an optimisation of the cathode/extraction mesh geometry and light focussing should bring us closer to the typical SEM spatial resolutions of order 100A.

References

1. J. A. Valdmanis, S. S. Pei be published in Picosecond Electronics and Optoelectronics, (Springer-Verlag,1987)
2. J. Bokor, A. M. Johnson, R. H. Storz and W. M. Simpson, Appl. Phys. Lett. 49, 226, (1986).
3. R. B. Marcus, A. M. Weiner, J. H. Abeles and P. S. D. Lin, Appl. Phys. Lett. 49, 357, (1986).
4. see, for instance, E. Menzel and E. Kubalek, Scanning 5, 103, (1983).
5. Gail A. Massey, IEEE J. Quantum Electron. QE-20, 103, (1984).

Photoemissive Testing of High-Speed Electrical Waveforms

A.M. Weiner, P.S.D. Lin, and R.B. Marcus

Bell Communications Research, 331 Newman Springs Road,
Red Bank, NJ 07701, USA

Photoemissive sampling was recently introduced as a new, contactless method for probing high-speed electrical waveforms on circuits and devices on any semiconductor [1-3]. Previously the best temporal resolution achieved by this technique was limited to 40 ps. In this paper we describe photoemissive sampling experiments which demonstrate a temporal resolution of 5 ps.

Photoemissive sampling, illustrated schematically in Fig. 1, utilizes ultrashort optical pulses to stimulate the emission of photoelectrons from metal interconnection lines on the surface of the device under test. A measurement of the potential at the emitting surface is derived from an energy analysis of the photoelectrons. By varying the delay of the optical probe beam, the complete electrical waveform is reconstructed.

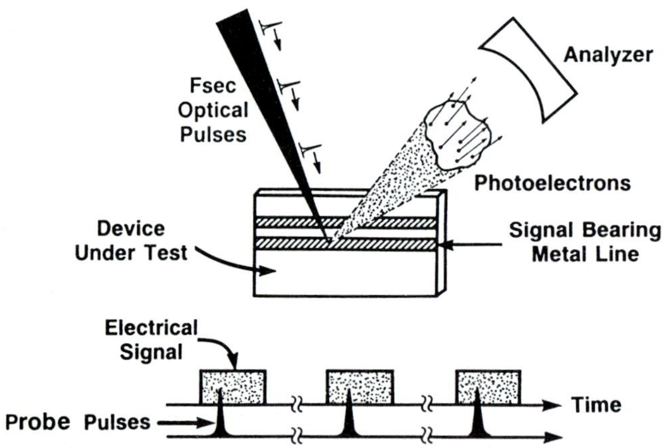

Fig. 1. A schematic illustration of photoemissive sampling

Our experiments were performed with 2-eV, 80-fs pulses from a colliding-pulse-modelocked (CPM) ring dye laser, which generates photoelectrons via a three-photon photoelectric effect. The test sample, a 5-μm, gold coplanar stripline on polished GaAs, is placed within a vacuum chamber held at approximately 10^{-7} Torr. Two separate optical beams are focused onto the test sample using a single 40x microscope objective lens. One beam is used for photoconductive excitation of a fast electrical signal; the second beam is the probe. A wire grid placed 320-μm above the sample provides an extraction field for the photoelectrons, which are subsequently detected using a microchannel plate detector. The energy analysis is performed by a retarding field analyzer mounted directly in front of the MCP. The present detection scheme retains linearity for signal voltage swings of several volts.

The sample geometry is shown schematically in Fig. 2. The center conductor is 6.25-μm wide, separated by 5-μm from ground planes on either side, for a design impedance of 50 Ω. A 100-mV electrical step function is generated photoconductively using the "sliding contact" geometry [4]. The probe beam is focused onto the center line at a spot 40-μm distant from the excitation position. The photoelectron current due to the probe beam exceeds the background photoelectron current generated unintentionally by the excitation beam by more than three orders of magnitude.

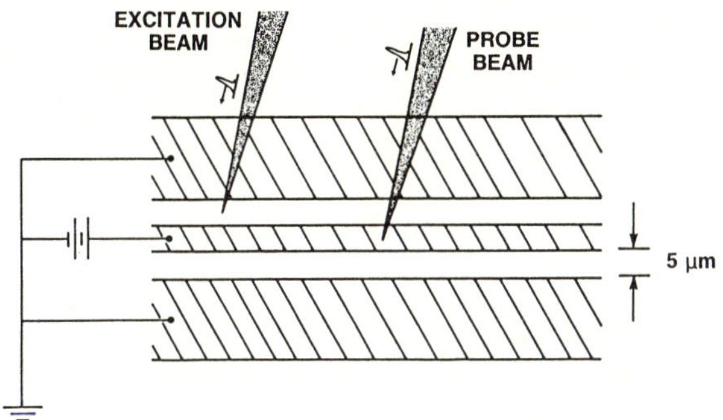

Fig. 2. Excitation and probe geometry for the 50−Ω, gold coplanar stripline on polished GaAs. The excitation and probe locations are separated by 40−μm.

The temporal response of photoemissive sampling is determined by the transit time of photoelectrons out of the region of time-varying, local electric fields due to the signal on the sample [1,2,5]. The transit time T is given by

$$T = \frac{\sqrt{2m}}{eE}\left[\sqrt{eES + U_0} - \sqrt{U_0}\right] , \tag{1}$$

where E, S and U_0 are respectively the extraction field, the spatial

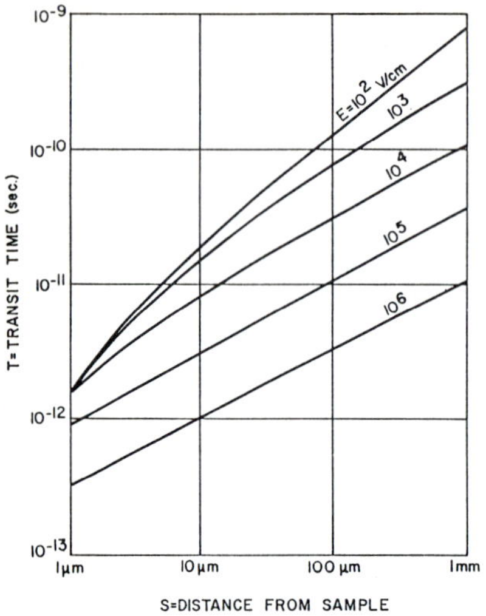

Fig. 3. Transit time as a function of distance from sample, calculated from Eq. (1) with $U_0 = 1$ eV

extent of the time-varying electric field and the initial kinetic energy in the direction of the extraction field; m and e are the electron mass and charge. Figure 3 shows a plot of the transit time for 1-eV electrons, as a function of E and S. In a real circuit, S depends on the widths and spacings of the metal lines being probed and may be much smaller than the distance to the extraction electrode [5,6]. For S = 5-μm, with E = 30 kV/cm and $U_0 = 1$ eV, the transit time is 3.4 ps.

The effect of the sample geometry on the transit distance S is illustrated by Fig. 4. Equipotential lines have been calculated for a hypothetical circuit which consists of two metal lines, each 1-μm wide and separated by 1-μm, placed in an extraction field of 100 kV/cm. The solid lines correspond to electrodes at 0 V and 1 V, as shown; the dashed lines correspond to both electrodes at 0 V. The effect of changing electrode voltages on the position of the equipotentials diminishes as the distance from the circuit increases. At a distance of 1-μm, the influence of a changing sample potential is reduced by more than 50%.

A time-resolved measurement of the electrical step generated on the 5-μm coplanar transmission line is shown in Fig. 5, using an extraction field of 85 kV/cm. The 10-90% risetime is 5.3-ps; the steepest slope present on the rising edge would correspond to an even faster risetime of 4.2 ps. With weaker extraction fields, the risetimes were slower; for example, at 30 kV/cm, the risetime was 7.5 ps. These risetimes are in

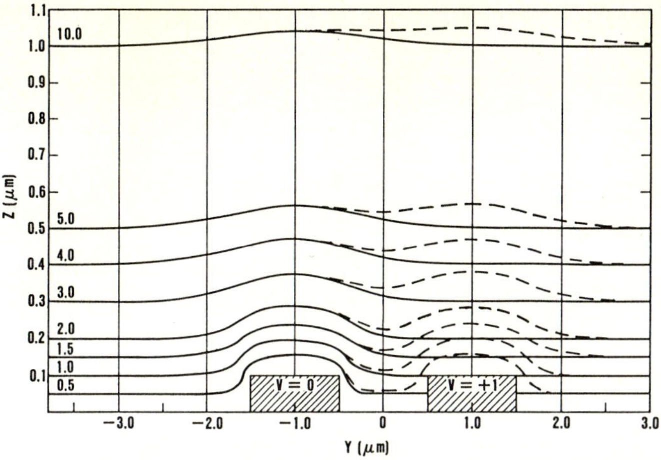

Fig. 4. Equipotential lines for 1− μm wide coplanar lines separated by 1− μm. The extraction field is 100 kV/cm. Solid curves correspond to electrodes at 0 V and 1 V, respectively; dashed curves correspond to both electrodes at 0 V.

reasonable agreement with the calculated transit times T, using S in the range 10-15 μm.

The signal to noise ratio of 30 evident in Fig. 5 corresponds to a voltage sensitivity of approximately 10 mV/$\sqrt{\text{Hz}}$. A shot-noise limited sensitivity in the range of several hundred μV/$\sqrt{\text{Hz}}$ should be possible by optimizing the analyzer and by operating with higher photoelectron current.

Our experiments demonstrate that photoemissive sampling is capable of probing 70 GHz signals with a temporal resolution of 5.3 ps. A temporal response in the 2-3 ps range seems achievable for circuits with 1-μm design rules.

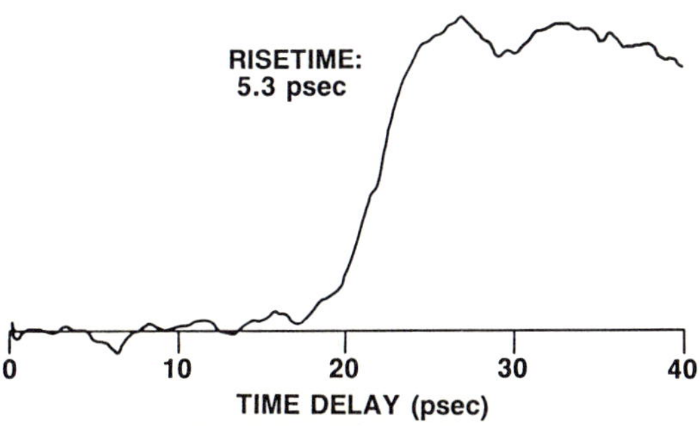

Fig. 5. Measured voltage waveform from the sample described in Fig. 2, showing 5.3 ps risetime

We are grateful to J. H. Abeles for design and fabrication of the 50-Ω coplanar transmission line and to J. Orloff for electron trajectory calculations.

REFERENCES

[1] R. B. Marcus, A. M. Weiner, J. H. Abeles and P. S. D. Lin, Appl. Phys. Lett. 49, 357 (1986).

[2] J. Bokor, A. M. Johnson, R. H. Storz and W. M. Simpson, Appl. Phys. Lett. 49, 226 (1986).

[3] R. Clauberg, H. K. Seitz, A. Blacha, J. A. Kash and H. Beha, Proceedings of the High-Speed Electronics Conference, Stockholm, Sweden (1986).

[4] M. B. Ketchen, D. Grischkowsky, T. C. Chen, C. -C. Chi, I. N. Duling III, N.J. Halas, J. -M. Halbout and G. P. Li, Appl. Phys. Lett. 48, 751 (1986).

[5] H. Fujioka, K. Nakamae and K. Ura, J. Phys. D. 18, 1019 (1985).

[6] A. M. Weiner, R. B. Marcus, P. S. D. Lin and J. H. Abeles, in Ultrafast Phenomena V, G. R. Fleming and A. E. Siegman, eds. (Springer-Verlag, Berlin, 1986), p. 127.

High-Speed Optical Logic Using GaAs

H.M. Gibbs[1], *N. Peyghambarian*[1], *Y.H. Lee*[1], *J.L. Jewell*[2], *A. Migus*[3], *A. Antonetti*[3], *D. Hulin*[4], *and A. Mysyrowicz*[4]

[1]Optical Sciences Center, University of Arizona, Tucson, AZ 85721, USA
[2]AT&T Bell Laboratories, Crawfords Corner Road, Holmdel, NJ 07733, USA
[3]Laboratoire d'Optique Appliqué, Ecole Polytechnique-ENSTA, F-91120 Palaiseau, France
[4]Groupe de Physique des Solides, Ecole Normale Supérieure, 2 Place Jussieu, Tour 13, F-75005 Paris, France

The status of bulk-GaAs etalons and their application to high-speed optical logic are summarized. A bulk-GaAs nonlinear etalon can make a decision in ≅1 ps and recover to make another in <100 ps, making possible high-speed optical logic. Several characteristics are of particular importance: time to make a logic decision ("switch-on time"), time required for the device to recover and be ready to make the next decision (recovery time), energy required per logic decision (switching energy), ability to switch the next device (cascadability) or perhaps several devices (fanout), etc.

Recently, research has centered on the use of a nonlinear etalon as a logic gate with the inputs at a wavelength having good absorption leading to index changes at a second probe wavelength.[1] With an appropriate initial detuning between the Fabry-Perot transmission peak and the probe wavelength, each of the logic operations can be performed. Several logic gates have been demonstrated using a high-finesse etalon containing GaAs-AlGaAs multiple quantum wells at room temperature.[2] Input pulses with energies less than 3 pJ focused to 5 to 10 μm on the device produced contrasts greater than 5:1 in the NOR gate. A recovery time of 5 ns permitted 82-MHz continuous operation. Thermal stability was demonstrated in the NOR gate even though the device was poorly heat sunk. The NOR gate has been operated using diode laser sources.[3]

Because the physical mechanisms leading to refractive index changes are extremely rapid, one expects a very short switch-on time. A Fabry-Perot peak was shifted in 1 to 3 ps; the etalon consisted of alternating layers of 152-Å GaAs and 104-Å AlGaAs with a total GaAs thickness of ≅1.5μm.[4] Less than 100 fJ/μm^2 energy per unit area incident on the device was used for the logic operation.

The recovery times of the first GaAs gates were typically 5 to 10 ns, determined by the lifetime of the carriers in the light beam. The recovery time was greatly reduced by enhancing surface recombination by removing one or both of the AlGaAs windows used to stop chemical etching and reduce surface recombination. A 1.5-μm-thick bulk-GaAs etalon without the top AlGaAs window and etched to form a 2 D array of 9μm x 9μm pixels with 20 μm center-to-center spacing was shown to have a recovery time of less than 200 ps with resolution limited by the input pulse duration.[5] Then 5- to 10-ps input pulses were used to study etalons with no AlGaAs windows and with no etching of pixels.[6] A thickness of 0.5 μm was found to be optimum for current etalon designs: complete recovery in 150 ps with 16-pJ input-pulse energy. A 0.3-μm-thick sample showed even faster recovery (down to about 40 ps) with 20-pJ input-pulse energy. A 1350-Å-thick sample of inferior quality showed recovery times as fast as 30 ps but required much higher energies. Bombardment by 2-MeV protons was less effective. The most heavily bombarded sample (2x10^{14} protons/cm^2) showed 270-ps complete-recovery time, but the contrast was severely degraded (2:1) even with more input-pulse energy (40-pJ).

These recovery times are probably quite adequate for parallel computations where thermal considerations will limit the product of pixel density and repetition rate. Answers emerging from one etalon can proceed to the next etalon, etc., so that many etalons can be used to avoid using any one too often. Where very high-speed response is needed, for example, in a waveguide directional coupler encrypting or multiplexing data in a pipeline mode, the newly discovered AC Stark shift may be the answer.[7,8] If the pump pulse is tuned far enough below the exciton frequency, in the transparency region of the material, the carrier lifetime is no longer a limitation since carriers are not excited. A complete shift/recovery cycle of the Fabry-Perot transmission peak has been accomplished in 1.5 ps.[9]

The gates described herein utilize different wavelengths for inputs and probe (output), making cascading difficult. Two-wavelength operation utilizing an isolated resonance is one possibility. We are reassessing one-wavelength operation (optical transistor or bistability operation) to see what price one must pay for ease of cascadability.

We gratefully acknowledge support for the Arizona research by the Air Force Office of Scientific Research, the Army Research Office, the National Science Foundation (Grant Nos. 8317410 and 8610170), and the Optical Circuitry Cooperative.

1. J. L. Jewell, M. C. Rushford, and H. M. Gibbs, Appl. Phys. Lett. *44*, 172 (1984).

2. J. L. Jewell, Y. H. Lee, M. Warren, H. M. Gibbs, N. Peyghambarian, A. C. Gossard, and W. Wiegmann, Appl. Phys. Lett. *46*, 918 (1985).

3. M. Ojima, A. Chavez-Pirson, Y. H. Lee, J. F. Morhange, H. M. Gibbs, N. Peyghambarian, F.-Y. Juang, P.K. Bhattacharya, and D. A. Weinberger, Appl. Opt. *25*, 2311 (1986).

4. A. Migus, A. Antonetti, D. Hulin, and A. Mysyrowicz, H. M. Gibbs, N. Peyghambarian, and J. L. Jewell, Appl. Phys. Lett. *46*, 70 (1985).

5. Y. H. Lee, M. Warren, G. R. Olbright, H. M. Gibbs, N. Peyghambarian, T. Venkatesan, J.S. Smith, and A. Yariv, Appl. Phys. Lett *48*, 754 (1986).

6. Y. H. Lee, H. M. Gibbs, J. L. Jewell, J. F. Duffy, T. Venkatesan, A. C. Gossard, W. Wiegmann, and J. H. English, "Speed and Effectiveness of Windowless GaAs Etalons as Optical Logic Gates," talk FK2, p. 364, Technical Digest of Conference on Lasers and Electro-Optics, June 9-13, 1986, and Appl. Phys. Lett. *49*, 486 (1986).

7. A. Mysyrowicz, D. Hulin, A. Antonetti, A. Migus, W. T. Masselink, and H. Morkoc, Phys. Rev. Lett. *56*, 2748 (1986).

8. A. Von Lehmen, J. E. Zucker, J. P. Heritage, and D. S. Chemla, postdeadline talk PD18, International Quantum Electronics Conference, June 9-13, 1986.

9. A. Migus, D. Hulin, A. Mysyrowicz, A. Antonetti, W. T. Masselink, H. Morkoc, H. M. Gibbs, and N. Peyghambarian, postdeadline talk ThU9 Conference on Lasers and Electro-Optics, June 9-13, 1986.

Comparison of Sampling Oscilloscopes with ~35 ps Transition Durations

J.R. Andrews

Picoseconds Pulse Labs, Inc., P.O. Box 44, Boulder, CO 80306, USA

The purpose of this paper is to compare transient responses of all of the various ultrafast sampling oscilloscopes. This study was limited to the fastest samplers with transition durations (10-90% risetimes) of <35 ps and corresponding bandwidths of >10GHz. It is very important for users of these instruments to realize that none of them will give the correct answer when measuring picosecond domain signals. The same pulse signal measured with different samplers will result in different pulse waveforms as displayed on the CRT's.

Six samplers from four manufacturers were compared. The manufacturers were: Hewlett-Packard, Colorado Springs, Colorado; Tektronix, Beaverton, Oregon; Picosecond Pulse Labs, Boulder, Colorado and Iwatsu, Tokyo, Japan. The HP samplers were the Models 1430A and 1430C used with both the 140 series (1411A & 1424A plug-ins) and 180 series (1811A plug-in) oscilloscopes. The Tek samplers were the S4 and S6 used with the 7000 series oscilloscopes and the 7S11/7T11 or 7S12 plug-ins. The S4 and S6 could also be used with the older 560 series scopes. The PSPL sampler was a Model S-1430D. PSPL does not manufacture a complete sampling oscilloscope. The S-1430D sampler is designed to be used interchangeably with either the HP 140 or 180 series or Tek 560 or 7000 series sampling oscilloscopes. The Iwatsu sampler was the SH-4B used with their new SAS-8130 digital sampling oscilloscope. It can also be used with their SAS-601B oscilloscope.

Two different pulse generators were used. The first was the fastest tunnel diode we could find. It was an HP1106A powered by a PSPL TD-1105 Bias Supply. It was specified by HP to have a nominal 20 ps transition duration. We estimate that the particular unit used had a 17 ps transition duration. For the nanosecond domain comparisons a PSPL 6100 Reference Flat Pulse Generator (RFPG) was used. It produces a very flat, clean step from 500 mV to 0 V with a 400 ps transition duration. The 6100 RFPG is based upon an NBS RFPG design.

The collection of photographs in Fig. 1, shows the picosecond domain transient responses of all of these samplers when driven by the same HP1106A tunnel diode. The vertical scale is 50 mv/div. The horizontal scale is 20 ps/div. for the top two rows and 200 ps/div. for the bottom two rows. Fig. 2 shows the nanosecond domain step responses of the samplers when driven by the PSPL 6100 RFPG. The scale factors are 100 mv/div. and 5 ns/div. From the picosecond domain photos of Fig. 1, transition duration estimates were made. The results are listed in Table I. The simple root sum of squares equation was used with 17 ps for the TD generator transition duration.

In the complete paper presented at the conference, detailed comments were made about each sampler. These comments and a list of references are published in an application note /1/ available from the author.

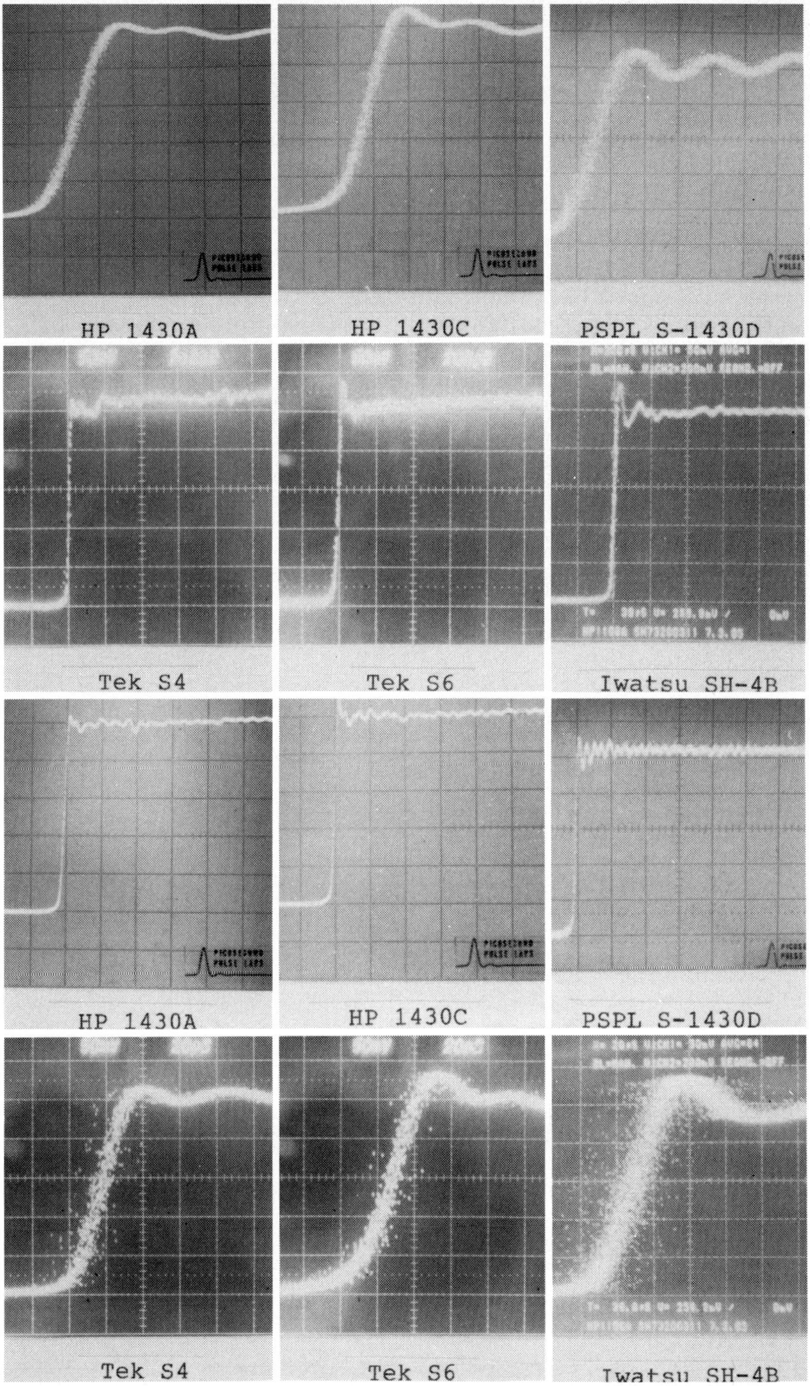

Fig. 1 Picosecond domain transient responses of 6 different sampling heads at 20 ps/div. and 200 ps/div.

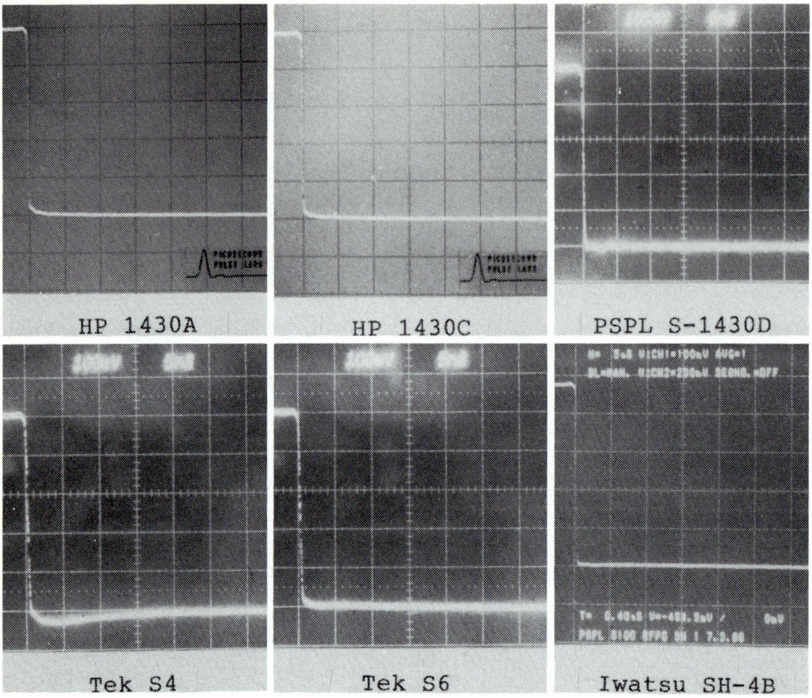

Fig. 2 Nanosecond domain step responses of 6 different sampling heads at 5 ns/div.

Table I Transition Durations of Various Samplers

Model	Observed	Calculated	Mfgr. Spec.
HP1430A	30 ps	25 ps	28 ps
HP1430C	28 ps	22 ps	20 ps
Tek S4	26 ps	20 ps	<25 ps
Tek S6	36 ps	31 ps	<30 ps
PSPL S-1430D	33 ps	28 ps	28 ps
Iwatsu SH-4B	30 ps	25 ps	30 ps max.

1. J.R. Andrews: Comparison of Broadband Sampling Oscilloscopes, AN-2, (Picosecond Pulse Labs, Inc., Boulder, Colorado, Aug. 1986).

Picosecond Optoelectronic Study of a Thin Film Transmission Line Structure

G. Arjavalingam, J.-M. Halbout, G.V. Kopcsay, and M.B. Ketchen

IBM T.J. Watson Research Center, P.O. Box 218,
Yorktown Heights, NY 10598, USA

1. Introduction

In several laboratories high-speed logic circuits which are capable of generating electrical pulses with risetimes in the order of ten picosecond have been demonstrated [1] . Consequently it is now important to be able to transmit these pulses without significant distortion and to be able to characterize the transmission line structures used for this purpose with picosecond time resolution. In addition to the delay and distortion of the pulses as they propagate between generating and receiving points, knowledge of the noise coupled onto adjacent lines is of critical importance to the workings of computers and communication systems. Here we present measurements of high-speed pulse propagation and coupling on an experimental thin film transmission line structure. Photoconductive switches are used for generating and sampling the ultrashort electrical pulses used in this study [2].

2. Experiment

A schematic of the experimental layout is shown in figure 1. The thin film structure shown consists of 8 μm wide and 5.35 μm thick copper microstrip transmission lines separated from a uniform ground plane by a 6.5 μm thick layer of polyimide. Center to center spacing of the lines is 25 μm . Photoconductive switches fabricated on polysilicon and which have been described elsewhere [3] were connected electrically to both ends of each of two adjacent lines using wirebonds. The switches are known to produce 2 ps wide electrical pulses when triggered by 1.5

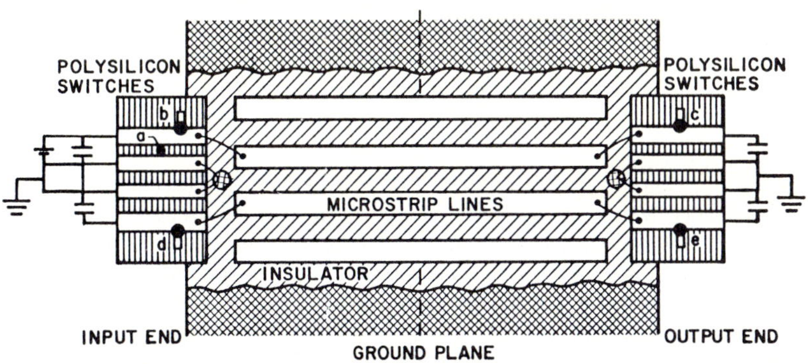

Figure 1. Experimental Layout:. The polysilicon switches are connected to the the transmission line structure usings wirebonds. Terminating capacitors are attached to the ends of the polysilicon switches as shown. The total length of the microstrip lines is 37.8 mm. The figure is not to scale.

ps wide optical pulses. It has been shown previously that there is no significant degradation of the ultrashort electrical pulses when they propagate through wirebonds of 25 μm diameter and 250 μm length [4] , and this was further confirmed for the 1 mm long bonds used here. Miniature capacitors were attached across the far ends of the lines of all four switching devices to provide known impedances and facilitate modelling.

The laser source used to drive the photoconductive switches is a Nd:YAG laser, acousto-optically mode-locked at 100 MHz repetition rate. Its 73 ps pulses are compressed by passage through 40 meters of single mode optical fiber followed by a double-pass optical delay line composed of a pair of grazing incidence gratings. The compressed pulses are frequency doubled in a 5.0 mm thick KTP crystal to yield pulses of autocorrelation width 2 ps and average power 350 mW. The pulses launched on the structure that is investigated are produced at point a (figure 1) in a sliding contact configuration [3] and characterized at point b with a side gap detector. A representative pulse is shown in figure 2a. The pulse transmitted by the 37.8 mm long microstrip line is coupled onto the second semiconductor structure and sampled at point c. The near and far end coupled noise are measured at points d and e respectively.

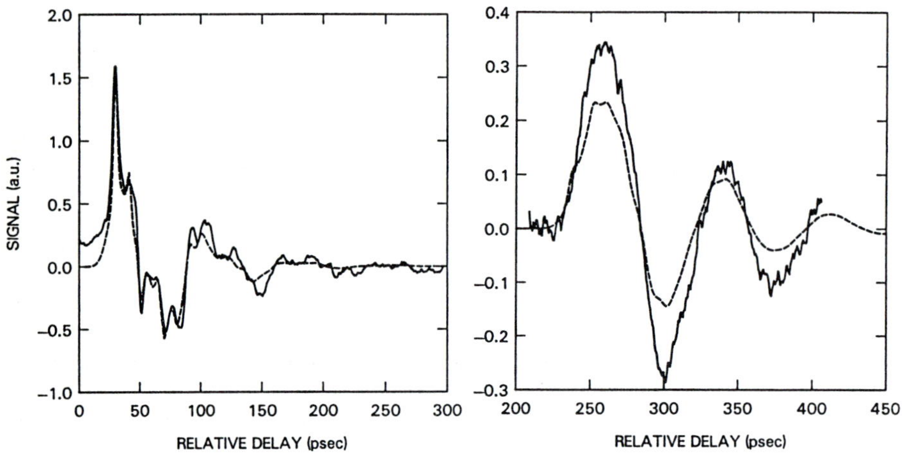

Figure 2. a. Signal measured at point b in fig. 1 on the input side of the transmission line structure. b. Signal measured at point c in fig. 1 after propagation through the microstrip line. The output of the circuit simulation model is shown as a dashed line.

3. Measurements and Modelling

For comparison with the experimental data, pulse propagation and coupling were modelled using the ASTAP transient circuit analysis program [5]. Quasi-static transmission line parameters of the coupled microstrip structure were obtained numerically using two dimensional capacitance analysis [6]. Frequency dependent skin effect loss was determined using the incremental inductance rule [7] at the high frequency limit, and by an network theory model [8] in the dc to mid-frequency range. A general lossy, frequency dependent, coupled transmission line model was used for the transient analysis [9]. It should be noted that, although propagation velocity is frequency dependent for mixed dielectric media [10], due to the small cross-sectional size of the structure used for our experiments the quasi-static model used here is accurate over the frequency range of interest. Models of the chips, capacitors and wire bonds were also included in the circuit analysis. A current source excitation consisting of the sum of two gaussians whose parameters were empirically matched to the shape of the optoelectronically generated pulse was used as input at the generating point a in figure 1, and the resulting time dependent voltages at the other nodes were calculated from the model. Figure 2a shows the signal measured at point b in figure 1 together with the voltage predicted by the model. The transient features on the measured waveform which are due to reflections from discontinuities in the experimental structure agree quite well with the model.

Figure 2b shows the signal measured at point c in figure 1 after propagation through the transmission line structure. A 4.6 ps wide input pulse results in an output about 40 ps wide, and the time delay between the peak of the input and output waveforms corresponds to a line delay of 5.07 ps/mm. The model quite accurately predicts the pulse dispersion and delay characteristics of the structure, as well as the shape of the measured waveform. However, there is an amplitude discrepancy of 30% between the measurement and the model calculations which may be due both to uncertainties in the detector sensitivities, and to inadequacies in the model.

The near and far-end coupled noise were measured at points d and e in figure 1. While there was fair agreement between the simulation and the measured far-end coupling, the measured near-end noise waveform had larger amplitude and more high-frequency structure than predicted. We have experimentally ruled out the possibility of local coupling on the chip and between the wire bonds. However, the additional coupling may be due to the finite impedance of the ground return connections in the experimental structure, and the end effects at the transition between the coplanar lines on the chip and the microstrip lines being measured. A related phenomenon called ΔI noise occurs in real computer systems.

4. Conclusions

In conclusion we have demonstrated the use of picosecond optoelectronics to characterize the properties of an experimental thin-film transmission line structure. The good agreement obtained between measurement and simulation of the propagation delay and dispersion illustrates the potential application of picosecond optoelectronic sampling for studying high-speed circuit interconnection structures.

5. Acknowledgements

The authors wish to acknowledge the IBM Yorktown Silicon Facility for making the photoconductive switches and V.Ranieri for expert technical assistance.

References

1. See for example R.A. Murphy, in Picosecond Electronics and Optoelectronics, ed. G.A. Mourou, D.M. Bloom, and C.H. Lee, (Springer Verlag) Berlin, 1985, and references therein.

2. D.H. Auston, in Picosecond Optoelectronic Devices,ed. by C.H. Lee, (Academic Press), London, 1984.

3. M.B. Ketchen, D. Grischkowsky, T.C. Chen, C-C. Chi, I.N. Duling III, J-M. Halbout, J.A. Kash and G.P. Li, Appl. Phys. Lett. 48, 751, (1986); see also D.R. Dykar, T.Y. Hsiang, and

G.A. Mourou, in Picosecond Electronics and Optoelectronics, ed. by G.A. Mourou, D.M. Bloom, and C.H. Lee (Springer Verlag), Berlin, 1985.

4. P.G. May, G.P. Li, J-M. Halbout, M.B. Ketchen, C.C. Chi, M. Scheuermann, I.N. Duling III, D. Grischkowsky and M. Smyth, in Proceedings of the Topical Meeting on Ultrafast Phenomena V, Optical Society of America, June 1896, Snowmass, Colorado.
5. W.T. Weeks, A.J. Jimenez, G.W. Mahoney, D. Mehta, H. Qassenzadeh and T.R. Scott, IEEE Transactions on Circuit Theory, CT-20 628, (1973).
6. W.T. Weeks, IEEE Transactions on Microwave Theory and Techniques, MTT-18 35, (1970).
7. H.A. Wheeler, Proceedings of the IRE, 30 412, (1942).
8. W.T. Weeks, L. L. Wu, M. F. McAllister and A. Singh, IBM Journal of R&D, 23 652, (1979).
9. A.J. Groudis and C. S. Chang, IBM Journal of R&D, 25 25, (1981).
10. G. Hasnain, G. Arjavalingam, A. Dienes and J.R. Whinnery, SPIE Proceedings, 439 159, (1983).

Part II

Transient Transport

Direct Subpicosecond Measurement of Carrier Mobility of Photoexcited Electrons in GaAs

M.C. Nuss and D.H. Auston

AT&T Bell Laboratories, 600 Mountain Avenue,
Murray Hill, NJ 07974, USA

1. INTRODUCTION

Hot carrier phenomena in semiconductors play an important role in high-speed devices like ultrafast photodetectors and hot-electron transistors, but they are also an intrinsic phenomena in submicron devices due to the large internal electric fields. Hot carrier effects have been investigated by direct heating of electrons by an applied electric field, and transport parameters like the drift velocity were determined by microwave time-of-flight techniques[1] or optical pump-and-probe techniques involving picosecond laser pulses[2]. Hot electrons can also be studied by photoinjection of carriers with excess energy, thereby creating hot electrons directly. The hot-electron dynamic can then be probed by a second ultrashort pulse[2]. Because of the availability of ultrashort laser pulses in the picosecond and even femtosecond time regime, the relaxation phenomena of hot carriers can be traced directly in time and extremely valuable information has been extracted from these experiments. On a femtosecond time scale, both absorption saturation[3,4] and time resolved photoluminescence[5] have been used to probe the ultrafast relaxation of injected carriers. All these studies probe the carrier distribution in the bands after injection. An important property for high speed devices is, however, the time dependent mobility or drift velocity. These transport parameters can be studied by microwave conductivity and time-of-flight techniques[1]. However, the time resolution of these techniques is not sufficient to observe transient transport properties on a picosecond or even femtosecond time scale as they occur in a semiconductor after photoinjection of hot carriers.

2. SUBPICOSECOND PROBING OF CARRIER MOBILITY

We have used a technique using both femtosecond optical and electrical pulses to measure the mobility of hot photoinjected carriers in GaAs on a femtosecond time scale. The technique uses the basic concepts of microwave conductivity measurements, however has a superior time resolution (290 fs) not only due to the availability of femtosecond electrical pulses to measure the mobility, but also greatly benefits from the synchronization of the optical carrier injection pulse to the electrical probing pulse. The measurement technique is noncontacting, does not require electrodes on the sample and yields absolute values of the carrier mobility.

Our technique uses femtosecond electrical pulses generated by optical rectification of femtosecond laser pulses in the electro-optic crystal $LiTaO_3$[6]. These pulses consist of only a single cycle of electromagnetic radiation with a duration as low as 300 fs and a spectrum that peaks in the far-infrared range of the electromagnetic spectrum (~1 THz). A schematic of the experimental setup is depicted in Fig.1. Optical pulses of 60-fs duration are generated in a colliding pulse mode locked (CPM) laser (λ = 625nm) and amplified by a copper vapor laser to ~ 1-μJ energy per pulse at 8 kHz repetition rate. A small portion of this beam (~4%) is split off to derive two beams for the generation and probing of the femtosecond electrical pulse. Both beams are focussed in parallel to a spot size of 8μm diameter into the 1 mm thick $LiTaO_3$-crystal, separated by 20μm (Fig.1). The electrical pulse evolves from the generating beam in the form of a conical Cherenkov shock wave[6]. The second, much weaker, probe beam serves to measure the propagating electrical pulse via electro-optic sampling. The complete waveform can be mapped out by changing the time delay t_{probe} between generating and probing beams.

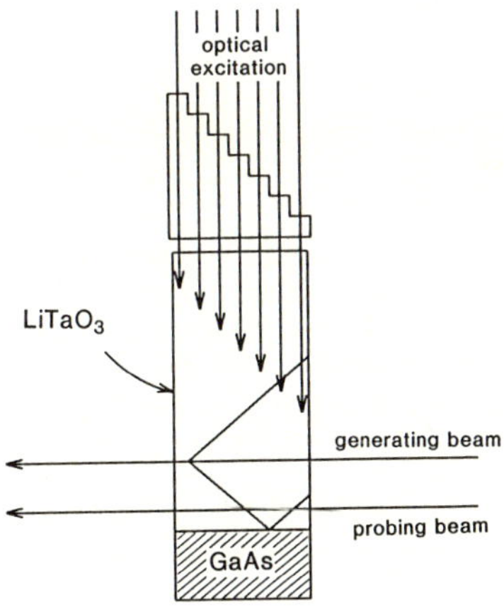

Fig.1
Schematic of the experimental arrangement

The GaAs sample is optically contacted to the polished crystal face at the bottom of the $LiTaO_3$ crystal. The intrinsic sample is grown by molecular beam epitaxy on a semi-insulating GaAs substrate and consist of a 9300 Å thick buffer layer of GaAs followed by a 1900 Å thick $Al_{0.52}Ga_{0.48}As$ layer, a 4600 Å thick GaAs, and a 105 Å thick $Al_{0.33}Ga_{0.67}As$ cap layer. The conical wave of the femtosecond electrical pulse sweeps across the surface of the sample, is reflected and crosses the probe beam again after a certain delay time. From the top of the $LiTaO_3$ crystal, opposite to the GaAs crystal, the remaining 96% of the 60 fs optical pulse illuminates the sample after passing through a transmission echelon. The echelon provides a tilted wavefront of the optical pumping beam in synchronism with the conical wave of the electrical pulse. The optical excitation pulse creates carriers with a density between $5\times10^{17}cm^{-3}$ and $1.2\times10^{19}cm^{-3}$ within the 2500 Å absorption depth with an excess energy of 500meV for the electrons and 70meV for the holes.

3. TRANSIENT REFLECTIVITY AND ELECTRON MOBILITY

We can now record the reflected femtosecond electrical waveform for each delay T_{ex} of the optical carrier excitation pulse. A series of traces obtained this way is plotted in Fig.2. Time zero of the delay T_{ex} and the time resolution of the plot can be

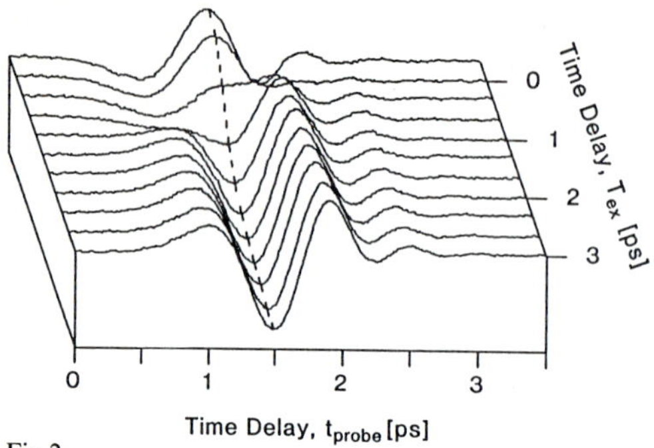

Fig.2
Reflected electrical pulses for different settings of the optical pump delay T_{ex}. The dashed line marks the probe delay at which the data of Fig.3 were taken.

determined by a crosscorrelation experiment of the optical excitation pulse and the optical probe pulse using the third order nonlinear susceptibility $\chi^{(3)}$ of the $LiTaO_3$ crystal. The crosscorrelation trace has a Lorentzian shape with a FWHH of 290 fs. The reflected electrical pulse shape in the very back is recorded without optical excitation and represents the dielectric response of the GaAs crystal without free carriers. It is essentially an attenuated replica of the incident pulse (not shown). As we go towards the front of the set, the phase as well as the amplitude changes dramatically. In particular, the phase changes by 180° as expected for an insulator-to-conductor transition.

The electric field of the reflected femtosecond electrical pulse E_r can be derived from the boundary condition at the thin conducting layer at the interface between $LiTaO_3$ and GaAs:

$$E_r(t) = \rho E_i(t) - \frac{1}{Y} \cdot J_s(t) \tag{1}$$

where ρ is the reflection coefficient of the interface without the presence of free carriers, J_s the sheet current in the thin conducting layer and $Y = \sqrt{\epsilon_0/\mu_0}\,(\sqrt{\epsilon_1}\cos\theta_i + \sqrt{\epsilon_2}\cos\theta_t)$ with θ_i, θ_t being the angle of incidence and transmission and ϵ_1, ϵ_2 the dielectric constants of $LiTaO_3$ and GaAs, respectively. As can be seen from Fig.2, only for the short time of carrier injection around $T_{ex}=0$ does the phase of the reflected electrical pulse change; it remains nearly constant during the subsequent cooling of the hot carriers. This means that the sheet current J_s follows the electric field E(t) directly without phase change, so that Ohm's law is instantaneous in time: $J_s(t)=\sigma(t)\cdot E(t)$. This behavior is also expected since the relaxation times of the hot electron gas (30-300 fs) are shorter than the cycle time of the femtosecond electrical pulse (1 ps) and the current follows the driving electric field without any integrating effects (no memory). This is the only approximation we make in our analysis and the lack of any phase shifts after the injection of carriers in Fig.2 justifies this approximation. Ohm's law is valid here even for hot carriers, since the electric field of the probing pulse is small (~100V/cm). $E(t)=E_i(t)+E_r(t)$ is the driving field of the carriers in the conducting sheet. We then gain an expression for the reflected electric field E_r:

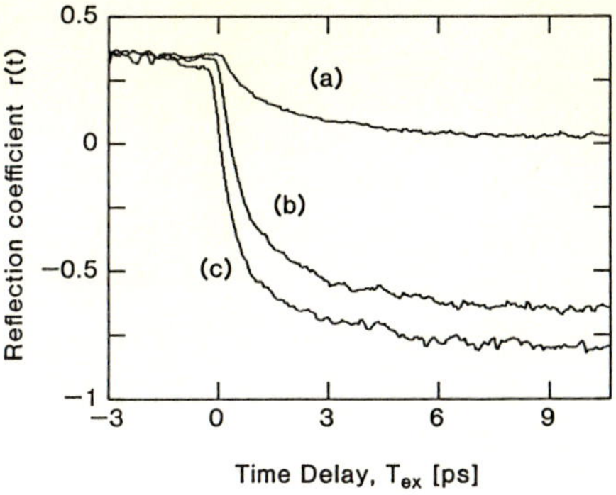

Fig.3
Reflection coefficient r(t) recorded along the dashed line in Fig.2. Traces (a)-(c) correspond to injected carrier densities of $5\times10^{17}cm^{-3}$, $5\times10^{18}cm^{-3}$ and $1.2\times10^{19}cm^{-3}$.

$$E_r = \left\{\rho - (1+\rho)\cdot\frac{\sigma_s}{Y+\sigma_s}\right\}\cdot E_i = r(t)\cdot E_i(t)\,. \tag{2}$$

The expression in parenthesis is the reflection coefficient r(t). The reflection coefficient at the fixed probe delay t_{probe} marked by the dashed line in Fig.2 is shown in Fig.3 as a function of excitation delay T_{ex} for three different carrier excitation densities.

The carrier densities were obtained by comparing the reflectivity at long delay times ($\gg$10 ps), when all carriers are in equilibrium with the lattice at 300K, with the Drude reflectivity of a semiconductor plasma[7]. Knowing the reflection coefficient, the conductivity $\sigma_s = Y(\rho - r)/(1+r)$ can be obtained from the expression of the reflection coefficient in Eq.2. As we know the injected carrier densities, except for the short time (~290-fs) during which the carriers are injected, the mobility of the electron gas can be derived:

$$\mu(t) = \frac{\sigma_s}{n_s\, e} = \frac{Y}{n_s\, e}\cdot\frac{\rho - r(t)}{1+r(t)} \tag{3}$$

where n_s is the sheet carrier density and e the electronic charge. The mobility $\mu(t)$ obtained according to Eq.3 from the reflection coefficient in Fig.3 is plotted in Fig.4.

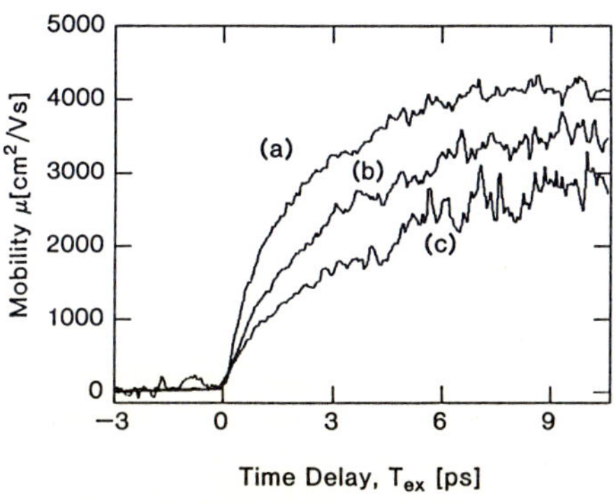

Fig.4
The r.h.s. of Eq.3 obtained from the reflection coefficient r(t) as function of delay time T_{ex} after carrier injection. For $T_{ex}>0.3$ps, this parameter is identical to the carrier mobility μ. (a): $n=5\times10^{17}cm^{-3}$, (b): $n=5\times10^{18}cm^{-3}$ and (c): $n=1.2\times10^{19}cm^{-3}$.

We would like to emphasize, that the time resolution of the reflectivity data (and therefore also of the mobility data in Fig.4) are limited in our experiment by the amount of synchronization of the optical excitation pulse with the electrical probing pulse. This was determined to be 290 fs. The ultimate time resolution would only be limited by the ability to resolve rapid changes within the reflected electrical pulse, being the duration of the probing optical pulse (60fs).

4. RESULTS AND DISCUSSION

Since the mobility for the holes is much lower than for the electrons, the data shown in Fig.4 represent essentially the electron mobility. The transient mobility data of Fig.4 show three extremely interesting features:

(i) immediately after the photoexcitation (T_{ex} ~300 fs) the carrier mobility is very low (<500 cm^2/Vs) and reaches a steadv state value onlv after ~ 10 ps. These low values can not be explained by the reduced mobility of the hot electron gas in the central (Γ) valley alone. Since the Γ-L intervalley separation of 310meV is less than the electron excess energy (500meV), scattering into the L-valley has to be taken into account. For excess energies of 500meV, the intervalley scattering rates are extremely high ($\tau_{\Gamma L}$ ~ 35 fs [11]), so that a substantial fraction of the injected electrons will scatter into the L-valley, having a very low mobility of ~250cm^2/Vs [12]. It has been estimated from Monte-Carlo simulations[13], that as much as 80% of the injected carriers reside in the L-valley under comparable conditions. However, the relaxation back to the Γ-valley by emission of a LO-phonon is estimated to take ~1-2 ps[14]. We think, that the rise time of the mobility following carrier injection (see (iii)) is largely the relaxation time from the L-valley back into the central valley. This relaxation process has not been resolved directly before. There is also a contribution from the fraction of the electrons that remains in the Γ-valley. This fraction evolves rapidly into a Boltzmann distribution with an electron temperature $T_e >> T_L$. The distribution cools down to the lattice temperature T_L by LO-phonon emission with an initial time constant of ~2 ps[2]. This time constant is in the same range as the relaxation time from the L-valley back to the Γ-valley and can hardly be resolved.

(ii) the maximum mobility for long delay times (>10 ps) decreases from 4200cm^2/Vs at a carrier density of $5\times10^{17}cm^{-3}$ to 3500cm^2/Vs at $n=5\times10^{18}cm^{-3}$ and to 2900cm^2/Vs at $n=1.2\times10^{19}cm^{-3}$ and is even at the lowest carrier density much less than the maximum electron mobility of ~7000cm^2/Vs of the sample grown by molecular beam epitaxy. Because of the sandwiching of the GaAs between two AlGaAs layers, surface scattering should be negligible. Polar optical scattering actually decreases with increasing electron density and yields mobilities at least two times higher than observed in our measurements[8]. However, photoexcitation injects both electrons *and* holes of equal density. Since the holes are much heavier than the electrons ($m_e=0.067m_0$, $m_h=0.57m_0$), they act as nearly elastic scattering centers for the electrons, thereby reducing the electron mobility[9]. At 300K the electron-hole (e-h) scattering can be adequately described by the equation for ionized impurity scattering with the electron mass replaced by the reduced mass[10]. The agreement of the calculated mobilities with our experimental data is good and shows, that e-h scattering is indeed important and responsible for the observed mobilities. Electron-electron scattering, however, is not effective in reducing the carrier mobility, since the average momentum of the electron gas is conserved during collisions.

(iii) The rise time of the electron mobility as the carriers cool to the lattice temperature T_L(300K) is strongly dependent on the number of injected carriers. At n = $5\times10^{17}cm^{-3}$, the (exponential) time constant is 1.4 ps (Fig.4a), which increases to 2.5 ps at n = $5\times10^{18}cm^{-3}$ (Fig.4b) and to 3.2 ps at n = $1.2\times10^{19}cm^{-3}$ (Fig.4c). The dependence of the mobility rise time on carrier density observed in our experiment can be explained with the generation of nonequilibrium LO-phonons during the cooling process. This effect has been shown to decrease the cooling rate of a highly excited semiconductor plasma considerably[2,15,16,17]. In our experiment, the density of states into which LO-phonons can be emitted is considerably less than the number of hot electrons that can exchange energy with the lattice by emission of LO-phonons. With the phonon lifetime being ~7 ps[18], the phonon occupation number can then deviate appreciably from the value at 300K if the emission rate of phonons from the hot electron gas is comparable to or larger than the decay rate. This quenches further emission of LO-phonons, thereby slowing down the cooling rate.

5. CONCLUSION

In conclusion, we have determined the mobility of hot photoinjected carriers in GaAs directly with <300 fs time resolution using femtosecond electrical pulses. We find very low electron mobilities (<500cm^2/Vs) immediately after injection. The mobility then increases with a time constant between 1.4 ps and 3.2 ps for injected carrier densities ranging between $5\times10^{17}cm^{-3}$ and $1.2\times10^{19}cm^{-3}$, respectively. There is evidence that this time constant is not simply due to the cooling of carriers in the central valley, but due to the relaxation of electrons from the rapidly populated L-valley back into the Γ-valley. We find the equilibrium mobility (after >10 ps) to decrease with increasing density and being limited by electron-hole scattering.

There are immediate implications of our results for high speed devices. Our data show, that a considerable delay in the response on the order of several picoseconds can be expected for high speed photodetectors and photoconductors, when the energy of the photons is well above the bandgap of the semiconductor. Our results also suggest, that picosecond time-of-flight and related experiments on a time scale of <10 ps have to be interpreted carefully to take the low electron mobility for the first few picoseconds into account.

We wish to thank F. Capasso and S. Schmitt-Rink for many stimulating discussions and suggestions and A. C. Gossard, J. H. English and J. M. Parsey for the expert growth of samples used in our experiments.

6. REFERENCES

1. see, e.g. C. Canali, F. Nava and L. Reggiani, in *Hot-Electron Transport in Semiconductors*, edited by L. Reggiani, Topics in Applied Physics Vol.58, (Springer, Berlin, 1985)

2. for a review, see J. Shah, IEEE J. Quantum Electron. *QE-22*, 1728 (1986)

3. J. L. Oudar, D. Hulin, A. Migus, A. Antonetti and F. Alexandre, Phys. Rev. Lett. *55*, 2074 (1985)

4. W. H. Knox, C. Hirlimann, D. A. B. Miller, J. Shah, D. S. Chemla and C. V. Shank, Phys. Rev. Lett. *56*, 1191 (1986)

5. D. Block, J. Shah and A. C. Gossard, Solid State Comm. *59*, 527 (1986)

6. D. H. Auston, K. P. Cheung, J. A. Valdmanis and D. A. Kleinman, Phys. Rev. Lett. *53*, 1555 (1984)

7. D. H. Auston and K. P. Cheung, J. Opt. Soc. Am. *B2*, 606 (1985)

8. K. Seeger, *Semiconductor Physics*, Springer Series in Solid State Sciences Vol.40, (Springer, Berlin, 1982)

9. R. A. Höpfel, J. Shah, D. Block and A. C. Gossard, Appl. Phys. Lett., *48*, 148 (1986)

10. T. P. McLean and E. G. S. Paige, J. Phys. Chem. Solids, *16*, 220 (1960)

11. F. Capasso, R. E. Nahory and M. A. Pollack, Solid State Electr., *22*, 977 (1979)

12. The mobility in the L-valley is obtained by scaling the mobility in the central valley by the ratio of the electron masses (m_Γ=0.067m_0, m_L=1.2m_0)

13. M. A. Osman, M. J. Kann, D. K. Ferry and P. Lugli, same proceedings.

14. E. M. Conwell, *High Field Transport in Semiconductors*, in Solid State Physics, Suppl.9, Eds. F. Seitz, D. Turnbull and H. Ehrenreich, (New York, 1967)

15. J. Shah, A. Pinczuk, A. C. Gossard, W. Wiegmann and K. Kash, Surf. Science *174*, 363 (1986)

16. J. Collet, A. Cornet, M. Pugnet and T. Amand, Solid State Comm., *42*, 883 (1982)

17. W. Cai, M. C. Marchetti and M. Lax, Phys. Rev. *B34*, 8573 (1986)

18. D. von der Linde, J. Kuhl and H. Klingenberg, Phys. Rev. Lett., *44*, 1505 (1980) ; J. A. Kash, J. C. Tsang and J. M. Hvam, Phys. Rev. Lett., *54*, 2151 (1985)

Nonstationary Transport in MODFETs and Heterojunction Devices

K. Hess

Coordinated Science Laboratory and Department of Electrical and Computer Engineering, University of Illinois, Urbana, IL 61801, USA

Self-consistent simulations of transient transport processes in single layer and single quantum-well modulation doped structures are reported .

The availability of vector and parallel machines has given enormous opportunities in recent years for the detailed understanding of these complicated (mostly nonlinear) transport problems. Monte Carlo simulations which have been too computer-time consuming in the past can now be accomplished with ease. Also, complex two-dimensional physical processes can be visualized by color-coded moving pictures. Computation therefore complements conventional theory and experimental research in new ways. This is demonstrated on the basis of a movie showing the two-dimensional dynamics of a high electron mobility transistor (HEMT).

Figure 1 represents a photograph out of the movie and shows the distribution of the electron energy typical for transistor operation. The interface between the GaAs and AlGaAs is perpendicular to the figure and intersects at the top boundary of the shown device cross section. The device is a 1 μ m gate, 3 μ m source-drain distance HEMT.

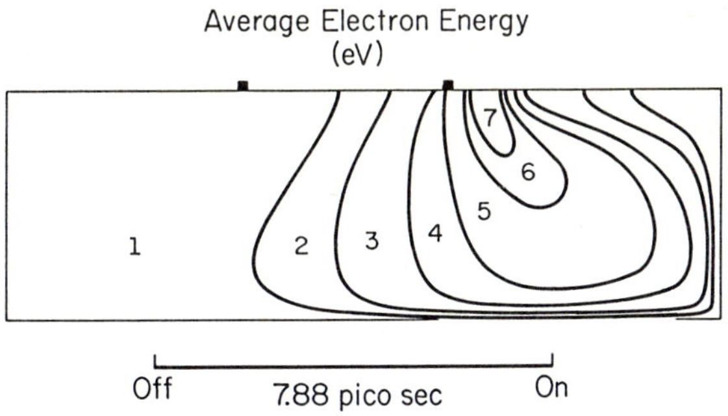

Fig. 1. Typical average electron energy for a high electron mobility transistor. The numbers denote the energy in electron volts: each higher number corresponds to an increase of $\approx$ 70 meV starting with 1 = 70 meV.

The movie [1] demonstrates clearly how conventional device modeling procedures need to be changed to account more precisely for the more complicated electronic transport processes. For example, the switch-on time of the HEMT is more than twice as long as the time necessary for switching it off because of geometrical considerations and the dynamics of overshoot effects.

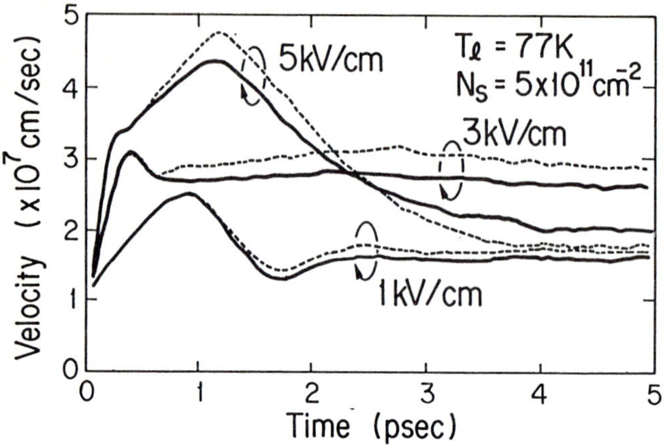

Fig. 2. Transient velocity characteristics at T = 77 K for different applied electric fields. Broken and solid lines indicate the results for one time solution of the Schroedinger equation and the *fully self-consistent method,* respectively.

To completely understand transport in HEMT structures one neeeds to take into account the dependence of the self-consistent interface potential (gate voltage, modulation doping) on the energy distribution of the electrons (drain voltage) [2]. By solving the Schroedinger equation repeatedly as the distribution function evolves, we show that overshoot effects can be reduced by the effects of the distribution function on the interface potential as shown in Fig. 2.

We have also applied our simulation code to studies of femtosecond thermalization after optical excitation [3] as has been reported by Knox and coworkers [4]. Although our program contains over 20 different scattering mechanisms, we show that electron-electron interactions are sufficient to explain the experimental results. It is important, however, to include the reduced dimensionality into the electron-electron scattering rate. The relaxation of the excited electrons toward a thermalized distribution function is shown in Fig. 3.

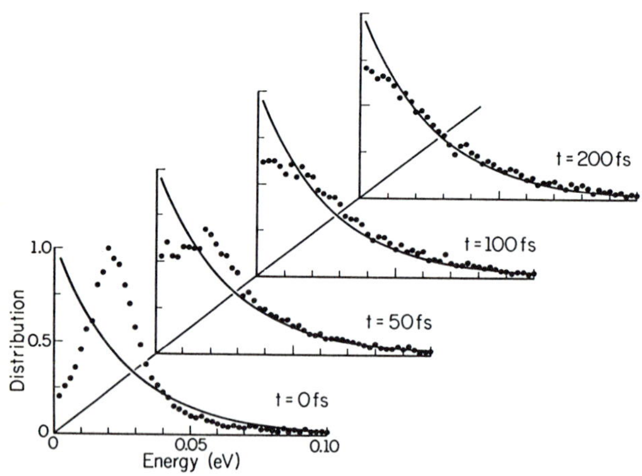

Fig. 3. Evolution of electron distribution with time (dots). The solid line represents a thermalized (Maxwellian-type) distribution function. These results agree with the experiments of Knox and coworkers [4].

Acknowledgement:

This work has been supported by the Army Research Office and the Office of Naval Research.

References: .

1. D. J. Widiger, I. C. Kizilyalli, K. Hess, J. J. Coleman: IEEE Transactions on Electron Devices, Vol. ED-32, 1092 (1985).

2. K. Yokoyama, K. Hess: J. Appl. Phys., Vol. 59, 3798-3802 (June 1986).

3. D. Bailey, C. Stanton, K. Hess: to be published.

4. W. H. Knox, C. Hirlimann, D. A. B. Miller, J. Shaw, D. S. Chemla, C. V. Shank: Phys. Rev. Lett. 56, 1191 (1986).

Monte Carlo Investigation of Hot Photoexcited Electron Relaxation in GaAs *

M.A. Osman[1], *H.L. Grubin*[1], *P. Lugli*[2], *M.J. Kann*[3], *and D.K. Ferry*[3]

[1]Scientific Research Associates, Inc., Box 1058, Glastonbury, CT 06033, USA
[2]Dipartimento di Fisica e Centro Interuniversitario di Struttura della Materia della Universita, Modena, Italy
[3]Center for Solid State Electronics Research, Arizona State University, Tempe, AZ 85281, USA

Recently, several authors [1-3] have investigated the dynamics of electron-hole plasma generated by picosecond and sub-picosecond laser pulses in GaAs. These measurements have indicated the importance of electron-hole interaction in the cooling process (especially in the presence of a cold hole plasma) and the existence of non-equilibrium phonon distributions as a result of cooling of photoexcited electrons and holes. Previous theoretical studies of the cooling process have ignored the presence of the upper valleys even when the energy of the excited electrons exceeds the energy separation between the central and the upper valley. Additionally, assumptions had to be imposed on the form of the distribution function of the electrons and phonons. Ensemble Monte Carlo (EMC) techniques avoid these assumptions and have been used to investigate the effects of electron-hole (e-h) interaction [4] and non-equilibrium phonons [5] on the relaxation rate of photoexcited electrons. In this study we present the first EMC calculation to account for both the e-h interaction and hot phonon effect.

The Monte Carlo model takes into account the e-h , carrier-phonon interactions and the non-equilibrium LO phonon-electron interactions. The band model consists of three nonparabolic valleys for the conduction band and a parabolic heavy-hole valence band presence. The parameters used in the EMC calculations are the same as those used in references [4] and [5]. Electron-electron, electron-plasmon and the screening of electron-phonon interactions have been negelected. Initially, to avoid the complications arising from the transfer of electrons to the upper valleys, we have examined the roles of hot phonons and electron-hole interaction for excitation energy of 1.8 eV. Under this situation, the energy of the excited electrons is 250 meV while that of the holes is 40 meV, so that the change in the LO phonon population is mainly due to the emission of LO phonons by the electrons. The time evolution of the LO-phonon distribution N(q) is calculated as a function of the wavevector q from the Monte Carlo simulation using the approach outlined in [5]. Instead of recalculating the scattering rates for the LO-phonons using the perturbed phonon distribution, we have used a Large value for N(q) and used the rejection technique to accept the final state after each scattering event involving an LO phonon.

The relaxation of electrons excited by a 1.8 eV laser pulse was examined for an excitation level of 5×10^{16} cm^{-3} to understand the

* Supported by the Office of Naval Research.

manner in which e-h interaction and hot phonons influence the cooling process. From Fig. 1, it is clear that when the e-h interaction is ignored, the hot phonon effect is significant only for times longer than 1 ps. Curve 1 in Fig. 2 shows that between 0.6 and 2 ps the energy-loss-rate through the emission of LO-phonons is reduced when the phonon heating is taken into account compared to the situation where hot phonons are neglected (curve 1). When the e-h interaction is taken into account, the cooling rate slows for times less than 2 ps as shown by curve 3 in Fig. 1. When both the e-h interaction and hot phonons are considered (curve 4), the cooling rate is slightly faster for times below 1 ps partly due the initial overestimation of the enhancement of the phonon emission process inherent in using a single large value for N(q). However, it is obvious that the cooling process for times larger than 2 ps is controlled by the hot phonons, which lead to a slower cooling rate. Additionally, no significant change in the energy of the photoexcited holes was noticed, when the electron-hot-phonon interaction was considered. The situation where hole-hot-phonon interaction is taken into account might lead to an increase in the hole energy as a result of the phonon reabsorption process. This work is underway and the results will be presented elsewhere.

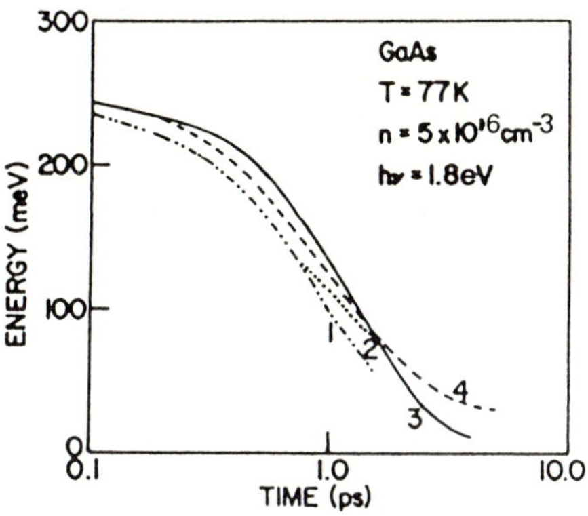

Fig. 1. Time evolution of the electron energy: 1) e-ph, 2) e-ph (hot), 3) e-ph + e-h, 4) e-ph (hot) + e-h.

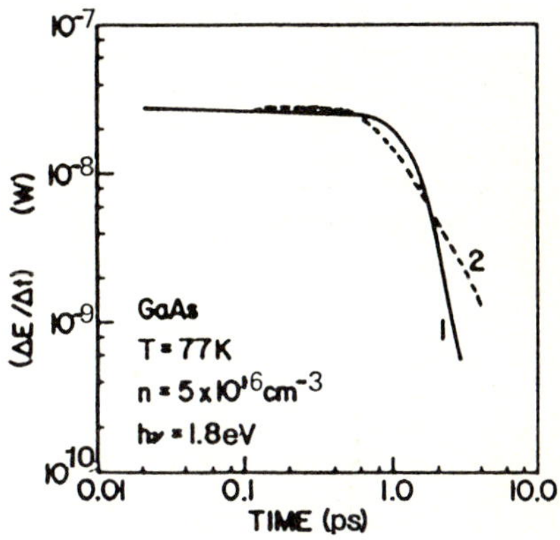

Fig. 2. Time dependence of the electron-energy-loss rate: 1) without hot phonons, 2) with hot phonons.

The situation where the electrons were transferred to the upper valleys was investigated for the case of excitation by a 588 nm laser pulse using EMC and taking into account the carrier-phonon interactions only. The fraction of electrons in the central valley is shown in Fig. 3 as a function of the time after the application of the pulse. From Fig. 3, it is obvious that the simple picture of electrons cascading down the central valley by emitting LO-phonons is not complete. The electrons rather transfer to the upper valleys and cool down in these valleys by emitting phonons through polar-optical and deformation potential interactions as shown in Fig. 4 (curves 2 and 4). The energy loss rate through LO phonon emission is dominated by the central valley electrons for times greater than 1 ps and reach a maximum after 2 ps. Note that after 2 ps about 80% of the electrons

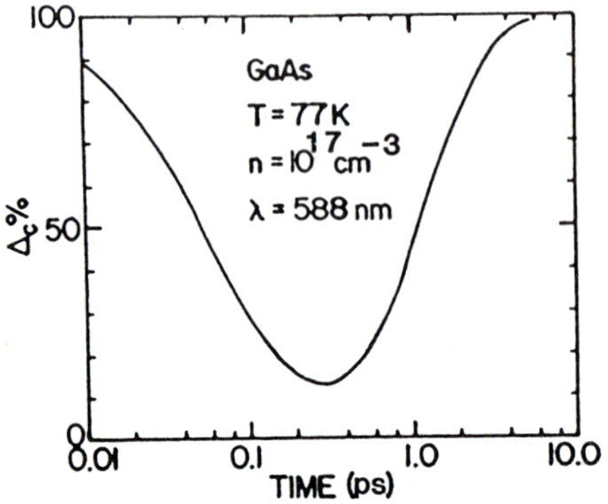

Fig. 3. Time evolution of the fraction of electrons in the central valley.

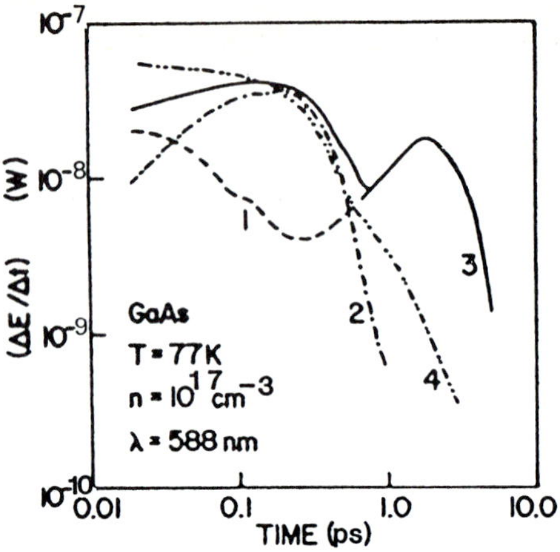

Fig. 4. Time dependence of the electron-energy-loss rate: 1) LO (Γ), 2) LO (L), 3) total LO 4) Intervalley scattering

are in the central valley. The energy of the electrons in the central valley (curve 1 in Fig. 5) remains above 100 meV up to 3 ps after the excitation and an increase in the average energy occurs at 2 ps due to the return of the electrons to the central valley. The initial rapid drop in the average kinetic energy of the electron ensemble (curve 2 in Fig. 5) is due to the transfer of a large fraction of the electrons to the upper valleys as can be seen in Fig. 3. Curve 3 in Fig. 5 indicates that the cooling of the electron in the L-valley is not negligible.

In conclusion, we have shown that the cooling process of the photoexcited electrons is controlled by e-h interaction for times less than 1 ps and by hot phonons for times greater by 2 ps. In addition, we have demonstrated that the cooling of the electrons in the upper valleys is important for high excitation energies.

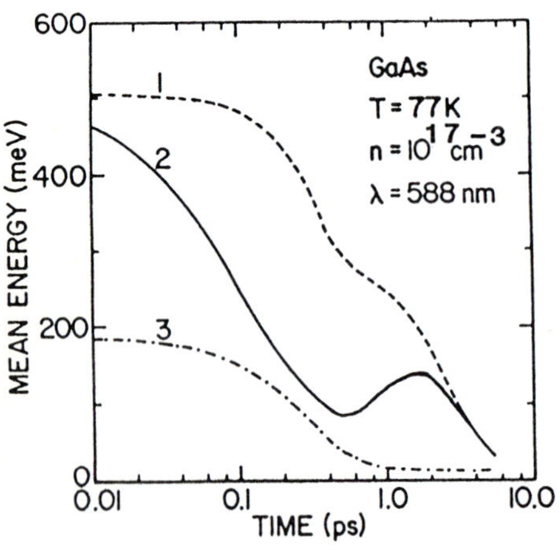

Fig. 5. Mean kinetic energy of the electrons in: 1) central valley, 2) central + L-valley, 3) L-valley.

References

1. J. A. Kash, J. C. Tsang, and J. M. Hvam, Phys. Rev. Lett. 54, 2151 (1985).
2. J. Shah, A. Pinczuk, A. C. Gossard, and W. Weigmann, Phys. Rev. Lett. 54, 2045 (1985).
3. C. L. Collins and P. Y. Yu, Phys. Rev. B30, 4501 (1984).
4. M. A. Osman, U. Ravaioli, R. Joshi, W. Potz, and D. K. Ferry, in Proc. 18th Intern. Conf. Phys. Semiconductors, Stockholm, 1986, Ed. by O. Engstrom, in press.
5. P. Lugli, C. Jacoboni, L. Reggiani, and P. Kocevar, submitted for publication.

Longitudinally Localized Optical Carrier Injection for Femtosecond Transport Studies

D.H. Reitze, T.R. Nuss, and M. Downer

Physics Department, University of Texas at Austin, Austin, TX 78712, USA

A new generation of microelectronic devices will be based on ballistic and quantum tunneling transport processes, which involve transit times on a femtosecond scale across submicron semiconductor layers. Fundamental experimental studies of these phenomena will require the use of femtosecond optical techniques to inject and probe carriers locally, just as analogous picosecond techniques have been used in studying slower transport processes in thicker layers [1,2]. However, because optical absorption (and therefore carrier injection) depths at near band gap wavelengths (typically ~1μm) greatly exceed the thickness of the semiconductor layers in which most femtosecond transport phenomena occur (typically 0.01 to 0.1 μm), adequate spatial resolution to observe these processes directly has not been achievable. We introduce a new technique for optically injecting carriers of any desired energy within the first few hundred angstroms of the surface of a semiconductor layer with femtosecond time resolution, so that the evolution of femtosecond transport processes can be resolved.

Our technique is based upon internal reflection (see Fig. 1). A light pulse passes through a high index (n) dielectric to the interface with a lower index (n') semiconductor sample at an incident angle Ø greater than the critical angle $\sin^{-1}(n'/n)$. Since the lower index semiconductor is slightly lossy, energy is absorbed from the evanescent wave in the form of photo-excited

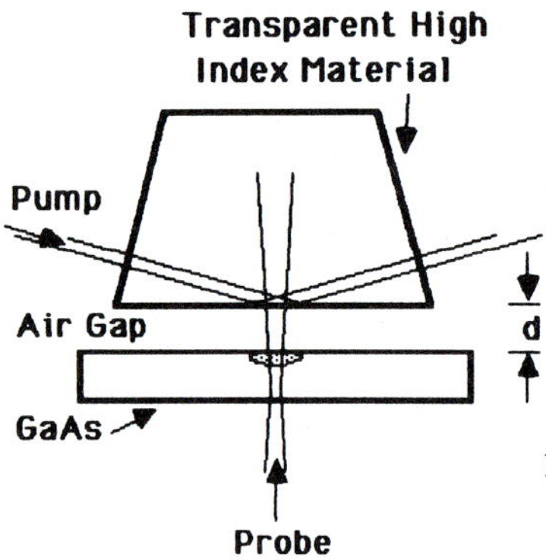

Fig. 1 -- Schematic experimental arrangement for time-resolved transport measurement using internal reflection carrier excitation.

carriers. Note that this absorption occurs within the penetration depth (approximately $\lambda/2\pi n'$) of the evanescent wave, typically only a few hundred angstroms for semiconductor materials and wavelengths of interest. The extremely small penetration depth of evanescent waves has been used extensively in the past as a tool in cw infrared spectroscopy of thin films and surfaces [3]. Actual contact is not necessary in order to apply the internal reflection method. The presence of a small air, or other low index, gap between the two materials, as shown in Fig. 1, alters neither the critical angle nor the penetration depth of the evanescent wave into the semiconductor, although the *amount* of absorption depends on the size of this gap, as shown in more detail below.

We have calculated absorption depths and absorbances for three practical cases of potential interest in experimental transport studies. These calculations use the thin film optics equations [4] with variable incident angle and complex refractive index $n_3 + ik_3$ in the third layer. The solid curve in Fig. 2 assumes an excitation wavelength of 0.83 μm (1.5 eV), slightly above the room temperature band gap of undoped GaAs. The excitation pulse enters through a prism of trigonal selenium, which has a band gap at 1.7 eV, and an extraordinary index higher by approximately 0.1 than GaAs in the 1.5 to 1.7 eV spectral region [5]. The evanescent wave is therefore absorbed by local electron-hole pair creation. Note that the penetration depth above the critical angle is twenty times smaller than the normal incidence absorption depth. The dashed curve in Fig. 1 assumes an excitation pulse at 3.0 μm (0.4 eV), incident through germanium upon GaAs n-doped to 10^{18} cm^{-3}. Here the absorption mechanism is the localized heating of conduction band electrons through free carrier absorption. The evanescent wave absorption is again localized within a few hundred angstroms, smaller by a factor of 10^5 than normal incidence absorption. The absorption depths in shown in Fig. 2 are independent of the size of the air gap.

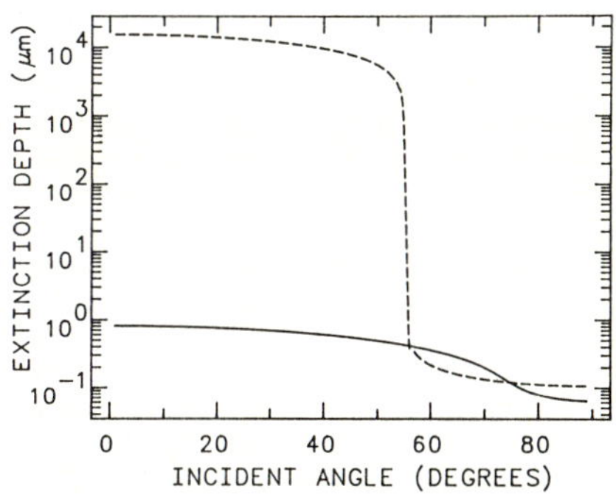

Fig. 2--Extinction depth for 0.83 μm pulses incident from trigonal selenium upon GaAs (solid curve) and for 3.0 μm pulses incident through Ge upon n-GaAs (dashed curve). Note sharp decreases above critical angles (75° for Se/GaAs, 55° for Ge/n-GaAs).

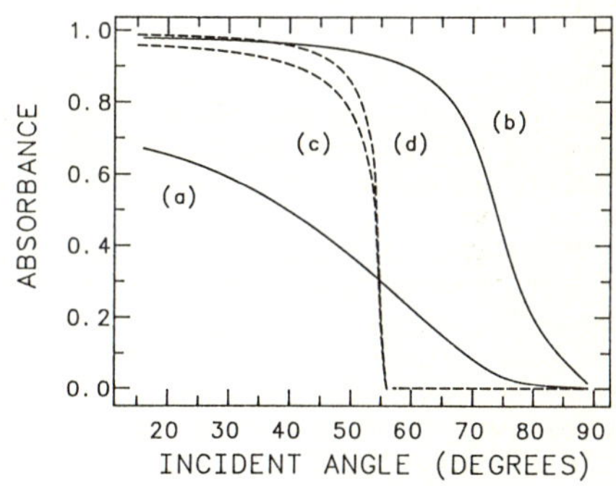

Fig. 3--Absorbance of GaAs layer through Se internal reflection element, λ = 0.83 μm, with air gap (a) 500Å and (b) 100Å; and for Ge internal reflection elenent, λ = 3.0 μm, with air gap (c) 500Å and (d) 100Å (absorbance is ~10^{-4} above critical angle for curves (c) and (d)).

Figure 3 shows the percent absorption which can be achieved with an s-polarized incident pulse for 100 Å and 500 Å air gaps. Note that absorbances as high as 30% are achievable even above the critical angle. We have calculated that the phase changes which occur upon internal reflection [7] will broaden 50 fsec excitation pulses by less than 5% in the first case. Thus carrier injection occurs essentially within the duration of the optical pulse.

Our preliminary measurements of the optical properties of selenium samples indicate that the real part of the extraordinary index of refraction n_e is 3.74 at 1.06 μm, 0.25 greater than GaAs, confirming earlier measurements [5]. Optical coupling of Se/GaAs interfaces has also been achieved at 1.06 μm, although experiments at the GaAs band edge have so far been hampered by below gap absorbance in our selenium samples significantly greater than literature values [8], evidently caused by a high defect concentration.

In the third case of interest, an excitation pulse of wavelength 1.06 μm from a mode-locked Nd:YAG laser enters a heterostructure consisting of GaAs (E_G =1.42 eV), $Al_{.8}Ga_{.2}As$ (E_G =2.56 eV), and $Al_{.4}Ga_{.6}As$ (E_G =1.92 eV). The excitation pulse can be compressed to 1.0 picosecond using a single mode optical fiber and grating pair [6] and amplified in Nd:glass. Because the band gaps of all three materials exceed the excitation pulse

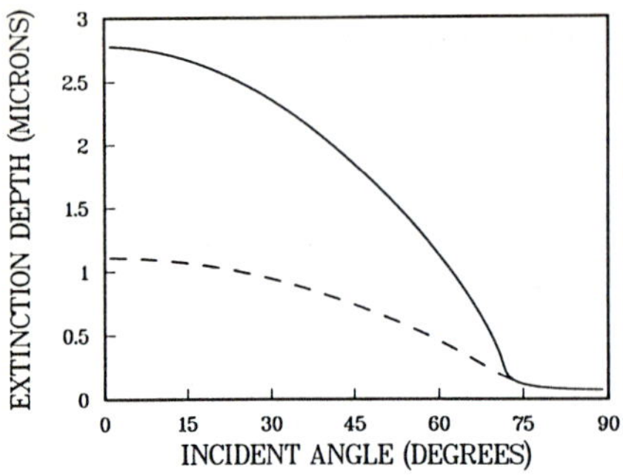

Fig. 4 --Extinction depth for 1.06 μm pulses incident upon $Al_{.4}Ga_{.6}As$ for intensities 200 GW/cm^2 (solid curve) and 500 GW/cm^2 (dashed curve). The critical angle for GaAs/ $Al_{.4}Ga_{.6}As$ is 70.5°.

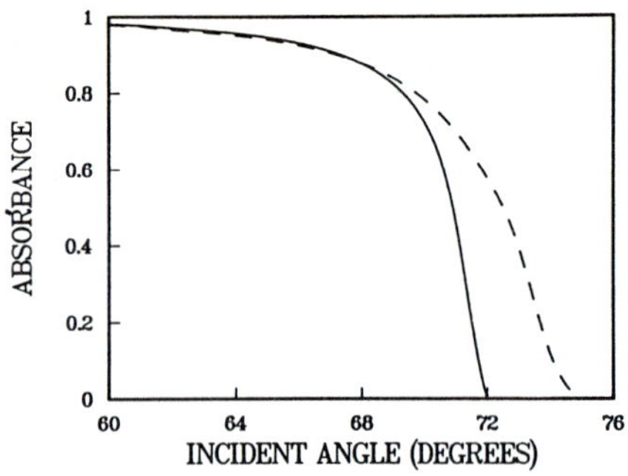

Fig. 5 -- Absorbance of $Al_{.4}Ga_{.6}As$ layer for 1.06 μm pulses for intensities 200 GW/cm^2(solid curve) and 500 GW/cm^2 (dashed curve). The barrier layer is 50 Å deep.

photon energy (1.17 eV), linear absorption is suppressed so that localized carrier excitation occurs in the $Al_{.4}Ga_{.6}As$ layer via a two-photon absorption process. The $Al_{.8}Ga_{.2}As$ layer in which no two-photon absorption occurs serves as a barrier preventing carriers excited in the GaAs layer from migrating to the $Al_{.4}Ga_{.6}As$ layer. Figure 4 shows the evanescent wave penetration depth assuming incident intensities of 200 and 500 GW/cm^2 at the GaAs/$Al_{.8}Ga_{.2}As$ interface and a 50 Å $Al_{.8}Ga_{.2}As$ barrier layer.

Figure 5 displays the amount of light absorbed in the third layer. The abrupt transition to zero absorption slightly above the critical angle is due the angular dependence of the imaginary part of the complex index of refraction k_3 in the $Al_{.4}Ga_{.6}As$ layer. As the angle of incidence approaches the critical angle, k_3 rapidly approaches zero. However, absorbances above 40% are possible even above the critical angle.

In order to conduct meaningful transport measurements, the internal reflection injection technique must be combined with optical probing methods possessing comparable depth resolution. An example is shown in Fig. 1, where a probe beam monitors band edge absorption, an appropriate technique for voltage-biased structures in which the Franz-Keldysh effect is modulated as moving carriers screen the applied electric field [1]. Optical reflectivity [2] provides an appropriate probe in other circumstances.

Dynamics of surface states and absorbates can also be studied with this technique. This work is supported by the Joint Services Electronics Program and the Texas Advanced Technology Research Program.

1. C.V. Shank, R.L. Fork, B.I. Greene, F.K. Reinhart, and R.A. Logan, Appl. Phys. Lett. *38*, 104 (1981).
2. B.F. Levine, C.G. Bethea, W.T. Tsang, F. Capasso, K.K. Thornber, R.C. Fulton, and D.A.Kleinman, Appl. Phys. Lett. *42*, 769 (1983).
3. N.J. Harrick, **Internal Reflection Spectroscopy** (Wiley, New York, 1967).
4. O. S. Heavens, **Optical Properties of Thin Solid Films** (Dover, New York,1985).
5. L. Gampel and F.M. Johnson, J. Opt. Soc. Am. *59*, 72 (1969); E.D. Palik, ed., **Handbook of Optical Constants of Solids**, (Academic Press, New York, 1985).
6. B. H. Kohler, D. M. Bloom, J. D. Kafka, and T. M. Baer: in **Ultrafast Phenomenon IV**, D. H. Auston and K. B. Eisenthal,ed. (Springer-Verlag, Berlin, 1984), p.19.
7. K.P. Cheung and D.H. Auston, Opt. Lett. *10*, 218 (1985).
8. W. Henrion: in **The Physics of Selenium and Tellurium**, W. C. Cooper, ed. (Pergamon Press, New York,1969)p. 75.

Femtosecond Nonequilibrium Electronic Heat Transport in Thin Gold Films

S.D. Brorson, J.G. Fujimoto, and E.P. Ippen

Department of Electrical Engineering and Computer Science and Research Laboratory of Electronics, Massachusetts Institute of Technology, Cambridge, MA 02139, USA

With the advent of femtosecond lasers has come the promise of directly measuring ultrafast electron transport effects in electronic materials and devices. Using a front pump/back probe[1] measurement of transient reflectivity changes, we have observed ultrafast electronic heat transport through thin films of gold. For film thicknesses of 500 - 2000 Å, we find thermal transport occurs on a timescale of 40 - 200 fs. In this regime, the transport of heat seems to be governed by nonequilibrium electronic motion, since equilibrium heat diffusion occurs over tens of picoseconds.

Our experiment is a variation of the usual transient thermomodulation technique.[2] An ultrashort laser pulse heats the surface of a noble metal. The resulting temperature increase changes the occupancy of electronic states near the Fermi level. This changes the reflectivity of the metal, particularly for frequencies near the d-band to Fermi level optical transition. Thus, monitoring the reflectivity of the sample with a delayed probe pulse gives information about the electron temperature as a function of time. Previous thermomodulation experiments demonstrated that the electrons are out of equilibrium with the lattice for times less than ~ 1 ps.[2-4] The decay of the reflectivity transient was attributed to nonequilibrium electron temperature relaxation via the electron-phonon interaction. By pumping the front of the sample and probing reflectivity changes on the back (Fig. 1), we have used this technique to study the propagation of heat through the sample.

The experiments were performed on films of gold deposited on sapphire. Sample thicknesses ranged from 500 to 2000 Å. All samples used were thicker than the optical skin depth (~ 150 Å). The laser source was a colliding pulse mode-locked (CPM) dye laser employing

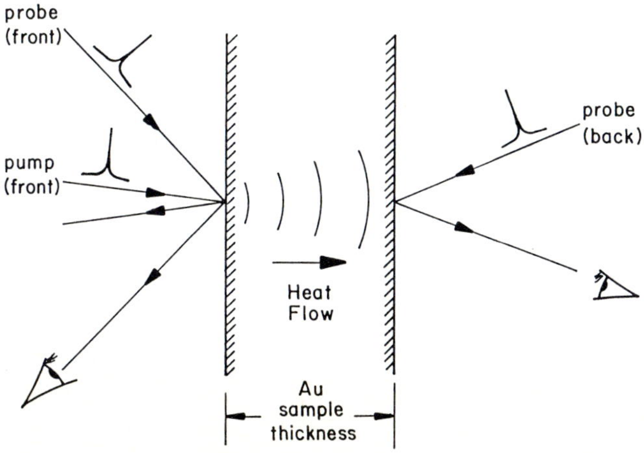

Figure 1. Simplified schematic of the experiment. Usually, the sample is pumped and probed on the front surface. By probing the back, we are able to measure heat transport across the sample.

4 prisms to control the cavity dispersion (λ_o = 630 nm, $t_p \sim 50 fs$ FWHM assuming a $sech^2$ pulse shape). The pump and probe beams were derived from a conventional modified interferometer arrangement. A motor-driven stepper stage with 0.1 μm resolution was employed to vary the delay between the pump and probe pulses. The pump beam was chopped and focused onto the front surface of the sample with a 40× microscope objective, while the probe was similarly focused onto the back of the sample. The reflected probe beam was monitored with a photodiode. The thermomodulation signal was detected with a lock-in amplifier tuned to the chopping frequency of the pump. The zero delay point was determined by reversing the role of pump and probe beams and repeating the experiment. Several examples of pump/probe data are shown in Fig. 2.

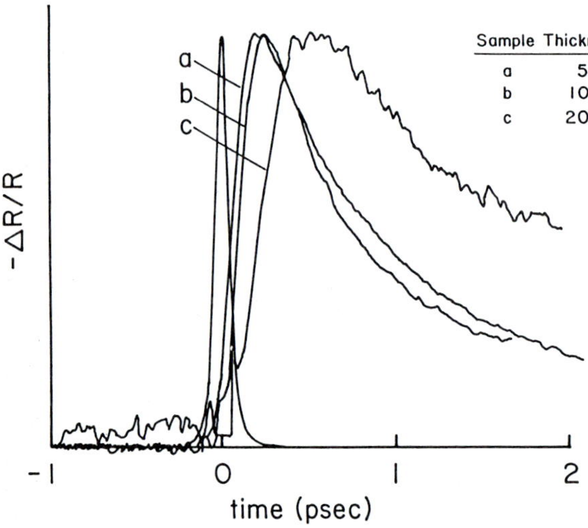

Figure 2. Back surface reflectivity change *vs.* time for various thickness gold films. As the film thickness increases, the delay of the rising edge of the signal also increases. An autocorrelation of the laser pulse is also plotted to indicate the zero delay position.

We have repeated this experiment for a variety of sample thicknesses. Numerical fits were performed after each experiment to determine the delay as a function of the sample thickness. These results are summarized in Fig. 3. Two features are noteworthy. First, the measured delays are much shorter than would be expected if the heat was carried by the diffusion of electrons in equilibrium with the lattice. This suggests that heat is

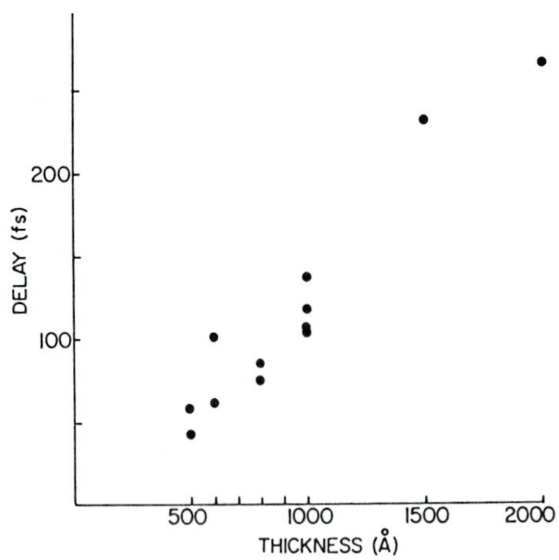

Figure 3. Delay *vs.* sample thickness for several gold films.

transported via the electron gas alone, and that the electrons are out of equilibrium with the lattice on this timescale. Second, referring to Fig. 3, we see that it takes $\sim$ 100 fs for heat to be transported $\sim$ 1000 Å. Thus, we may extract a heat transport 'velocity' of $\sim 10^8$ cm/sec. Interestingly, this is of the same order of magnitude as the Fermi velocity of electrons in Au, 1.4×10^8 cm/sec. This supports the hypothesis that heat transport occurs via nonequilibrium electrons.

Rapid heat transport is also seen in the results of front pump/front probe measurements on films of varying thickness. Shown in Fig. 4 is thermomodulation data on samples of thickness 500, 800, and 1000 Å.

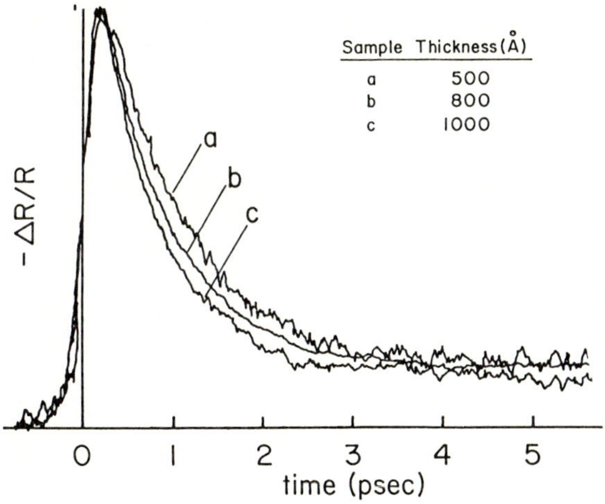

Figure 4. Front surface reflectivity change *vs.* time for various thickness gold films. As the thickness of the film increases, the decay time constant decreases. This can be interpreted as due to the effect of rapid electronic heat transport.

The initial relaxation time decreases slightly with increasing film thickness. This effect is consistent with transport of heat out of the optically active region on a femtosecond time scale. For very thin films, relaxation dominates over transport and the reflectivity decay time is determined by the electron-lattice equilibration process. For thicker films, both relaxation and nonequilibrium transport occur simultaneously. Hence, the measured reflectivity transient decays more quickly. Since this effect is noticeable for all films studied, heat transport is apparently occurring on a femtosecond time scale.

In summary, we have used the transient thermomodulation technique to study the ultrafast transport of heat in thin gold films. Using a front pump/back probe technique, we observed nonequilibrium heat transport occurring on the same time scale as electronic motion. This rapid transport is confirmed by front pump/front probe measurements which show that the reflectivity transient decay time depends on the gold film thickness. Extensions of this technique to other structures may allow for ultrafast probing of transport in other materials and devices.

The authors would like to acknowledge helpful discussions with R. W. Schoenlein, W. Z. Lin, G. L. Eesley, and C. A. Paddock. SDB acknowledges the support of a doctoral fellowship granted by the IBM Corporation. JGF acknowledges support from the National Science Foundation Presidential Young Investigators Program contract number 8552701-ECS. This work was supported in part by the AFOSR Grant 85-0213 and the Joint Services Electronics Program under contract number DAAL03-86-K-0002.

References

1. This technique is similar to that used by Levine, *et al.* to study carrier transport in $Al_xGa_{1-x}As$. B. F. Levine, W. T. Tsang, C. G. Bethea, and F. Capasso, *Appl. Phys. Lett.* **41**, 470 (1982).

2. G. L. Eesley, *Phys. Rev. Lett.* **51**, 2140 (1983); G. L. Eesley, *Phys. Rev. B* **33**, 2144 (1986).

3. H. Elsayed-Ali, M. Pessot, T. Norris, and G. Mourou, in **Ultrafast Phenomena V**, Ed. by G. R. Fleming and A. E. Siegman (Springer-Verlag, Berlin, 1986), p. 264.

4. R. W. Schoenlein, W. Z. Lin, J. G. Fujimoto, and G. L. Eesley, **Ultrafast Phenomena V**, Ed. by G. R. Fleming and A. E. Siegman (Springer-Verlag, Berlin, 1986), p. 260.

Part III

Ballistic Transport and Resonant Tunneling

Ballistic Transport and Energy Spectroscopy of Hot Electrons in THETA Devices *

M. Heiblum

IBM T.J. Watson Research Center, P.O. Box 218,
Yorktown Heights, NY 10598, USA

In 1928, FELIX BLOCH predicted, theoretically, that electrons in an ideally periodic crystal will move ballistically, in other words, freely without collisions, ignoring the periodic dense array of atoms that make the crystal. The underlying matrix of atoms determine the ultimate velocity the electrons can acquire and their 'effective mass' in the crystal.

Unfortunately, however, a purely periodic crystal does not exist in nature. Random impurities, lattice vibrations (phonons), and mutual repulsion among the electrons destroy the ballistic motion of the electrons. A statistical averaging is then generally used to find the average velocity as a function of the applied electric field and the mean free time between collisions (resulting in a mean free path or mfp).

When the size of the crystal is comparable to the electrons' mfp, a substantial number of electrons is expected to traverse the crystal ballistically, while the rest are expected to scatter at least once [1]. Recently, this anticipated phenomenon had been observed, in an unambiguous way, by HEIBLUM et al. [2] in a new type of device: Tunnelling Hot Electron Transfer Amplifier (THETA), a device proposed in 1980 [3] . The existence of ballistic electrons in THETA devices enabled the observation of a variety of phenomena associated with small structures. Since most of these effects were very recently reported in the literature, I'll review them here briefly and refer the reader to the original sources.

1. The THETA Device

The history of hot electron devices was described in some detail before [3,4] . The THETA device was proposed originally in the GaAs-AlGaAs system due to the expected long mfp in GaAs and the maturity of crystal growth techniques, such as Molecular Beam Epitaxy. In this device a quasi-monoenergetic hot electron beam, with an adjustable energy, is injected (by an emitter, see Fig. 1) into a thin, doped transport region (base). The base is terminated at its other end by a potential barrier that allows only those electrons that remain sufficiently hot to cross it into the collector. The electrons that lose more energy or those that are in thermal equilibrium with the lattice remain in the base. The injector (emitter) is composed of a tunnel junction formed by a thin, undoped, AlGaAs layer sandwiched between two doped GaAs layers. The applied voltage across the tunnel barrier determines the injection energy of the electrons. The other potential barrier that terminates

* This paper is given also in the 'Fourth International Workshop on Future Electron Devices', Tokyo, Japan (1987).

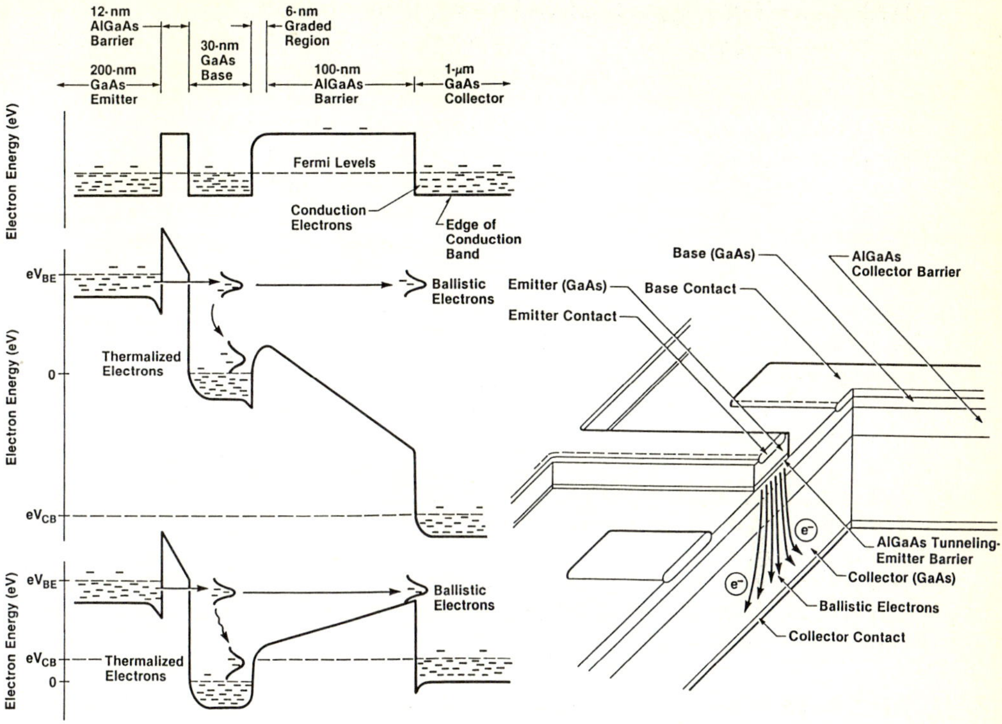

Fig. 1. On the left, an energy band diagram for Γ electrons in the heterojunction - THETA device. Top - unbiased device, center - device biased as an amplifier, bottom - device biased as a spectrometer. On the right, a cut through a schematic representation of a fabricated device.

the base and precedes the collector layer is made of a relatively thick, undoped, AlGaAs thus preventing tunnelling (Fig. 1).

If most of the injected current that can be modulated to carry information arrives at the collector, power or voltage gain can be accomplished by having the total output resistance considerably greater than the input resistance. Since a minimum level of injection energy is needed for the ballistic electron current to surmount the collector barrier height, the device can be operated as a switch with a threshold voltage equal to the collector barrier height. The intrinsic speed of the device will be limited by the transit time of the hot electrons through the structure which can be as short as 0.2 pS.

By varying the potential height of the collector barrier with an applied voltage between base and collector, V_{CB}, the arriving current distribution can be partly blocked from entering the collector. In this way, the electron energy distribution can be mapped as V_{CB} is changed causing the top of the barrier to scan through the distribution. This is the known 'solid-state energy spectroscopy' technique. [3, 5 – 9]

2. Observation Of Ballistic Transport

In the initial phases of the work on hot electron devices, fabricated devices did not exhibit ballistic transport [7,9]. This was because the energy of the injected carriers was very high, thus allowing them to scatter into upper satellite valleys (the L and X valleys in GaAs) and because of the poor quality of the AlGaAs barrier serving as a spectrometer. However, when the injection energies were reduced to below 0.3 eV (which is the approximate Γ to L valley energy splitting) and the material quality was improved, ballistic transport was observed, via the above mentioned spectroscopy technique, in devices with base thickness as wide as 80 nm and collector barriers 100 nm wide.

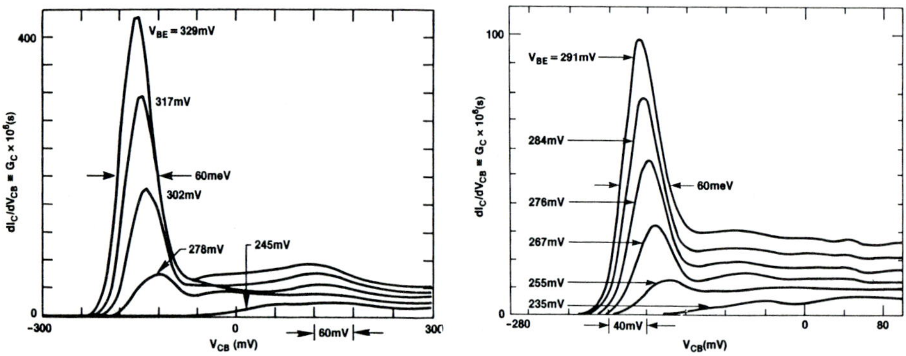

Fig. 2. Measured electron momentum distributions as a function of energy of the collected ballistic electrons for 30 nm (left) and 80 nm (right) base widths. The widths of the distributions are similar, however, about 30% and 15% of the injected currents are ballistic for the 30 nm and 80 nm bases, respectively. The devices have slightly different collector barrier heights.

The mapped ballistic electron distribution had a full width half-height of about 60 meV and a peak energy equal to the injection energy. The 60 meV width is expected also to be generated by the tunnel emitter at the input to the base. The 60 meV width of the distribution was the same for different lengths of the base, however, the number of ballistic electrons in the distribution was smaller for devices with longer bases (Fig. 2). These rather surprising results clearly prove that the ballistic portion of the distribution progresses through the device without any scattering events (which will tend to widen the distribution that traverses longer bases), while the electrons that scatter lose most of their energy, thermalize, and vanish from the detected energy distribution. In devices with low collector barrier heights (about 0.14 eV) and thus wider energy windows for ballistic transport ($E_L - \Phi_C \simeq 0.16$ eV, where E_L is the energy minima of the the L valleys and Φ_C is the collector barrier height), at least 75% of the injected electrons traversed the device ballistically. In these devices a current gain, $\alpha = I_C/I_E = 0.9$ was measured. In some devices, where quantum size effects were prominent (see section 4), the differential $\alpha(= dI_C/dI_E)$ was modulated with peak values exceeding 0.9 and the current gain in a common emitter configuration, $\beta = dI_C/dI_B$, approached 20 at limited ranges of injection energy (Fig. 3).

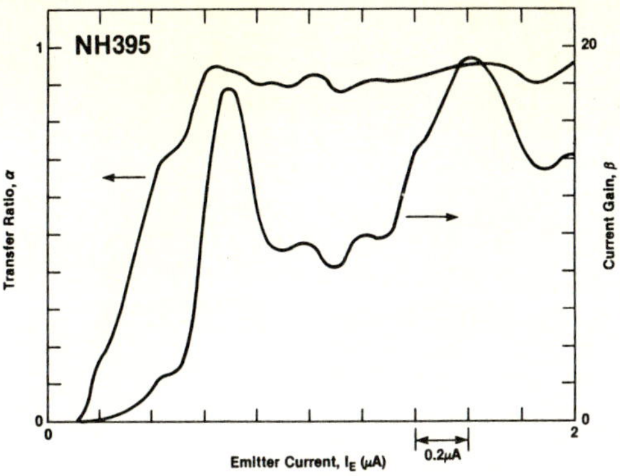

Fig. 3. Differential α and β as a function of emitter current in a device that exhibit strong quantization effects. The base width is 25 nm.

3. Measuring Band Discontinuities

Since the THETA device is a threshold type device, an abrupt onset in the collector current is observed when $eV_{BE} + \zeta = \Phi_C$, where V_{BE} is the injection voltage and ζ is the Fermi level in the base. In AlGaAs that is free of impurities, the barrier height is equal to the conduction band discontinuity, ΔE_C, between the GaAs and the AlGaAs. In these cases, our experimental results can be expressed as $\Delta E_C \simeq 9x$, where ΔE_C is in meV and x is the Al mole percent in the AlGaAs alloy (x<40%). [10] This corresponds to a ratio Q = ΔE_C:$\Delta E_V \simeq$70:30, where ΔE_V is the valence band discontinuity.

4. Quantum Size Effects

When the length of the active region (base) is smaller than about 80 nm, the separation among the quasi two-dimensional electronic energy levels in the base, starts being comparable to the energy width of the ballistic electron beam. Then, the current injected into the base will be enhanced each time the injection energy coincides with a successive energy level. This effect was clearly observed in all structures that carried a substantial fraction of their current ballistically (Fig. 4). [11] Moreover, even when the injection energy exceeded the potential height of the collector barrier, 'resonant' (or 'virtual') states in the energy continuum were clearly observed via modulation effects in the tunnelling currents from the emitter, the collected currents in the collector, and the differential current gain α (Fig. 5).

As long as the base width exceeds about 50 nm the quantum size effects do not affect significantly the charge distribution in the base. Thus, the classical potential calculated via Poison equation is accurate enough to explain the position of the observed modulation peaks (Fig. 4). However, when the base width is on the order of 30 nm, only three levels or so are occupied and the true charge distribution deviates markedly from its classical estimation. Then, a self-consistent, iterative, solution of Poisson and Schrödinger equations is necessary for the calculation of the exact potential and the energy levels in the base. The resultant potential and charge density calculated in the 'classical' and 'self-consistent' approaches are compared in Fig. 6. The bottom figure in Fig. 4 shows the difference between the calculated tunnelling transmissions into the base in the potentials calculated with the two approaches. As we see,the agreement between the experimental and the

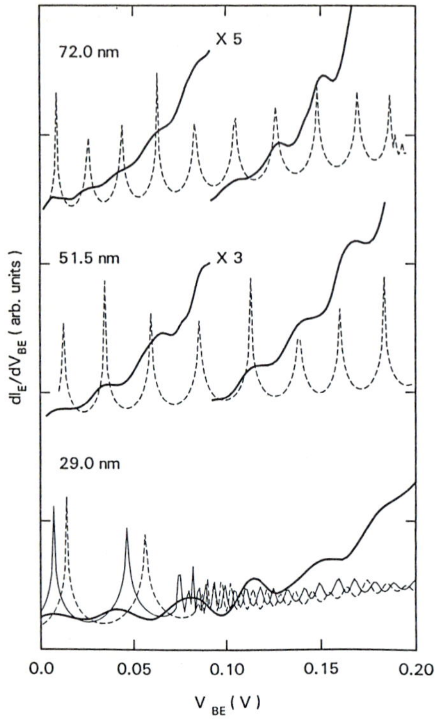

Fig. 4. The experimental derivatives of the injected currents as a function of the injection voltage for three different THETA devices with base widths 72, 51.5, and 29 nm (heavy, solid lines). The dotted lines are the calculated tunnelling transmissions of electrons at the Fermi level of the emitter with the 'classical' potential in the base. The thin, solid line for the 29 nm device, is the transmission calculated with the 'self-consistent' potential.

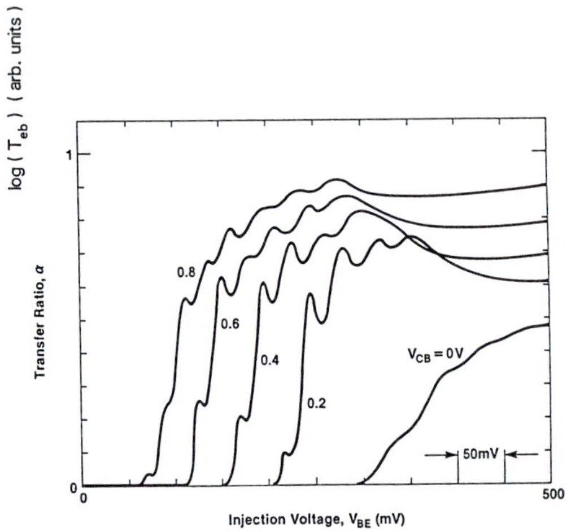

Fig. 5. The differential current gain, α, for different base-collector voltages, exhibits the existence of 'virtual states' in the base. The device has a 30 nm wide base and doping of 1×10^{18} cm^{-3} . Note the disappearance of the modulation for $V_{BE} > 0.3$ V.

Fig. 6. Charge densities and potential distributions calculated in the 'classical' and 'self-consistent' ways for the device with base width of 29 nm.

'self-consistent' results are good. When we integrate over the total electron distribution in the emitter (introducing broadening), the agreement with experiments is excellent. [11]

An important outcome of these experiments is a rather exact determination of the 'appropriate effective mass' at an energy range where non-parabolicity effects are significant. Since only the ballistic electrons (coherent and phase maintaining) are responsible for the observed effects (and not these electrons that have transferred to the satellite valleys or those that have thermalized), the electron mass that is determined is that of the Γ ballistic electrons. The 'appropriate effective mass' used in these calculations is the 'energy effective mass', m, defined via the expression $E(k) = \hbar^2k^2/2m$, where E is the total energy, k the wave vector, and $\hbar$ is the reduced Planck constant. We find that $m(E) = 0.067m_0(1 - \gamma E)$ where $\gamma = -0.834\ (\text{eV})^{-1}$. This mass and not the 'optical mass' or the 'curvature mass' should also be used in all the calculations of enegy levels in quantum wells and superlattices. [11,12]

5. Transfer Into L valleys.

When the injection energy exceeds the energy of the L (or X) valleys, the ballistic electrons will have a certain probability to scatter to these valleys (via optical phonon or impurity scattering). Electrons scattered to the L valleys will randomize in k-space and move in a diffusive motion toward the collector, 'trapped' at the bottom of the L valleys since scattering times back to the Γ valley are much longer than the transit time. Most of the 'L electrons' will not be collected due to an 'L related' potential barrier at the collector barrier, and will be missing from α. This phenomenon is observed clearly in devices that carry a large number of ballistic electrons, [13] and have a relatively thick base. It was estimated that about 25% of the otherwise ballistic electrons scatter via longitudinal optical phonons in a 30 nm long base in one pass that lasts about 30 fS.

The intervalley scattering was verified by applying hydrostatic pressure on the devices and observing the change in the onset of scattering to the L valleys. Due to the decrease in the Γ to L energy splitting with increased hydrostatic pressure (about -5.5 meV/kbar), the onset of scattering was observed to occur at lower energies as the pressure increased, as shown for one device in Fig. 7. Note in the

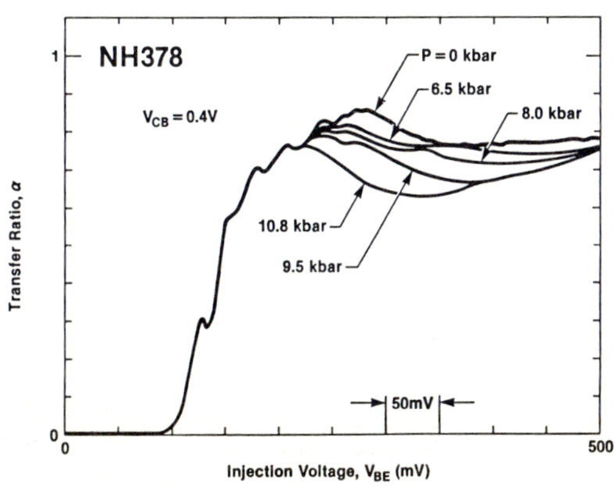

Fig. 7. The differential $\alpha = dI_C/dI_E$ as a function of the injection voltage, at different applied hydrostatic pressures. The onset of transfer into the L valleys is indicated by the drop in α that is shifted to lower energies as the applied pressure increases.

figure, that after the scattering of a sizable fraction of the ballistic electrons to the L valleys (for $eV_{BE} + \zeta > E_L - E_\Gamma$), the modulation of α (due to the 'virtual states') disappears.

6. Where Are We Headed From Here?

Even at this early and immature stage in the development of the THETA device it has proved to be very useful as a 'mini-laboratory' for studying transport of hot electrons. A thorough understanding of the different scattering mechanisms, including the quantum effects is necessary. This should be compared with results from a variety of devices with different base widths and doping concentrations. Spectroscopy should be pursued with sharper electron distributions to look for the existence of weak elastic scattering events. In addition, the effects of different doping methods in the base, such as 'selective doping' or 'planar doping' should be investigated. Indium-based materials might provide a wider window for ballistic transport, a lower base resistance, and an easier way to form ohmic contacts. Since the ultimate speed of the devices is directly related to the current density the devices can support, different types of emitters such as thermionic and resonant tunnelling emitters should be compared with the conventional tunnel emitter for their current driving capabilities. The current gain, α, should be at least 0.95 to make the device practical. This could be achieved most probably by decreasing the base width to about 20 nm or less, grading the collector barrier to minimize quantum mechanical reflections, and invoking one of the doping techniques mentioned above.

The final form of the THETA device, its usefulness as a discrete device or as one of many in a digital circuit, its ultimate speed, and its highest gain are not known yet, but the initial results certainly merit an aggressive research program in this field.

Acknowledgments I wish to thank my collaborators in the past two years: D. C. Thomas, I. M. Anderson, G. C. Wilson, C. M. Knoedler, and J. W. Mitchell in the fabrication process, L. Osterling in material growth, M. V. Fischetti, W. P. Dumke, D. J. Frank, and P. J. Price in the theoretical analysis, and M. I. Nathan and E. Calleja in the experiments.

References

1. S. Tiwari, Int. J. Electronics, 61,(1986), to be published.
2. M. Heiblum, M. I. Nathan, D. C. Thomas, and C. M. Knoedler, Phys. Rev. Lett. 55, 2200 (1985).
3. M. Heiblum, US patent no. 4,286,275, filed 1980; M. Heiblum, Solid-State Electron. 24, 343 (1981).
4. M. Heiblum, in High - Speed Electronics, Eds. B. Källbäck and H. Beneking, Springer Series in Electronics and Photonics, 22, 11 (1986).
5. D. J. Barterlink, J. L. Moll, and N. I. Meyer, Phys. Rev. 130, 972 (1963).
6. P. Hesto, J-F. Pone, and R. Castagne, Appl. Phys. Lett. 40, 405 (1982).
7. N. Yokoyama, K Imamura, T. Ohshima, H. Nishi, S. Muto, K. Kondo, and S. Hiyamizu, IEEE Electron Device Meeting Proceedings, San Francisco (1984).

8. J.R. Hayes, A.F.J. Levi, and W. Wiegmann, Electron. Lett. 20, 851 (1984); Phys. Rev. Lett. 54, 1570 (1985).

9. M. Heiblum, D. C. Thomas, C. M. Knoedler, and M. I. Nathan, Appl. Phys. Lett. 47, 1105 (1985).

10. M. Heiblum, I. M. Anderson, and C. M. Knoedler, Appl. Phys. Lett. 49, 207(1986).

11. M. Heiblum, M. V. Fischetti, W. P. Dumke, D. J. Frank, I. M. Anderson, C. M. Knoedler, and L. Osterling, Phys. Rev. Lett., Submitted.

12. T. Hiroshima and R. Lang, Appl. Phys. Lett. 49 , 456 (1986).

13. M. Heiblum, E. Calleja, I. M. Anderson, W. P. Dumke, C. M. Knoedler, and L. Osterling, Phys. Rev. Lett. 56, 2854 (1986).

Microwave and Millimeter-Wave Resonant Tunneling Diodes

T.C.L.G. Sollner, E.R. Brown, and W.D. Goodhue

Lincoln Laboratory, Massachusetts Institute of Technology, Lexington, MA 02173, USA

Resonant tunneling through double-barrier heterostructures has attracted increasing interest recently, largely because of the fast charge transport [1] it provides. In addition, the negative differential resistance regions which exist in the current-voltage (I-V) curve (peak-to-valley ratios of 3.5:1 at room temperature [2-4] and nearly 10:1 at 77 K have been measured) suggest that high-speed devices based on the peculiarities of the I-V curve should be possible. For example, the negative differential resistance region is capable of providing the gain necessary for high-frequency oscillations [5]. In our laboratory we have been attempting to increase the frequency and power of these oscillators [6], and others have worked toward a better understanding of the equivalent circuit of the device [7] and the underlying processes responsible for the frequency response [8-10]. Three-terminal devices using resonant tunneling in various ways have also been proposed and fabricated [11-13]. In this paper we will describe our most recent results for oscillators as well as some new resonant-tunneling devices that have application in the millimeter and submillimeter-wave spectrum.

1. Resonant Tunneling Oscillators

Our recent room-temperature, millimeter-wave oscillator results are summarized in Fig. 1. The initial experiments at 20 GHz were performed in a coaxial circuit, and the oscillations

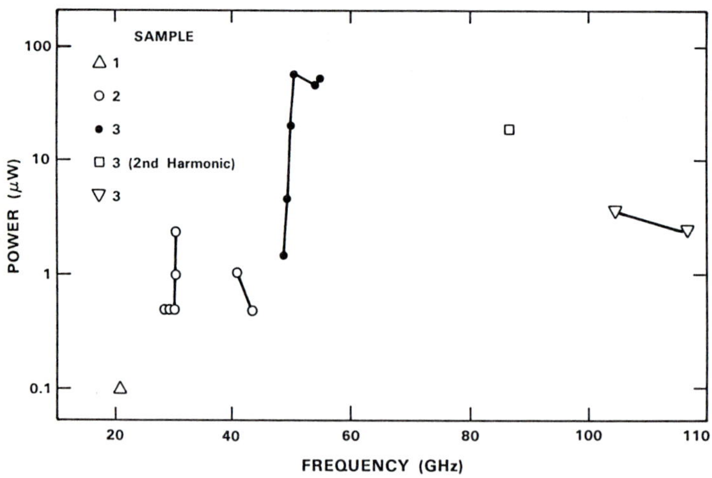

Fig. 1. Output power as a function of frequency for recent resonant-tunneling oscillators. The connected dots indicate continuous tunability in a particular waveguide structure. All measurements were made at room temperature.

around 30 and 40 GHz were achieved in WR-22 and WR-15 waveguide resonators, respectively [14]. A significant improvement in the quality of the devices, especially the use of thin AlAs barriers in place of AlGaAs barriers, resulted in oscillations near 56 GHz in the WR-15 resonator [6]. The oscillations between 104 and 108 GHz were obtained with the same AlAs-barrier material in a WR-6 resonator. We believe that this material is capable of oscillations approaching 200 GHz, and that optimized material should be capable of fundamental oscillations as high as perhaps 1 THz. More details on material parameters can be found in References [6] and [15].

Since the resonant-tunneling I-V curve is nonlinear, especially near the negative differential resistance region, the same device can act as both a mixer and an oscillator. Figure 2 shows the difference frequency between two oscillators, one of which is a resonant-tunneling quantum-well oscillator. The oscillator output power of this early device operating at 20 K is rather low; nevertheless, it appears to be an efficient mixer as well as a compact oscillator. The best DC bias point for mixing was found to be at either side of the negative differential resistance region, rather than in the center where the oscillator output power is maximized, to take advantage of the larger nonlinearity of the I-V curve there.

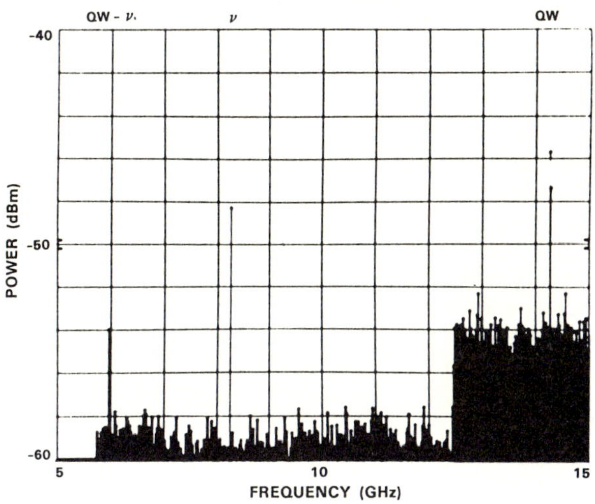

Fig. 2. Measured power spectrum of a resonant tunneling oscillator (marked QW) mixing with an injected signal (marked ν) to produce a difference-frequency output (marked QW-ν). This oscillator was cooled to 20 K.

2. Absolute Negative Resistance

If a resonant tunneling structure is appropriately pumped at high frequency, a negative resistance can appear with no DC bias applied, *i.e.* an absolute negative resistance exists. The dashed line in Fig. 3 shows a resonant-tunneling I-V curve typical of our early structures at 20 K. Ordinarily, the negative differential resistance regions occur at about ± 0.35 V. When an AC waveform (*i.e.*, the pump) is applied to the terminals with an amplitude of about 0.35 V, the negative differential resistance region is sampled for some fraction of the pump period, resulting in a negative resistance appearing near $V_{DC} = 0$. The solid

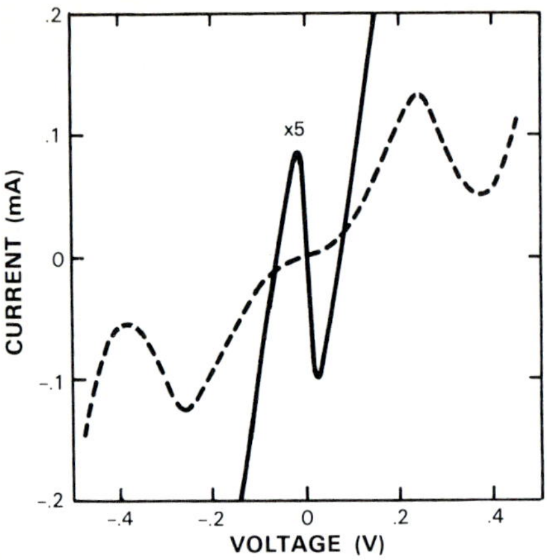

Fig. 3. Current-voltage curves for one of our early resonant-tunneling structures operating at 20 K: dashed line is measured DC characteristics; the full line results when a 1 GHz pump with an amplitude of about 0.4 V has been applied. The negative resistance in the region near V = 0 can be varied by changing the amplitude of the pump.

curve of Fig. 3 shows the DC I-V curve that results when a sinusoidal pump wave at 1 GHz with an amplitude of about 0.4 V is applied. Note that the value of this absolute negative resistance is nearly the same as the differential negative resistance in the unpumped DC curve, and the voltage region spanned by the absolute negative resistance is about half of that for the differential negative resistance.

This phenomenon can be understood qualitatively from Fig. 4. Approximating the pump waveform as a square wave and the unpumped I-V curve by straight lines, absolute negative resistance occurs at DC when the pump amplitude is greater than the peak voltage, but not greater than the valley voltage. In the actual experiment, the positive resistance regions are weighted more heavily than in the case of a square-wave pump, but the principle is the same as shown in Fig. 4.

From the explanation given above, one would expect to find absolute negative resistance whenever a material with negative differential conductance and an I-V curve that is antisymmetric (the current is an odd function of the voltage) is driven with a pump of the right amplitude and frequency. In fact, this effect has been observed by POZHELA [16] with a sample of bulk GaAs. In that case the negative differential mobility for electric fields above approximately 1 kV/cm is the basis for the absolute negative resistance.

To understand more quantitatively the origin of the absolute negative resistance, and to investigate the frequency dependence, the I-V curve can be represented by a polynomial. A reasonable fit to the I-V curve shown in Fig. 3 is obtained with the expansion

$$i = \sum_{n=1}^{4} a_{2n-1}\, e^{2n-1} \qquad (e < 0.5\ V) \qquad \text{and with}$$

$$a_1 = 0.32\ \text{mA/V}$$

$$a_3 = 8.2\ \text{mA/V}^3$$

$$a_5 = -124\ \text{mA/V}^5$$

$$a_7 = 403\ \text{mA/V}^7.$$

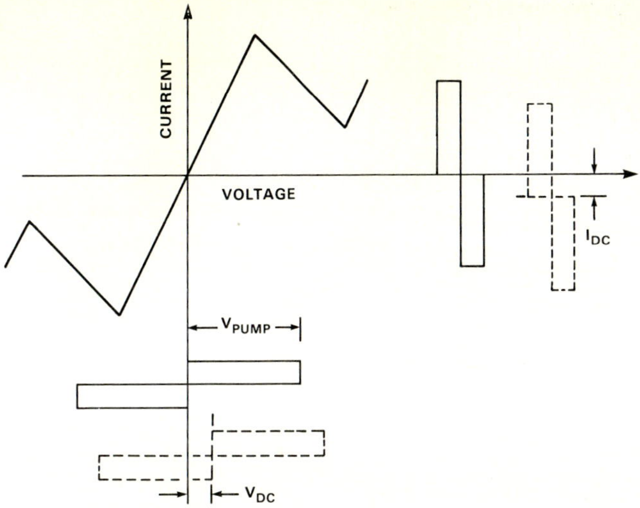

Fig. 4. Origin of the absolute negative resistance observed when a device with antisymmetric differential negative resistance regions is pumped with a waveform that has the amplitude necessary to reach these regions. A positive change in the DC value of the voltage waveform produces a negative change in the DC value of the current waveform.

Only odd powers of the voltage e appear because of the point symmetry of the I-V curve. Suppose the voltage is time-dependent with a large pump amplitude e_1 at frequency ω_1, and a small signal amplitude e_2 at frequency ω_2, *i.e.*,

$$e = e_1 \cos\omega_1 t + e_2 \cos\omega_2 t.$$

There will, of course, be harmonics of the pump generated by the nonlinearity of the I-V curve, and to do this analysis properly they should be calculated in a self-consistent manner (see, for example, KERR [17]). Higher harmonics will be neglected in the driving voltage, in spite of the fact that they appear in the conductance, to obtain a simple physical picture of the device behavior. Some quantitative information will be lost by this approximation. Keeping terms to first order in e_2 yields

$$e^{2n-1} \cong (e_1 \cos\omega_1 t)^{2n-1} + (2n-1)\, e_2 \cos\omega_2 t\, (e_1 \cos\omega_1 t)^{2n-2}.$$

Then the signal conductance is

$$\frac{di}{de_2} = \sum_{n=1}^{4} \left\{ a_{2n-1} (2n-1)(e_1 \cos\omega_1 t)^{2n-2} \right\} \cos\omega_2 t.$$

Unless $\omega_2 = (n-1)\omega_1$, the only terms inside the braces that contribute to the conductance at frequency ω_2 are the constant terms. These terms have been plotted in Fig. 5.

The foregoing asserts that at any frequency, even those well above the pump frequency ($\omega_2 \gg \omega_1$), the conductance of the pumped device will be negative at the signal frequency when the pump amplitude is within the region shown in Fig. 5. Furthermore, the magnitude of the negative resistance can be adjusted by varying the pump amplitude, a very useful circuit property. The fact that no DC connection need be made to the sample

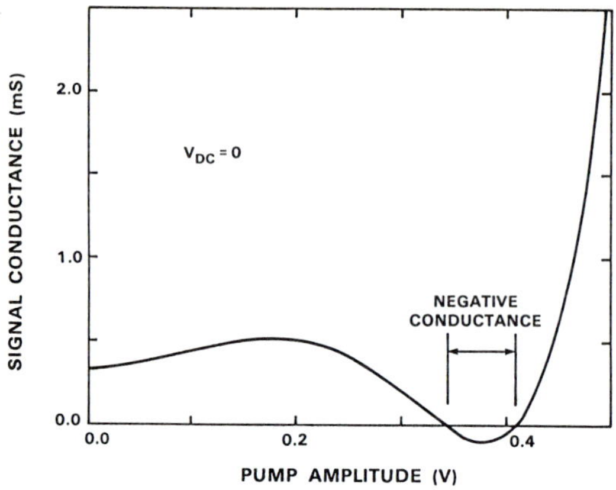

Fig. 5. Calculated conductance at the signal frequency as a function of the pump amplitude for I-V curve shown in Fig. 3. The DC bias voltage is zero. For small signal amplitudes the signal conductance can be made negative and is variable in the region near 0.4 V pump amplitude as marked. The signal frequency can be larger than the pump frequency, as discussed in the text.

simplifies the problem of stabilizing a two-terminal negative-resistance device, because the low-frequency circuit at which unwanted oscillations could occur is eliminated. In addition, further analysis shows that when $\omega_2 = 2\omega_1$ there is a slight increase in the negative conductance above that calculated in Fig. 5. This increase in negative conductance at a specified frequency could simplify frequency selection for an oscillator based on this effect. Finally, it should be noted that we have gained the advantage of operating with zero DC bias voltage with little sacrifice in negative conductance or dynamic range.

It is implicit in the representation of the $\cos^{2n}\theta$ by constant and time-variable terms that some form of averaging over the pump period is available. In the case of an oscillator based on the absolute negative resistance, this may take the form of a resonant cavity to store the signal energy during a pump period. The cavity Q gives an estimate of the number of signal periods that can be stored, so as long as $\omega_2 < Q\omega_1$ oscillation can occur. Thus an oscillation frequency of several hundred gigahertz with a pump of 10 GHz would require a cavity Q of only 100 or so.

It appears that the variable absolute negative resistance recently observed in double-barrier diodes can be used as the basis for oscillators up to the cutoff frequency of the diode. Also, this phenomenon is quite generally applicable to any material which has a point symmetric I-V relation with negative differential resistance. It could be useful with materials such as superlattices, which form domains when biased into their regions of negative differential resistance for long times; it may be possible to sweep them into that region for a time short compared to their domain-formation time with a high-frequency pump, so that the negative resistance is retained.

3. A Quintupler

The undulations of the unpumped I-V curve of Fig. 3 also suggest that there should be large harmonic content to the current waveform, leading to efficient harmonic multiplication. Figure 6 shows the calculated voltage and current waveforms for this same device

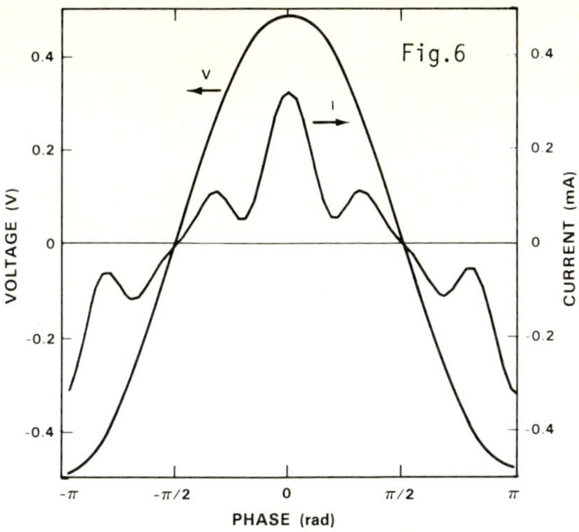

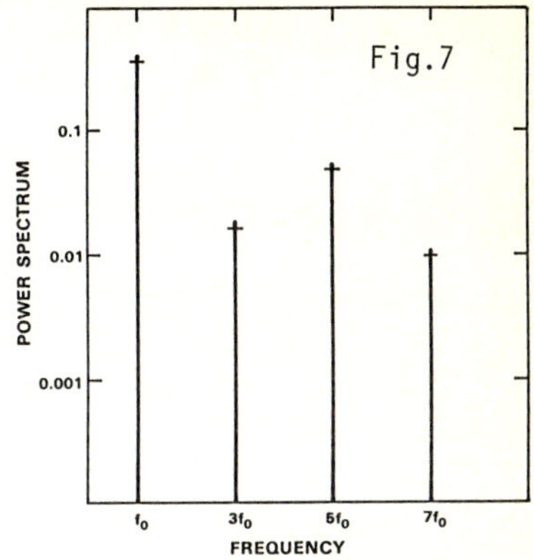

Fig. 6. Calculated voltage and current waveforms for the DC I-V curve of Fig. 3 when driven by a pump with a 50 Ω source impedance.

Fig. 7. Power spectrum calculated from the current waveform in Fig. 4. The even harmonics are absent because of the symmetry of the I-V curve, while the large 5th harmonic content derives from the occurrence of the two negative resistance regions.

when driven by a 50-Ω source with an amplitude of about 0.5 V, and the power spectrum of the current waveform is shown in Fig. 7.

There are several interesting features of this spectrum. One of the most difficult problems in harmonic multiplier design is to reactively terminate all the harmonics lower than the desired output. Since these harmonics are usually of larger amplitude than the higher-order ones, the reactive termination is necessary to avoid unwanted dissipation of power. In the present case the even harmonics are entirely missing because of the symmetry of the I-V curve, thus halving the number of harmonics which need to be considered. Also, for the pump power considered, the largest component after the fundamental is the 5th harmonic, so even if the power in the other harmonics is resistively terminated one could expect as high as 10% conversion efficiency to the 5th harmonic. So far we have observed about 1% conversion efficiency from 2 to 10 GHz, about the same as current state-of-the-art Schottky diode quintuplers. Optimizing the multiplier structure to provide the proper impedances should raise the efficiency significantly. The probable ultimate limit for the output frequency of a resonant-tunneling quintupler is the same as for the oscillator, on the order of 1 THz.

We have described a few interesting and useful high-speed devices based on resonant tunneling, and there are undoubtedly many more that await discovery. Optical interactions with these devices have not been discussed here, but this is a fruitful area of future research, combining the high temporal resolution of optical pulses with the high speed of resonant tunneling. We expect the general area of resonant tunneling to provide a fertile ground for both solid state physics and active device investigations for some time to come.

Acknowledgments

We wish to thank G.D. Johnson, K.M. Molvar, and N. Usiak for assistance in fabrication and packaging our devices. We are also indebted to C.L. Chen, P.J. Daniels, and P.E. Tannenwald for helpful discussions, and to C.A. Correa and C.D. Parker for help in making the measurements. This work was supported by the U.S. Army Research Office and by NASA.

4. References

1. T.C.L.G. Sollner, W.D. Goodhue, P.E. Tannenwald, C.D. Parker, and D.D. Peck, Appl. Phys. Lett. **43**, 588 (1983).
2. M. Tsuchiya and H. Sakaki, IEEE Int. Electron Devices Meeting, Washington, DC, 1985, p. 662.
3. T.J. Shewchuk, P.C. Chapin, P.D. Coleman, W. Kopp, R. Fischer, and H. Morkoç, Appl. Phys. Lett. **46**, 508 (1985).
4. W.D. Goodhue, T.C.L.G. Sollner, H.Q. Lee, E.R. Brown and B.A. Vojak, Appl. Phys. Lett. **49**, 1086 (1986).
5. T.C.L.G. Sollner, P.E. Tannenwald, D.D. Peck, and W.D. Goodhue, Appl. Phys. Lett. **45**, 1319 (1984).
6. E.R. Brown, T.C.L.G. Sollner, W.D. Goodhue, and C.D. Parker, Appl. Phys. Lett. **50**, 83 (1987).
7. P.D. Coleman, S. Goedeke, T.J. Shewchuk, and P.C. Chapin, J.M. Gering, and H. Morkoç, Appl. Phys. Lett. **48**, 422 (1986).
8. S. Luryi, Appl. Phys. Lett. **47**, 490 (1985).
9. D.D. Coon and H.C. Liu, Appl. Phys. Lett. **49**, 94 (1986).
10. W. Frensley, IEEE Int. Electron Devices Meeting Tech. Digest, Los Angeles, 1986, paper 25.5.
11. T.C.L.G. Sollner, H.Q. Le, C.A. Correa, and W.D. Goodhue, IEEE/Cornell Conf. Advanced Concepts in High Speed Semicond. Devices and Circuits, Ithaca, 1985, p. 252.
12. N. Yokoyama, K. Imamura, S. Muto, S. Hiyamizu, and H. Nishii, Jpn. J. Appl. Phys. **24**, L583 (1985).
13. F. Capasso, S. Sen, A.C. Gossard, A.L. Hutchinson, and J.E. English, IEEE Electron Device Lett. **EDL-7**, 573 (1986).
14. E.R. Brown, T.C.L.G. Sollner, W.D. Goodhue, B.J. Clifton, and P.E. Tannenwald, Device Research Conference, Amherst, 1986, paper IVB-1.
15. T.C.L.G. Sollner, E.R. Brown, W.D. Goodhue, and H.Q. Le, submitted to Appl. Phys. Lett.
16. J. Pozhela, *Plasma and Current Instabilities in Semiconductors*, Int. Ser. Solid State, vol. 18 (Pergamon, Oxford 1981).
17. A.R. Kerr, IEEE Trans. Microwave Theory Tech. **MTT-23**, 54 (1977).

Quantum Transport Calculation of Resonant-Tunneling Response Time

W.R. Frensley

Texas Instruments Incorporated, P.O. Box 655936, M.S. 154, Dallas, TX 75265, USA

The progress of semiconductor fabrication technology, particularly the hetero-epitaxial technology, has permitted the fabrication of devices whose behavior is dominated by quantum-interference effects. The most widely studied example of a quantum size-effect device is the resonant tunneling diode (RTD) [1,2]. This device exhibits interesting properties in the form of a negative-resistance region of its characteristic curve, and there are indications that its response time might be very short [2]. In the present work the RTD was taken as a prototype quantum device for the development of techniques for the analysis of such devices.

In the present model [3] the device is considered to be a finite region of semiconductor, characterized by a potential that includes the effects of applied voltages and of heterojunction band offsets. The boundaries of the device are taken to be interfaces to particle reservoirs, by which the terminals of the device are modeled. The internal state of the device is represented by the Wigner distribution function [4], which is the quantum analog of the classical distribution function $f(x,v)$ where x the is position and v is the velocity.

Regarding the contacts as particle reservoirs gives a well-defined model of the open-system nature of the device. The interaction between the reservoir and the device is simply described: Electrons in the device which impinge upon the reservoir pass into the reservoir without reflection and the distribution of electrons entering the device from the reservoir is given by the equilibrium distribution of the reservoir. These boundary conditions are a crucial aspect of the model, because they permit the existence of steady-state solutions under applied bias and they lead to a stable approach to those steady-state solutions after the bias voltage is changed.

The time evolution of the Wigner function is described by the Liouville equation

$$\frac{\partial f}{\partial t} = -\frac{\hbar k}{m}\frac{\partial f}{\partial x} - \frac{1}{\hbar}\int_{-\infty}^{\infty}\frac{dk'}{2\pi}\,V(x,k'-k)\,f(x,k') \tag{1}$$

where the kernel of the potential operator is given by

$$V(x,k) = 2\int_{0}^{\infty} dy\,\sin(ky)\,[v(x+\tfrac{1}{2}y) - v(x-\tfrac{1}{2}y)]\,. \tag{2}$$

The open-system boundary conditions are

$$\begin{aligned} f(0,k) &= F(\mu_l, T_l) \quad k>0 \\ f(l,k) &= F(\mu_r, T_r) \quad k<0 \end{aligned} \tag{3}$$

where F is the Fermi distribution function (integrated over the transverse momenta), $\mu_{l,r}$ are the Fermi levels, and $T_{l,r}$ are the temperatures of the respective contacts.

Equation (1) is discretized on a uniform mesh in the phase space (x,k). The boundary conditions lead to a natural discretization of the gradient term with a left-hand difference for $k>0$ and a right-hand difference for $k<0$. This is an "upwind" difference and is the means by which the boundary conditions stabilize the solutions of the Liouville equation. The dc characteristics of the RTD are obtained by solving the Liouville equation for $df/dt=0$. Figure 1 shows the Wigner distribution function which results from this procedure.

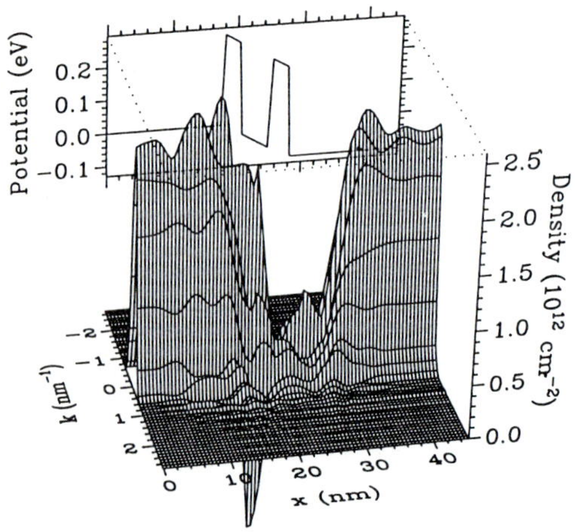

Figure 1. Wigner function for a resonant-tunneling diode biased at the peak current.

The steady-state calculation is repeated for a number of values of bias voltage to obtain the I(V) curve. Figure 2 shows the results of the I(V) calculation using the Wigner function and using a more conventional scattering-state model. The Wigner function predicts a higher current density than does the scattering theory. In particular, the valley current is significantly higher, leading to a lower predicted peak-to-valley ratio. Because the phonon scattering processes which are responsible for most of the temperature dependence of the I(V) curve are not included in the present

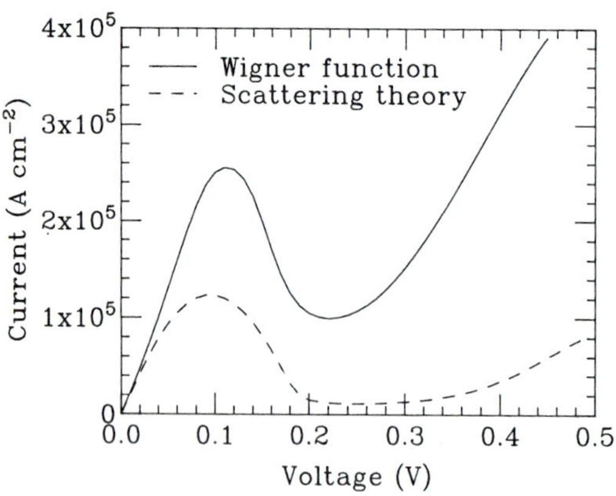

Figure 2. I(V) characteristics derived from the Wigner function calculation and from a conventional scattering theory calculation. The assumed device structure had 2.8 nm barriers of $Al_{0.3}Ga_{0.7}As$ around a 4.5 nm quantum well. The temperature was 300 K.

model, the I(V) curve calculated for lower temperatures shows a similar peak-to-valley ratio. At these lower temperatures, the Wigner function calculations in their present form significantly underestimate the peak-to-valley ratio as compared to the better experimental devices [5].

To obtain the transient response, one must integrate the Liouville equation with respect to time. The linear system of differential equations which results from the discretization in phase space is "stiff" in that the eigenvalues of the Liouville operator span a large range of magnitudes. To assure the stability of the solution, a fully implicit or backward-difference approach was used for the time integration. The steady-state solution for an appropriate bias voltage was used as the initial value for the Wigner function.

The calculated transient response of the RTD is shown in Fig. 3. For this calculation, the device was assumed to be initially in the steady state at the peak of the I(V) curve. The bias voltage was suddenly switched at t=0 to the value at the bottom of the I(V) curve. The device shows positive resistance behavior for a short time, as the current increases. There is a particularly prominent peak in the current at the downstream side of the quantum well after 5 fs. This is due to the electrons in the well piling up against the barrier. The current has essentially reached its steady-state value by 200 fs.

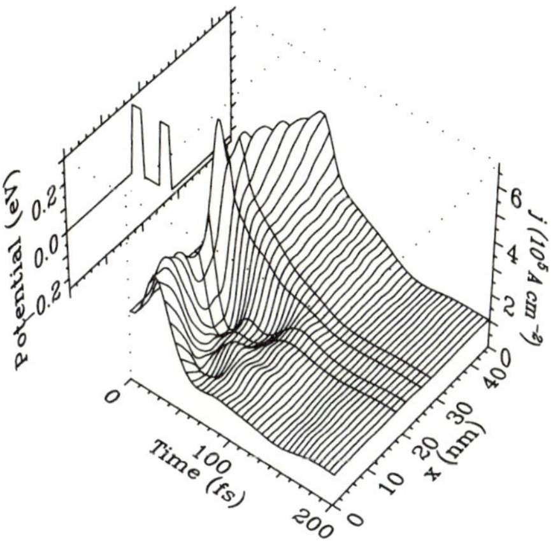

Figure 3. Transient response of the resonant-tunneling diode. Current density is plotted as a function of position within the device and as a function of time after a sudden change in bias voltage. At t=0 the bias voltage was changed from 0.11 V (corresponding to the peak current) to 0.22 V (the bottom of the valley).

The characteristics of the RTD are quite sensitive to the width of the barrier layers. As shown in Fig. 4, the peak current density varies exponentially with the width of the barrier. The peak electron density in the quantum well also varies, but much less sensitively. A simple estimate of the response time of the device may be obtained from $\tau = qn/j$, where q is the electron charge, n is the electron density in the well, and j is the current density. This estimate of the response time is also plotted in Figure 4. The response time estimated in this way is of the same order of magnitude as that which is obtained from the detailed transient response calculations described above. It varies approximately exponentially with the barrier width.

In summary, a model of an open quantum system has been presented. Irreversibility is introduced into the model through boundary conditions on the Wigner function which couple the system to particle reservoirs. The model was applied to the resonant-tunneling diode and the irreversibility permitted

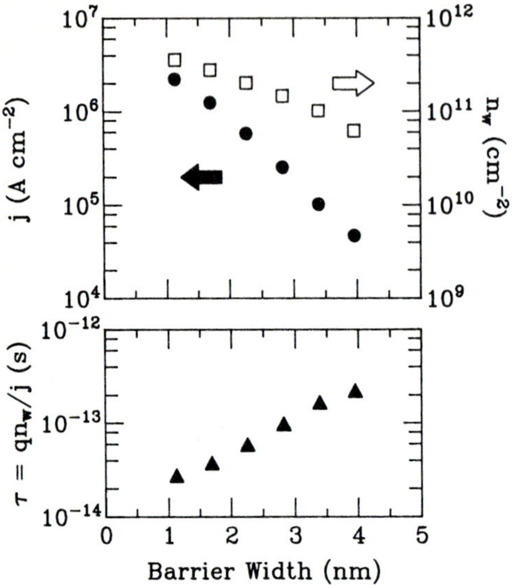

Figure 4. Peak values of current density (solid circles) and electron density in the quantum well (open circles) as a function of barrier width for a 4.5 nm quantum well. The simple estimate of response time $\tau = qn/j$ is plotted in the lower graph.

a calculation of the approach to steady state of the tunneling current after a change in the applied voltage. This transient response is a feature of tunneling devices which has not been addressed by the conventional models based upon scattering theory[6].

This work was supported in part by the Office of Naval Research and the Army Research Office.

1. L.L. Chang, L. Esaki, and R. Tsu: Appl. Phys. Lett. 24, 593 (1974).
2. T.C.L.G. Sollner, W.D. Goodhue, P.E. Tannenwald, C.D. Parker, and D.D. Peck: Appl. Phys. Lett. 43, 588 (1983).
3. W.R. Frensley, Phys. Rev. Lett. 57, 2853 (1986).
4. E. Wigner: Phys. Rev. 40, 749 (1932).
5. M.A. Reed, J.W. Lee, R.K. Aldert, and A.E. Wetsel: J. Mater. Res. 1, 337 (1986).
6. B. Ricco and M.Ya. Azbel: Phys. Rev. B29, 1970 (1984).

Resonant Tunneling Electron Spectroscopy

F. Capasso, S. Sen, A.Y. Cho, and A.L. Hutchinson*

AT&T Bell Laboratories, 600 Mountain Avenue,
Murray Hill, NJ 07974, USA

In this paper we demonstrate a new electron spectroscopy technique based on resonant tunneling. The key difference compared to conventional hot electron spectroscopy[1] is the use of a resonant tunneling double barrier in the collector of the structure (Fig. 1). The advantage of this new feature is that it allows one to obtain information on the electron momentum distribution $n(p_\perp)$ (or energy distribution $n(E_\perp)$) perpendicular to the layers *directly from the measured resonant tunneling collector current, without requiring the use of derivative techniques.* Figure 1 illustrates the band diagrams of two structures for resonant tunneling electron spectroscopy. The first one (Fig. 1a), realized by us in the present experiment, consists of a reverse biased pn heterojunction and can be used to investigate hot minority carrier transport. Incident light is strongly absorbed in the wide-gap p^+ layer. Photo-generated minority carrier

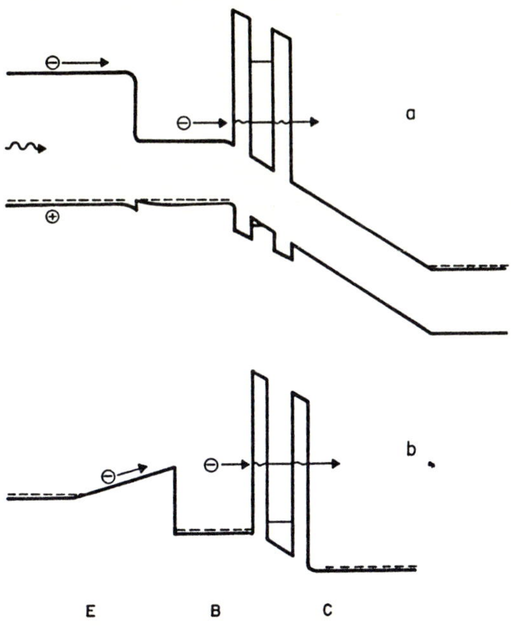

Fig. 1 **a.** Band diagrams of structure used for resonant tunneling spectroscopy (RTS) of hot minority carrier electrons.

b. Unipolar transistor structure for RTS of hot majority carrier electrons in the base layer.

* On leave from the Institute of Radio Physics and Electronics. University of Calcutta, Calcutta-700009, India.

electrons diffuse to an adjacent low-gap layer. Upon entering this region electrons are ballistically accelerated by the abrupt potential step and gain a kinetic energy $\cong \Delta E_c$. Collisions in the low gap layer tend to randomize the injected, nearly mono-energetic distribution, making it "hot". Hot electrons subsequently impinge on the double barrier in the collector. From simple considerations of energy and lateral momentum $p_\perp$ in the tunneling process[2] it can be shown that only those electrons with a perpendicular energy $p_\perp^2/2m_e^*$ equal (within the resonance width) to the energy of the bottom of one of the subbands (i.e., a resonance) of the quantum well, resonantly tunnel through the quantum well and give rise to a current. Thus by varying the applied bias (i.e., changing the energy difference between the resonance of the quantum well and the bottom of the conduction band in the low gap p^+ layer) and measuring the current, one directly probes the electron energy distribution $n(E_\perp)$ or equivalently the momentum distribution $n(p_\perp)$. Thermionic currents can be minimized by appropriate choice of the double barrier and low temperature. Identical arguments apply to the unipolar transistor structure of Fig. 1b. The structures were grown by MBE and consist of p^+n heterojunctions. Their band diagram is shown in Fig. 1a at a given reverse bias. The growth starts with a buffer layer followed by an undoped 5000 Å GaAs layer and an AlAs/GaAs/AlAs double barrier, with barrier and well thicknesses 20 Å and 80 Å respectively. The parameters of the double barrier were chosen

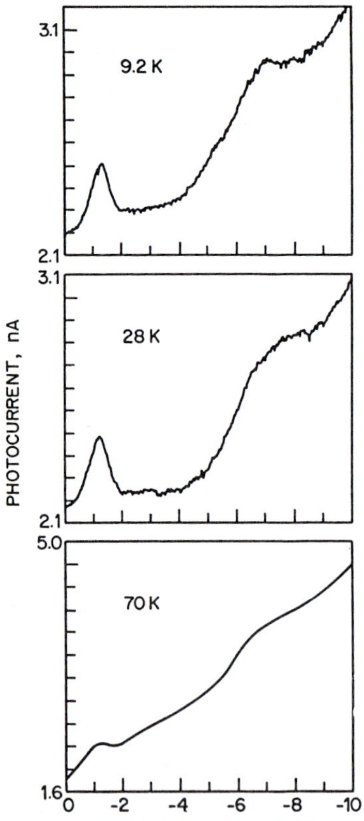

Fig. 2 **Photocurrent vs reverse bias voltage of the structure of Fig. 1a with a low gap GaAs p^+ layer of thickness = 500 Å. The wavelength of the incident light is 6328 Å.**

in such a way that over the applied voltage range (0-10V) the electron energy distribution is probed by only one resonance at a time. In this way at any given bias the current is due to resonant tunneling through a single resonance. For our double barrier the first resonance is at $E_1 = 60$ meV from the well bottom and the second at $E_2 = 260$ meV. A 20 Å undoped GaAs spacer layer separates the double barrier from the p^+ ($= 3x10^{18}cm^{-3}$) GaAs layer. Different thicknesses were used for this region (250 Å , 500 Å , 1000 Å) while keeping everything else the same. The last layer consists of 2 μm thick $Al_{0.3}Ga_{0.7}As$ doped to $p = 3x10^{18}cm^{-3}$. Fig. 2 illustrates the measured photocurrent for a p^+ GaAs layer thickness of 500 Å . Two distinct features are present at 1.2 V and 7 V respectively. One can easily show that the first peak corresponds to electrons with perpendicular energy of a few tens of meV ($\simeq$ 17 meV) that have resonantly tunneled through the first resonance of the quantum well. The second peak, clearly visible only at the lowest temperature, is much broader and corresponds to incident electrons with energy $E_\perp \simeq 122$ meV which have resonantly tunneled through the second resonance of the well. It is therefore clear that the energy distribution of the electrons in the p^+ GaAs layer, following high energy injection (0.233 eV, i.e. the conduction band discontinuity between GaAs and the $Al_{0.37}Ga_{0.7}As$) consists of two parts. One is strongly thermalized, with energies close to the bottom of the conduction band, while the other one is much hotter. Thus the distribution is strongly non-maxwellian similar to what has been found, by conventional electron spectroscopy, in the case of majority carrier electrons in the base of hot electron transistors[1]. Detailed analysis of the data for different p^+ GaAs layer thickness also show that the scattering time for energetic electrons ($\simeq$ 0.2 eV) injected in p^+ ($3x10^{18}cm^{-3}$) GaAs is very short ($\simeq 10^{-14}$ sec). This implies that it will be extremely difficult to implement a ballistic bipolar transistor.

REFERENCES

1. For a review see J. R. Hayes and A. F. J. Levi, IEEE J. Quantum Electron. *QE-22,* 1744 (1986).
2. S. Luryi, Appl. Phys. Lett. *47,* 490 (1983).

Analysis of Transit Time Effects due to Spacer Layers in Quantum Well Oscillators

V.P. Kesan, T.D. Linton, P.A. Blakey*, D.P. Neikirk, and B.G. Streetman*

Microelectronics Research Center, Department of Electrical and Computer Engineering, University of Texas at Austin, Austin, TX 78712, USA

1. Introduction

Quantum well oscillators [1-9] are the subject of much research activity, as potential candidates for millimeter-wave sources. Resonant tunneling through quantum wells is inherently a very fast transport mechanism, and hence very high cut-off frequencies (in the order of hundreds of GHz) have been suggested. The most common heterostructure used in quantum well oscillators has been the $Al_xGa_{1-x}As$/ GaAs system. However, the output power obtained from pure quantum well oscillators has been very low due to low dc bias voltages and area and impedance limitations associated with the capacitance of very narrow junctions. In order to improve room-temperature operation of quantum wells undoped spacer layers on either side of the quantum well have been used to prevent dopant migration from the heavily doped contact layers into the quantum well region. The thicknesses of these nominally undoped spacer layers vary greatly in thickness from 50Å to 3000Å [8-10]. Previously no transit time effects due to these spacer layers have been considered in the analysis of the rf performance of quantum wells when used as an oscillator.

We propose an optimized quantum well structure considering transit time and dopant migration effects using asymmetric spacer layers. A thin (~20Å) spacer layer is introduced before the quantum well to prevent dopant migration and a thicker unintentionally doped layer is used after the quantum well to form the drift region for transit time operation. We will show that the thickness of the spacer layer after the quantum well, now forming a drift region, when appropriately designed using transit time considerations for a particular frequency of operation, can significantly improve the rf performance of the device over pure quantum well oscillators. The presence of this drift region increases both the device impedance and bias voltage so higher output power can be obtained. Since this structure uses quantum well injection and transit time effects, we call it a QWITT diode.

2. Device Operation

The physical structure of this device, consistent with a typical molecular beam epitaxy growth sequence, is shown in Fig. 1. The device structure consists of a single GaAs quantum well between two $Al_xGa_{1-x}As$ barriers, with a thin region of undoped GaAs before the quantum well and a much thicker undoped GaAs layer after the quantum well. This structure is then placed between two n^+-GaAs regions to form contacts. The energy band diagram of the device, when no dc bias is applied, is shown in Fig. 1. For an isolated quantum well, when the voltage across the well equals $2E_1/q$, a large peak in the current through the device occurs. In the QWITT diode, an idealized picture of operation of this device can be explained using Figs. 2&3. If an appropriate bias is applied so that the Fermi level in the n^+-GaAs is above the energy level E_1 by an amount V_{RF} (Fig. 2), and an ac signal $V_{RF}\sin\omega t$ is superimposed on this bias, then resonant tunneling of electrons through the well will peak at an angle of $3\pi/2$ in the ac cycle (Fig. 3). The electrons should then traverse through the drift

*Regular address: Microelectronics and Computer Technology Corporation (MCC), VLSI-CAD Program, 3500 W. Balcones Center Dr., Austin, Texas 78759.

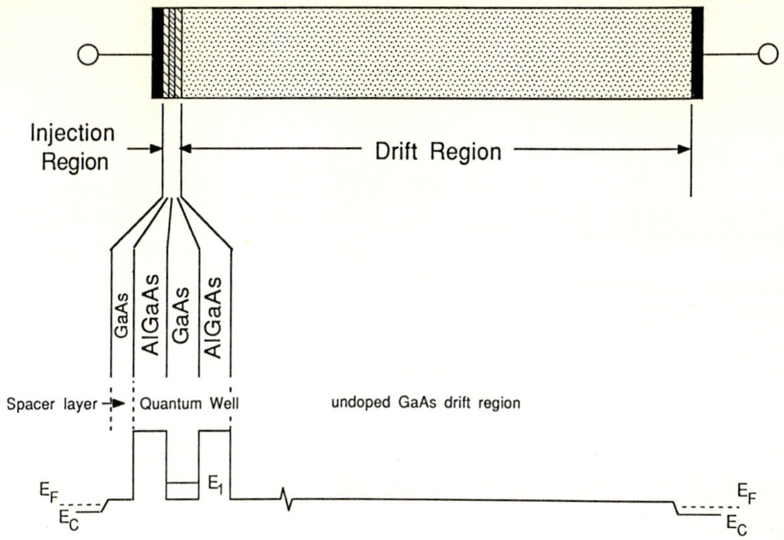

Fig. 1: Physical structure of the QWITT diode and energy band diagram of the device when no bias is applied

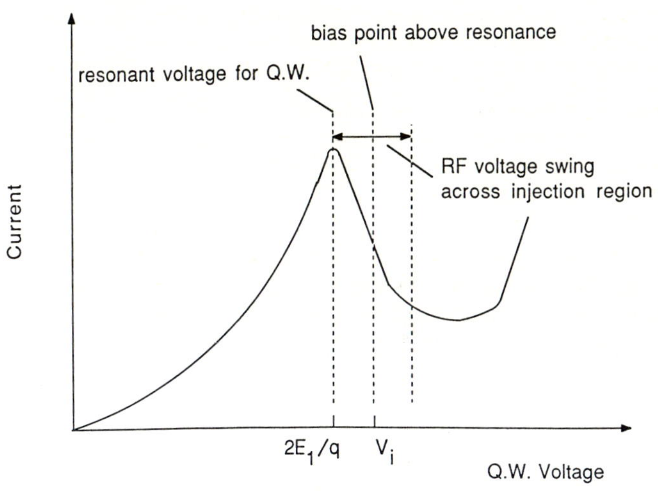

Fig. 2: The I-V curve for the quantum well injection region; the bias point is chosen above resonance such that the bias voltage across the quantum well is less than the resonant voltage; the figure also shows the rf voltage swing across the quantum well region

region at their saturation velocity v_s to produce a phase delay of about $\pi/2$, giving rise to a total delay of approximately 2π. The combination of the delay due to the injection and drift angles produces a dynamic negative resistance. Since the thickness of the drift region following the quantum well is chosen such that the electrons traverse this region for $\pi/2$ of the ac cycle (i.e. a time fraction of 1/4f), the thickness of this transit time region W would be given by

$$W = (v_s/4f). \tag{1}$$

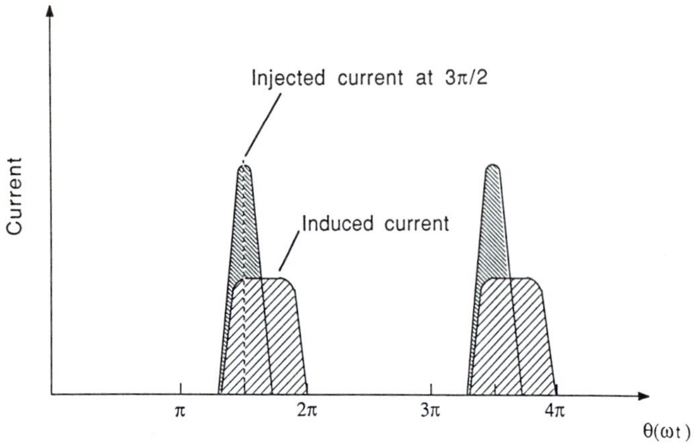

Fig. 3: Current pulse injected into the drift region and induced particle current in the circuit as a function of phase angle

Note that this case corresponds to biasing the quantum well above resonance into the negative conductance portion of its I-V curve (Fig. 2). Assuming a constant electric field throughout the injection and drift regions, the ratio of the voltage drops across the two regions is simply the ratio of their thicknesses. A typical quantum well structure is about 0.015 μm thick, and typical drift region thicknesses are 0.2 to 5.0 μm, depending on the desired frequency of operation. The voltage across the quantum well will thus be much less than the total voltage across the entire device. The dc bias across the device can also be appropriately adjusted so that the Fermi level in the n^+ GaAs is below the energy E_1 in the quantum well causing current injection to peak π/2 in th ac cycle. This mode of operation, which involves biasing the quantum well below resonance in the positive conductance regime of its I-V characteristic, has been discussed in detail elsewhere [11].

3. Small Signal Analysis

We have performed a small-signal transit time analysis [12, 13] of the QWITT diode, assuming instantaneous transport through the quantum well. Experimental dc I-V curves of the quantum well are used to determine the injection region conductance for the analysis. The small-signal negative resistance as a function of frequency for a particular drift region length and quantum well bias can then be computed. Figure 4 shows the results of such an analysis assuming a bias point in the negative conductance regime of the quantum well I-V characteristic [10]. The particular quantum well structure used in the analysis shown in Fig. 4 consists of a 70Å GaAs well with 14Å AlAs barriers together with a spacer layer of 15Å before the quantum well [10]. For frequencies less than 150 GHz, we can see that as the length of the drift region is increased from 0.001μm to 0.1μm, the magnitude of negative resistance increases, clearly showing that even when using 0.05-0.1μm spacer layers, transit time effects must be considered. An examination of these results indicates that the optimum drift angle is not frequency independent (as assumed in (1)), but varies between about π/3 and π/2. However, for any particular frequency of operation the optimum value of the drift region can be determined from Fig. 4. Hence, for a device designed to operate at 30GHz a transit time region of 0.3μm is desirable, while at 56GHz it would be more desirable to use a 0.2μm thick layer. In addition, for a given drift region length, the negative resistance decreases dramatically when the drift angle exceeds about 3π/4. This yields an approximate cut-off frequency for the QWITT diode of

$$f_{cut\text{-}off} \cong (3v_s/8W) \tag{2}$$

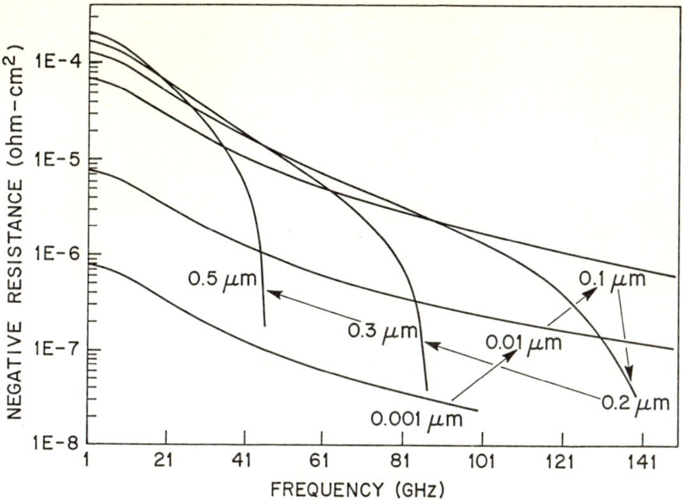

Fig. 4: Small-signal negative resistance for various drift region lengths plotted as a function of frequency for a 70Å GaAs quantum well with 14Å AlAs barriers

when biased in the negative conductance portion of the quantum well I-V characteristic. For example, when a spacer layer of 0.3μm is used there is a sharp drop in the value of negative resistance around 75GHz. This value of cut-off frequency is considerably lower than other estimates [9] made for a similar structure using a tunnel diode model. Based on this analysis an optimized structure that fully utilizes the improvement in rf performance obtained by using a thick spacer layer after the quantum well would involve asymmetric spacer layers, with the thickness of the layer following the quantum well obtained from transit time considerations. Such a structure would retain the improvement gained in dc and rf voltages and impedance matching by using a thicker structure, and also optimize the transit time effect in the structure to improve rf performance.

4. Conclusion

The transit time effects of spacer layers in quantum well oscillators have been considered. A small-signal analysis has been performed based on experimental dc I-V characteristics of the quantum well structure used. We have shown that the transit time effects due to the spacer layer after the quantum well significantly improves the magnitude of negative resistance obtained compared to pure quantum well oscillators. Hence, such a device would utilize the advantages of a thicker structure, and permit maximization of rf performance using transit time effects. This analysis also shows that there is a sharp cut-off frequency for a particular quantum well structure with a fixed spacer layer thickness.

This work was supported by the Texas Advanced Technology Research Program and the Joint Services Electronics Program under contract number AFOSR F 49620-86-C-0045.

1. T.C.L.G. Sollner, W.D. Goodhue, P.E. Tannenwald, C.D. Parker, and D.D. Peck, Appl. Phys. Lett., 43(6), 588 (1983).
2. T.C.L.G. Sollner, P.E. Tannenwald, D.D. Peck, and W.D. Goodhue, Appl. Phys. Lett., 45(12), 1319 (1984).
3. T.J. Shewchuk, P.C. Chapin, P.D. Coleman, W. Kopp, R. Fisher, and H. Morkoç, Appl. Phys. Lett., 46(5), 508 (1985).

4. S. Luryi, Appl. Phys. Lett., 47(5), 490 (1985).
5. T.J. Shewchuk, J.M. Gering, P.C. Chapin, P.D. Coleman, W. Kopp, C.K. Peng, and H. Morkoç, Appl. Phys. Lett., 47(9), 986 (1985).
6. P.D. Coleman, S. Goedeke, T.J. Shewchuk, P.C. Chapin, J.M. Gering, and H. Morkoç, Appl. Phys. Lett., 48(6), 422 (1986).
7. B. Jogai, K.L. Wang, and K.W. Brown, Appl. Phys. Lett., 48(15), 1003 (1986).
8. S. Ray, P. Ruden, V. Sokolov, R. Kolbas, T. Boonstra, and J. Williams, Appl. Phys. Lett., 48(24), 1666 (1986).
9. E.R. Brown, T.C.L.G. Sollner, W.D. Goodhue, B.J. Clifton, and P.E. Tannenwald, IEEE Dev. Res. Conf., Amherst, Mass. (1986).
10. M. Tsuchiya and H. Sakaki, IEEE Tech. Dig. Int. Electron Device. Meet., 662 (1985).
11. V.P. Kesan, D.P. Neikirk, B.G. Streetman, and P.A. Blakey, submitted to IEEE Elect. Dev. Lett. (1986).
12. J.L. Chu and S.M. Sze, Solid-State Electronics, 16, 85 (1973).
13. S.M. Sze, Physics of Semiconductor Devices, 2nd.ed. (John-Wiley, New York 1981).

Part IV

Quantum Wells

High-Speed Phenomena in GaAs Multiple-Quantum-Well Structures

A. Mysyrowicz[1*], *D. Hulin*[1*], *A. Migus*[1], *A. Antonetti*[1], *H.M. Gibbs*[2], *N. Peyghambarian*[2], *and H. Morkoc*[3]

[1]ENSTA, LOA, Ecole Polytechnique, F-91120 Palaiseau, France
[2]Optical Sciences Center, University of Arizona, Tucson, AZ 85721, USA
[3]Coordinated Science Laboratory, University of Illinois, Urbana, IL 61801, USA

We discuss different optical phenomena occurring within a few picoseconds after irradiation of GaAs multiple-quantum-well structures with femtosecond laser pulses.

In recent years an increasing effort has been devoted to the study of optically induced nonlinearities in quantum-well-structures, consisting of alternating ultrathin (~100Å) layers of two semiconductors with different bandgap energies. These structures appear to be promising candidates to play the role of active devices in ultrafast all-optical systems. Because of the forced confinement of electrons and holes in the same layer, pronounced excitonic absorption lines are observed in weakly excited samples up to room temperature, even in materials with small bulk exciton binding energy. The reinforcement of excitons is an attractive feature, since excitons can be rather easily erased by external radiation, leading thereby to the large and very rapid nonlinear response required in device applications. We describe here some of the underlying physical processes. The discussion is restricted to GaAs quantum wells, with the accent put on those effects affecting the lowest-lying heavy-hole exciton on a picosecond or subpicosecond time-scale.

Up to now, most of the effort has concentrated on the study of so-called dynamical optical nonlinearities. Here, the changes of the optical properties are induced by the presence of a real population of electron-hole pairs, the latter being generated by light falling in an absorption region of the material. The advantage is the high efficiency in this process (each incident photon can be absorbed and be therefore effective). However, it is power consuming, since energy is transferred from the incident light beam to the medium. It has also a response which may be too slow for many applications, since the switch-off time is limited by the excited carriers' lifetime.

In order to understand in detail the kinetics of dynamical nonlinearities, it is necessary to first recognize the quasi-equilibrium ground state of the excited system at low temperature, close to T = 0. Although the presence of a large population of electron-hole pairs is per se a highly nonequilibrium situation, a quasi-equilibrium regime can still be defined, in which thermalization amongst the particles has been achieved via interparticle collisions, allowing an effective electronic temperature to be established. As a further requirement for quasi-equilibrium the electronic temperature should be close to the lattice temperature, with $\Delta T < \hbar w_0$, where $\hbar w_0$ is the optical phonon energy, so that both the particle density and mean energy evolve at a slow rate on a picosecond timescale.

---

*Also at Ecole Normale Superieure, Paris, France

A good description of the quasi-equilibrium ground state of highly excited bulk GaAs has been given in terms of a two-component Fermi plasma [1]. A spectroscopic signature of such a plasma is the disappearance of exciton structures, due to the mutual screening of the Coulomb attraction between electrons and holes. A significant feature is the fact that the absolute energy of the exciton resonance stays constant, as the exciton oscillation strength gradually diminishes upon increase of injected free carrier density. This is attributed to an almost perfect cancellation between two effects acting in opposite directions: a reduction of exciton binding energy due to screening (leading to a shift of the exciton lines to the ionization continuum at higher energies) and a reduction of the band gap energy, due to the presence of electrons in the conduction and holes in the valence band (band gap renormalization).

There is growing experimental evidence that such a plasma state with strong screening does not represent the quasi-equilibrium ground state of a two-dimensional GaAs at low temperatures. Instead, the high density electron-hole pairs seem to maintain a strong Coulomb correlation. KELDYSH and KOZLOV [2], and NOZIERES and COMTE [3] have treated the 3-D case of e-h systems where strong pairing subsists in the dense regime. Schmitt-Rink et al. [4] have given several plausible arguments for a reduction of screening effects in a 2-D system and have calculated the expected properties of a highly Coulomb correlated 2-D system.

In that case, exciton bleaching also occurs, but for a different reason, having its root in the Pauli exclusion principle. It is due to an exhaustion of the Bloch states available to form new excitons (saturation in K-space). Another expected effect is a high energy shift of the excitons in the dense regime. This translates the predominantly repulsive potential between excitons, due to the Fermi nature of the constituents. The short-range repulsion is no more compensated by an equivalent red shift from screening effects which have the form of van der Waals attraction in a high density exciton system [3]. At higher temperatures, the usual screened plasma form the quasi-ground state, as in three-dimensional GaAs.

Experimental evidence for pairing is shown in fig. 1, where the subpicosecond time-resolved absorption of a GaAs MQWS (GaAs well size = 50 Å) at T = 15 K is shown at different delays following optical injection of a high density of carriers ($n \sim 10^{18} cm^{-3}$), using a pump beam tuned at 1.7 eV. Initially, the electron-hole system is hot, because of the large excess energy of the pump photons, and therefore, the plasma description applies. Complete exciton bleaching via screening is achieved just after the end of the pump pulse (Δt = 1 ps). Note that as it progressively bleaches during the pump pulse, the exciton maintains a constant energy value (see curve at Δt = 0.1 ps). At later times ($\Delta t \geqslant 10$ ps), the exciton oscillator strength partially recovers, in contrast to bulk GaAs where complete bleaching due to screening would persist for $\Delta t > 10^{-9}$s [6]. As it reappears, the exciton is shifted to higher energies as expected for a high density exciton gas, with bleaching now due to saturation of Bloch states instead of screening by free carriers.

According to the model of pairing, there should be a definite correlation between blue shift and bleaching from saturation of Bloch states, since both effects are due to the Pauli exclusion principle [4]. Figure 2 shows this to be the case, in remarkable quantitative agreement (within a factor 2) with theory. A study as a function of well thickness shows that the well size is a crucial factor for the reduction of screening effects [7].

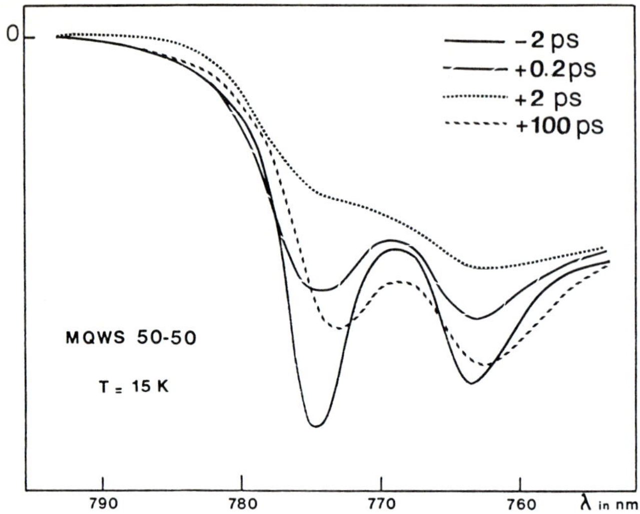

Figure 1. Subpicosecond time-resolved spectra of GaAs MQWS recorded at various delays following irradiation with a subpicosecond pump pulse tuned at λ = 740 m (see text)

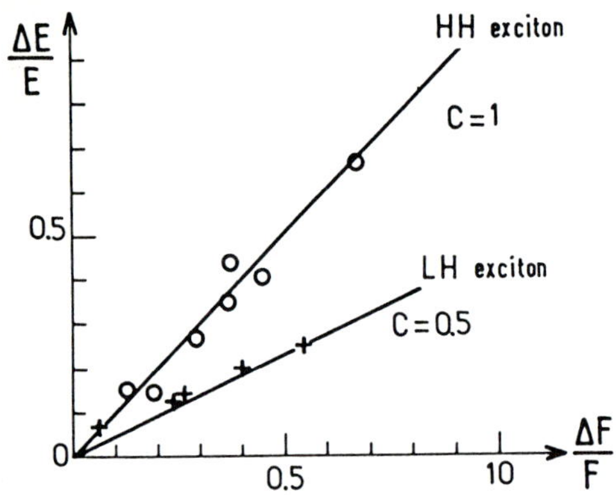

Figure 2. Exciton blue shift (normalized to exciton binding energy) as a function of fractional decrease of oscillation strength, as observed in a 50 [Å] GaAs MQWS at T = 15 [K], both for the heavy hole (HH) and light hole exciton (LH)

Having recognized the nature of the quasi-equilibrium state of highly excited GaAs MQWS, it is easy to understand some of the unusual features observed as fast transient. For instance, KNOX et al. [8] have observed in GaAs MQWS at T = 300 K an increased initial bleaching of excitons during ~ 300 fs if the sample is pumped with a laser tuned to the lowest exciton resonance. This observation runs against the conventional wisdom that free carriers should be more effective than excitons for bleaching exciton resonances. However, this effect is understood by noting that the comparison is made between bleaching from a population of cold excitons (before these optically injected particles thermalize to the lattice temperature), and from a distribution of electron-hole pairs at T = 300 K (after thermalization). Transient saturation of Bloch states is achieved for the initial cold excitons, with increased bleaching as a consequence, in agreement with calculations [4]. In the same spirit, a transient exciton blue shift has been observed in a 50 Å GaAs MQWS at temperatures close to the ambiant, before the injected cold excitons dissociate into free carriers to attain thermodynamic equilibrium [5].

The second type of optical nonlinearities arises from the nonlinear response of the system to an intense electromagnetic field. It corresponds to virtual interband transitions, and consequently has an inherent fast response time. Also it does not involve energy dissipation, since the sample is left in the ground state.

Figure 3 shows the effect of an intense femtosecond laser pulse tuned to the low-energy side of the exciton, in the transparency region of the GaAs MQWS at T = 15K. A strong blue shift together with bleaching is observed, but only during the presence of the pump pulse. This effect occurs in a sample with 100 Å GaAs well thickness, in which no exciton blue shift is observed under resonant or band-to-band pumping. The high energy shift seen here is attributed to an optical Stark effect [9]. Such an effect is well known in atomic spectrosocpy, but was not observed in excitonic transitions until very recently [9,10,11].

The "dressed atom" model can be applied to excitons in order to interpret in a simple physical way the induced exciton light-shift. A more elaborate model, with the many-body interactions between excitons treated from first principles has been derived recently [12]. It provides an explanation for the reduction of exciton oscillator strength which is clearly apparent in fig. 3. It is worth

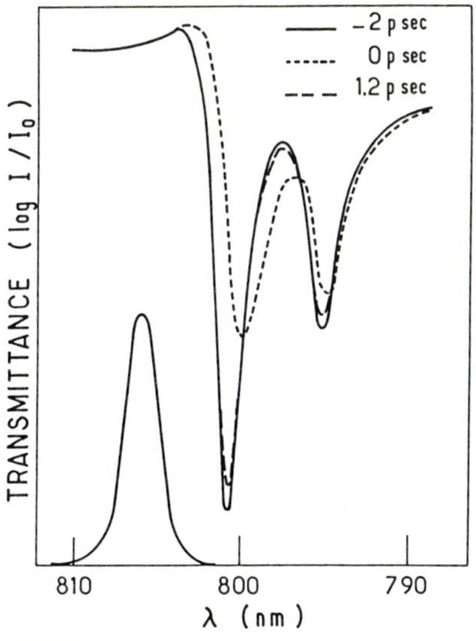

Figure 3. Subpicosecond time-resolved absorption spectra of a 100 [Å] GaAs MQWS at T = 15 [K], recorded before, during and immediately after an intense pump pulse tuned at λ = 806 nm, in the transparency region of the sample.

mentioning that contrary to other effects described above, the optical Stark effect of excitons is not tied to the reduced dimensionality of MQWS and has been in fact also observed in bulk GaAs [9].

The optical Stark effect has been exploited recently to demonstrate the operation of an optical logic gate with both switch-on and switch-off times in the subpicosecond range [13]. The device consisted of a 100 Å GaAs between two semitransparent gold layers, acting as a Fabry-Perot resonator. An incident pump pulse ($I_o \sim 10^9 W/cm^2 J$), adjusted to overlap a transmission mode of the Fabry-Perot, below the exciton in the transparent region of the device, induces a blue shift of the exciton. The corresponding reduction of the refractive index at lower photon energies leads to an energy displacement of the Fabry-Perot mode, thereby performing the logic operation.

In conclusion, GaAs MQWS exhibit fast nonlinear responses to incident optical pulses which can be useful for device applications. Some of the effects are tied to the reduced dimensionality of quantum wells, while others, such as the optical Stark effect are expected more generally in a variety of bulk semiconductors.

<u>References</u>

1. For a review see for instance C. Klingshirn, H. Haug: *Phys. Rep.* **70**, 315 (1981).
2. L. V. Keldysh, A. N. Kozlov: *Sov. Phys., JETP*, **27**, 571 (1968).
3. P. Nozieres, C. Comte: *J. de Physique* **43**, 1083 (1982).
4. S. Schmitt-Rink, D. S. Chemla, D. A. B. Miller: *Phys. Rev.* **B32**, 6601 (1985).
5. N. Peyghambarian, H. M. Gibbs, J. L. Jewell, A. Antonetti, A. Migus, D. Hulin, A. Mysyrowicz: *Phys. Rev. Lett.* **53**, 2433 (1984).
6. C. V. Shank, R. L. Fork, R. F. Leheny, J. Shah: *Phys. Rev. Lett.* **42**, 112 (1979).
7. D. Hulin, A. Mysyrowicz, A. Antonetti, A. Migus, W. T. Masselink, H. Morkoc, H. M. Gibbs, N. Peyghambarian: *Phys. Rev.* **B33**, 4389 (1986).
8. W. H. Knox, R. L. Fork, M. C. Downer, D. A. B. Miller, D. S. Ch emla, C. V. Shank, A. C. Gossard, W. Wiegmann: *Phys. Rev. Lett.* **54**, 1306 (1985).
9. A. Mysyrowicz, D. Hulin, A. Antonetti, A. Migus, W. T. Masselink, H. Morkoc, *Phys. Rev. Lett.* **56**, 2748 (1986).
10. A. von Lehmen, S. S. Chemla, J. E. Zucker, J. P. Heritage: *Optics Lett.* **11**, 609 (1986).
11. D. Fröhlich, A. Nöhte, K. Reimann: *Phys. Rev. Lett.* **55**, 1335 (1985).
12. S. Schmitt-Rink, D. S. Chemla: *Phys. Rev. Lett.* **57**, 2752 (1986).
13. D. Hulin, A. Mysyrowicz, A. Antonetti, A. Migus, W. T. Masselink, H. Morkoc, H. N. Gibbs, N. Peyghambarian: *Appl. Phys. Lett.* **49**, 749 (1986).

Picosecond Carrier Transport in GaAs Quantum Wells

J. Shah[1] *and R.A. Höpfel*[2]*

[1]AT&T Bell Laboratories, Crawford Hill Laboratory,
Holmdel, NJ 07733, USA
[2]Institute of Experimental Physics, University of Innsbruck,
Innsbruck, Austria

Picosecond techniques, in conjunction with applied electric fields, provide valuable information about carrier transport phenomena of interest in ultrasmall, ultrafast semiconductor devices. We review here our recent measurements of minority carrier drift velocities in GaAs quantum wells which, in combination with luminescence measurements, provide information about high field phenomena such as real-space-transfer and electron-hole energy exchange. We also demonstrate that, under appropriate conditions, minority carriers in GaAs quantum wells exhibit negative absolute mobility i.e. the minority electrons drift towards the negative electrode and the minority holes drift towards the positive electrode. We show that this unusual phenomenon results from momentum scattering between electrons and holes and deduce a momentum relaxation time of ~ 80 fs for electrons in a hole plasma and ~ 4 ps for holes in an electron plasma.

I. INTRODUCTION

High field transport phenomena in semiconductors are of fundamental interest and an understanding of these phenomena is essential in optimizing ultrasmall, ultrafast devices which usually operate at high electric fields. Picosecond techniques can play an important role in elucidating the nature of scattering processes which determine high field transport in semiconductors. We review here three recent experiments [1-5] in which we have used picosecond optical techniques to investigate quantities relevant to high field transport in semiconductors. We have determined the minority electron drift velocity in GaAs quantum wells by picosecond time-of-flight technique. We show that the negative differential mobility observed in these measurements results from real-space-transfer of electrons into the AlGaAs barriers. In addition, our measurements have allowed us to obtain information about energy as well as momentum relaxation by scattering between electrons and holes.

II. DRIFT VELOCITY BY TIME-OF-FLIGHT TECHNIQUE

Time-of-flight techniques provide a direct means of determining carrier drift velocities in semiconductors. We have determined the drift velocity of minority electrons in p-modulation doped quantum wells as a function of electric field up to 14 kV/cm [1,2]. The experimental arrangement is schematically shown in Fig. 1.

* Work done while at AT&T Bell Laboratories, Holmdel, N.J.

Fig. 1. Schematic diagram of a picosecond opto-electrical time- of-flight measurement.

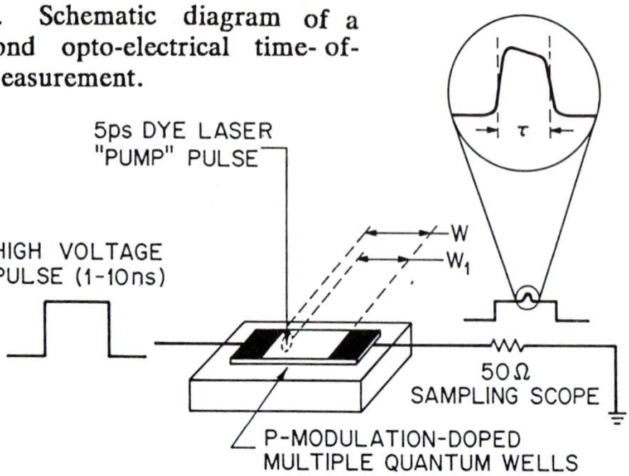

The presence of majority holes prevents the measurement problems related to space charge effects. Minority electrons were injected with a 5 ps pulse from a synchronously pumped, cavity-dumped dye laser operating at 6,200 Å. High fields were applied parallel to the layers of the quantum wells by ohmic p- type contacts on the multi-quantum well sample. Contact separation was typically 50 μm and the laser was focused to 2 μm width. The sample was mounted on a microwave stripline to permit fast measurements (system response < 50 ps). The photoinjected minority electrons gave essentially a square-topped photocurrent pulse (Fig. 1) from which the electron drift velocity was determined directly.

The drift velocity of electrons determined in this manner is plotted in Fig. 2. The most interesting feature of the data is that the drift velocity reaches a peak

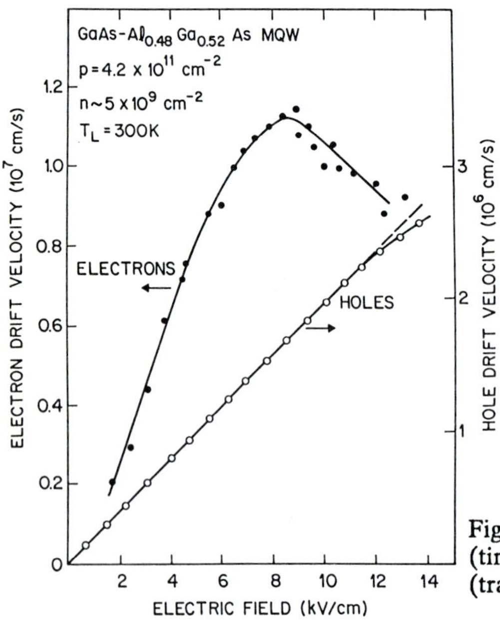

Fig. 2. Drift velocities of electrons (time-of-flight experiments) and holes (transport data) vs. electric field.

(1.1 × 10^7 cm/s) and then decreases; i.e. it shows a distinct negative differential mobility. Secondly, the low-field drift mobility is approximately 1,500 cm²/Vs and is much smaller than the typical values of > 7,000 cm²/Vs observed for electrons in n- modulation doped quantum wells. We attribute the lower value observed to momentum scattering between photoexcited electrons and the majority holes in our sample. We will come back to the question of momentum scattering between electrons and holes in Sec. IV. Fig. 2 also shows the drift velocity of majority holes deduced from the measured I-V curves.

III. SIMULTANEOUS LUMINESCENCE MEASUREMENTS: ELECTRON-HOLE ENERGY EXCHANGE RATE

The observation that there is a region of negative differential mobility immediately indicates that the electrons are becoming sufficiently hot to transfer to either the barrier layers or the higher conduction band valleys where the mobility is lower. We have made a direct determination of the electron temperature by measuring the luminescence spectra (inset of Fig. 3) under the influence of applied electric field. We find [2] that the electron temperature increases with electric field (Fig. 3) to a value of about 650 K and remains approximately constant in the region of negative differential mobility. We have argued that this temperature is too low for transfer to higher valleys in GaAs and that the observed negative differential mobility arises as a result of real-space-transfer to the barrier layers [2].

Figure 3 also shows the hole temperature estimated [3] on the basis of the measured I-V curve as well as from the known [6] hole energy loss rates to the lattice. We see that the holes are at a much lower temperature than the electrons. Since the electron and hole energy loss rates to the lattice are different [6], one expects to find different temperatures for electrons and holes in the absence of

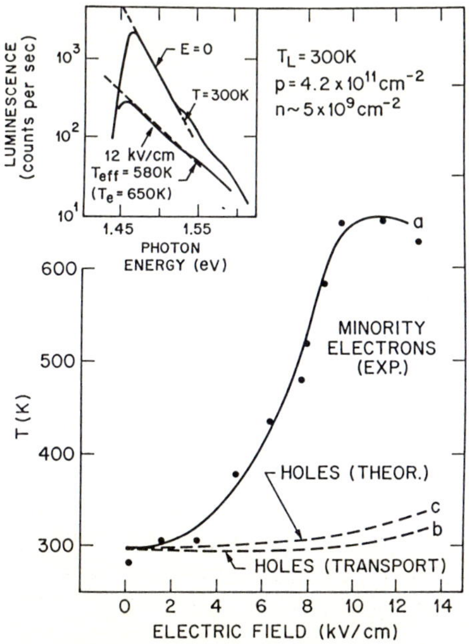

Fig. 3 Electron temperatures (from luminescence spectra) and hole temperatures (from transport data (b) and energy loss theory (c)) vs. electric field.

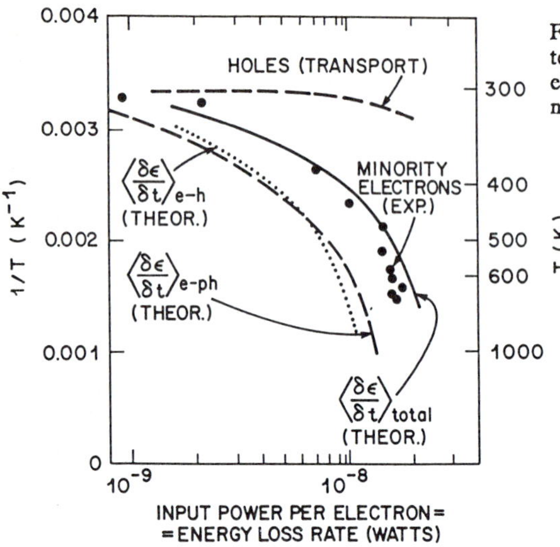

Fig. 4. Inverse of electron temperature vs. energy loss rate for electrons for various scattering mechanisms.

electron-hole scattering. Electron-hole scattering will tend to reduce this difference between the two temperatures. When the two temperatures are different, we expect a net flow of energy from electrons to the holes.

We plot in Fig. 4 the inverse of the electron temperature as a function of power loss into the lattice (which, in this steady state condition is equal to the power input from the electric field to the electrons). We compare [3] this with the expected energy loss rate from electrons into the lattice and find that (Fig. 4) the measured energy loss rate is larger than that expected by about a factor of 2. We attribute this difference to energy exchange between electrons and holes. This is supported by a calculation of energy exchange rate between electrons and holes from a simple gas plasma model [3]. While more elaborate calculations more appropriate to the semiconductor case are needed, these measurements provide the first estimate of electron-hole energy exchange rates.

IV. NEGATIVE ABSOLUTE MOBILITY OF MINORITY CARRIERS

We have seen in Sec. II that the low field mobility of electrons in a majority hole plasma is about 1,500 cm^2/Vs, a factor of about 5 smaller than the value of electron mobility in samples in which no holes are present. This is a result of electron-hole momentum scattering due to their Coulomb interaction. We describe in this section an unusual manifestation of momentum scattering between electrons and holes in a semiconductor [4,5].

Electric field was applied parallel to the plane of the layers of a p-modulation doped multiple quantum well sample. Minority electrons were photoexcited with a tightly focused (diameter 3 micron) picosecond laser (same as in Section II). The sample was mounted in variable temperature cryostat (lowest temperature 15 K). Since luminescence occurs only where the minority electrons are present, the motion of the electrons under the applied electric field can be monitored by measuring where the luminescence originates. We have done this by imaging the sample on a plane and scanning a narrow slit across this plane to measure the image of the time-integrated luminescence (Fig. 5).

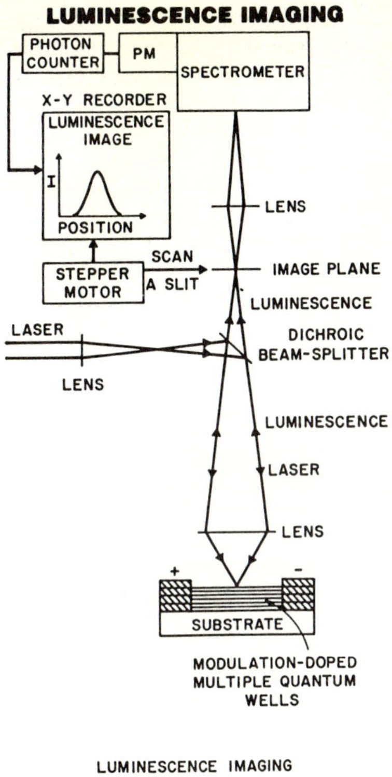

Fig. 5. Schematic diagram for the luminescence imaging experiment.

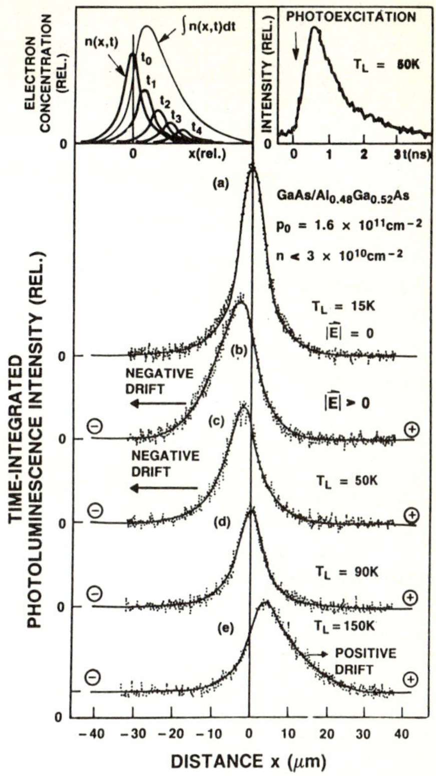

Fig. 6. Time-integrated luminescence images for different lattice temperatures: (a) 15 K; F=0 (b) 15 K; F=20 V/cm (c) 50 K; F=120 V/cm (d) 90 K; F=280 V/cm and (e) 150 K; F=600 V/cm. Curves (b,c) show negative drift, curve (e) shows positive drift and curve (d) shows no drift. The insets show the time integration of the drifting minority carriers (top left) and the time dependence of the luminescence intensity (top right).

Several typical scans are shown in Fig. 6. The top trace, taken at 15 K without any electric field, shows a symmetric distribution of electrons. All other traces were obtained in the presence of applied electric field at different lattice temperatures. At 15 K, the luminescence shifts towards the negative electrode; i.e. the minority electrons have a negative absolute mobility at low temperatures! With increasing temperature (Fig. 6), the negative drift becomes smaller, then becomes zero and finally reverses sign and becomes positive at temperatures above 90 K. We have obtained the electron drift velocity from the observed luminescence image [3] by assuming that the minority electrons drift without any appreciable broadening and at the same time decay according to the measured decay time. The electron drift velocity is the only adjustable parameter in this fitting procedure. The fits obtained by such a calculation are shown as solid curves in Fig. 6; also indicated in the figure are the drift velocities deduced from the fit.

The negative drift of electrons is a consequence of the momentum scattering between electrons and holes via Coulomb interaction. The holes in our high mobility samples interact strongly with electrons and drag the electrons with them to the negative electrode. Such an effect was first predicted for the case of InSb by McLean and Paige [7] but has not been observed in any semiconductor until now to our knowledge.

We have used a relaxation time approximation to the Boltzman equation to analyze this case and found that, for extreme minority injection ($\Delta n << P_o$) and negligible electron-lattice scattering [4], the electron mobility is given by subtracting the hole mobility from the mobility of electrons due only to electron-hole scattering. The latter quantity can be approximated by mobility due to electron-ionized impurity scattering. In Fig. 7, we have plotted the measured electron mobility as well as the negative of the measured hole mobility as a function of the lattice temperature. The difference between the two gives the electron mobility due only to electron-hole scattering and is in good agreement with the electron-ionized impurity scattering at appropriate impurity density [4].

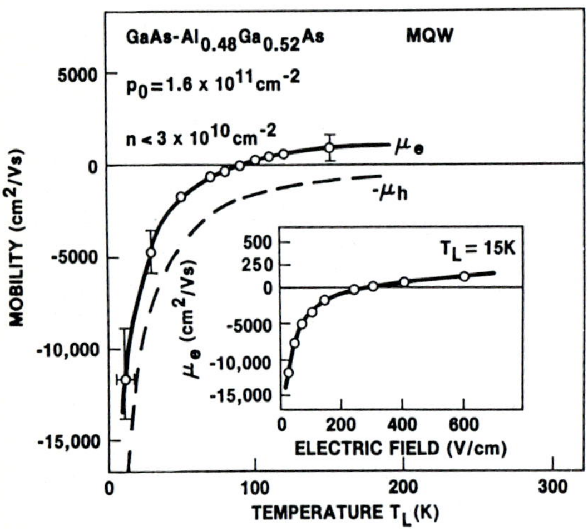

Fig. 7. The minority electron mobility vs. temperature; also shown is (-)hole mobility. The difference between the two is a measure of the electron mobility due to electron-hole interaction. The inset shows minority electron mobility vs. electric field.

Fig. 7 also shows the variation of the negative mobility of electrons as a function of applied electric field. The behavior with increasing field is similar to that with increasing temperature. This is to be expected because increasing electric field leads to the heating, and hence reduced mobility, of the majority hole gas.

Recently, we have also observed negative absolute mobility for minority holes in n-modulation doped samples [5], in spite of the fact that the holes are much heavier than electrons. We show the measured hole mobility as a function of temperature in Fig. 8. The negative mobility is high at low temperatures and

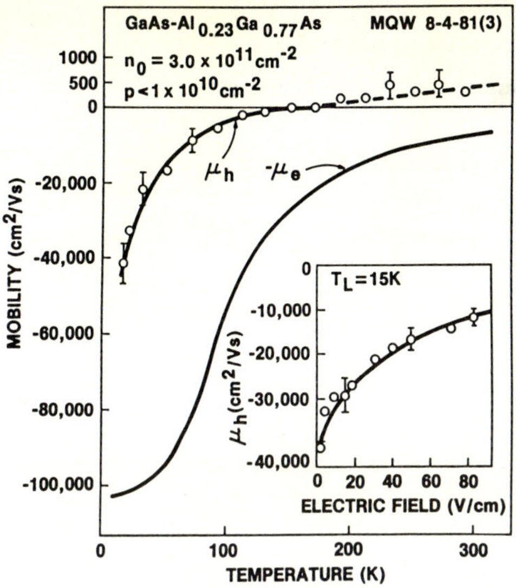

Fig. 8. The minority hole mobility vs. temperature; also shown is the (-)electron mobility and the minority hole mobility vs. electric field, in the inset.

approaches zero with increasing temperature, in a manner qualitatively similar to the behavior of the negative mobility of electrons in a majority hole plasma.

Knowing the minority carrier mobility from our measurements and the majority carrier mobility from Hall measurements, we have deduced the minority carrier mobility that results only from the interaction with the majority carriers. From this we have determined the momentum relaxation times for electrons in the hole gas and the holes in the electron gas. We find that the electron momentum relaxation time in a hole plasma is about 80 fs whereas the hole momentum relaxation time in an electron plasma is about 4 ps. About a factor of 6 of this large difference is due to the ratio of hole to electron effective masses. The remaining difference is very likely a result of the fact that the majority electrons are degenerate whereas the majority holes are non-degenerate. This is more fully explored in a forthcoming publication [8].

V. SUMMARY

In summary, time- and spatially resolved luminescence, as well as picosecond photoconductivity, in high electric fields provide excellent tools for studying carrier drift velocities and energy distribution functions in materials and systems relevant for high-speed electronics. Using these techniques, we have investigated drift velocities, distribution functions and electron-hole (momentum and energy) scattering rates in GaAs quantum wells.

ACKNOWLEDGMENT

We acknowledge collaboration with D. Block, A. E. DiGiovanni, A. C. Gossard, P. A. Wolff and W. Wiegmann.

[1] R. A. Höpfel, Jagdeep Shah, A. C. Gossard and W. Wiegmann, Physica 134B, 509-513 (1985).

[2] R. A. Höpfel, Jagdeep Shah, D. Block and A. C. Gossard, Appl. Phys. Lett. 48, 148 (1986).

[3] R. A. Höpfel, Jagdeep Shah and A. C. Gossard, Phys. Rev. Lett. 56, 765-768 (1986).

[4] R. A. Höpfel, Jagdeep Shah, P. A. Wolff and A. C. Gossard, Phys. Rev. Lett. 56, 2736-2739 (1986).

[5] R. A. Höpfel, Jagdeep Shah, P. A. Wolff and A. C. Gossard, Appl. Phys. Lett.49, 572-574 (1986).

[6] Jagdeep Shah, A. Pinczuk, A. C. Gossard, and W. Wiegmann, Phys. Rev. Lett. 54, 2045-2048 (1985).

[7] T. P. McLean, and E. G. S. Paige, J. Phys. Chem. Solids 16, 220 (1960).

[8] R. A. Höpfel, Jagdeep Shah, P. A. Wolff and A. C. Gossard, to be published.

Quantum-Confined Stark Effect in InGaAs/InP Quantum Wells Grown by Metal-Organic Chemical Vapor Deposition

I. Bar-Joseph, C. Klingshirn, D.A.B. Miller, D.S. Chemla, U. Koren, and B.I. Miller*

AT&T Bell Laboratories, Crawfords Corner Road, Holmdel, NJ 07733, USA

Large shifts of absorption spectrum in InGaAs/InP multiple quantum wells with applied electric fields are reported. Applications for light modulators are discussed.

Multiple quantum well (MQW), composed of alternate very thin layers of two different semiconductors, show unusual optical properties at room temperature. One effect of particular recent interest is the quantum-confined Stark effect (QCSE)[1]. In the QCSE, electric fields applied perpendicular to the MQW layers can shift the optical absorption edge to lower photon energies with the exciton absorption peaks remaining clearly resolved. This electroabsorptive effect has been applied to make small, high-speed optical modulators[2] and optical switching and signal-processing devices[3] in GaAs/GaAlAs MQW.

Perpendicular-field electroabsorption at longer wavelengths (i.e. $\lambda \sim 1.5\mu m$) has recently been studied in InGaAs and in GaSb MQW. The GaSb wells show relatively clear shifts of the exciton peaks with field[4], but the substrate is opaque at the wavelengths of interest for optical communication. The electroabsorption spectra of InGaAs MQW so far reported[5] only show a large broadening of the peaks with field, and relatively little clearly resolved shift. Such broadened peaks are of limited interest for applications.

In this paper we report the first observation of the QCSE in InGaAs quantum wells with InP barriers, in material grown by metal-organic chemical vapor deposition (MOCVD). In contrast to previous InGaAs MQW results, the exciton peaks remain well resolved up to very large fields at room temperature.

The sample structure is shown in the insert of Fig. 1. The MQW form the intrinsic region of a p-i-n diode, and

* Permanent address: Physikalisches Institut der Universität D-6000 Frankfurt am Main

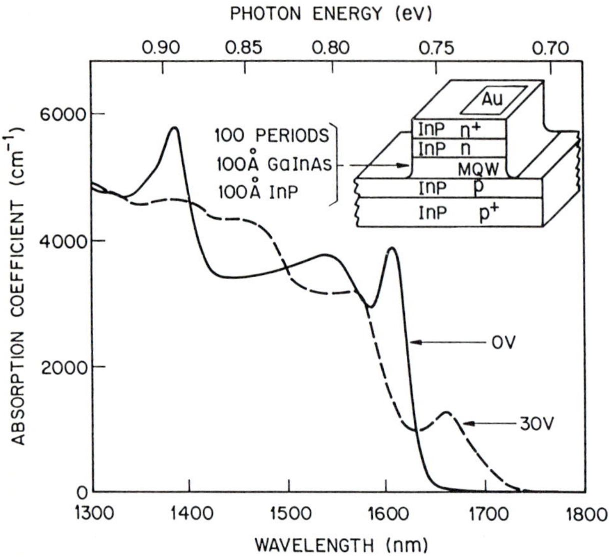

Figure 1: Room-temperature absorption spectra at 0V (solid line) and 30V (dashed line). The sample structure is shown in the insert.

consist of 100 periods of alternate InGaAs wells and InP barriers, each 100 Å thick. The sample is processed to form rectangular mesas with gold contacts as shown in the figure. Room - temperature absorption spectra at 0V and 30V are shown in Fig. 1. In the absence of applied voltage one can clearly distinguish the n_z=1 heavy (hh) and light (lh) hole exciton peaks as well as a strong resonance at the n_z=2 transition edge. The 30V spectrum shows the dramatic changes produced by the electric field. The n_z=1 hh-exciton is strongly red-shifted while the n_z=1 lh and n_z=2 hh transitions present less pronounced shifts. A new feature appears around 0.84 eV. It is located close to the calculated n_z=3 hh to n_z=1 electron "forbidden" transition, which becomes allowed in the presence of the field.

In Fig. 2a we present the absorption and in Fig. 2b the photocurrent spectra close to the n_z=1 transitions, for four applied voltages. This is the region of interest for applications to optical modulation. The photocurrent spectra closely follow that of the absorption, although the features are less pronounced. The shift of the n_z=1 hh absorption peak was compared with the theory[1], and very good agreement was found.

The large changes in the absorption spectrum reported here clearly indicate that light modulators made of InGaAs/InP MQW can be constructed. Waveguide modulators

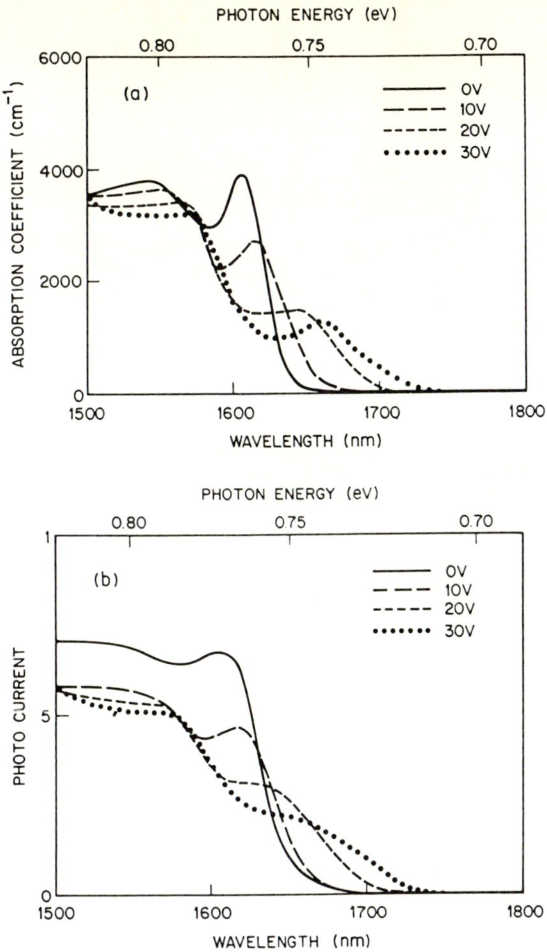

Figure 2: Absorption (a) and photocurrent (b) spectra in the vicinity of the $n_z=1$.

with only a few quantum wells may give increased modulation depth and reduce the driving voltage[5,6]. Self electro-optic effect device operation based on photocurrent feedback is also likely with this material system, as in GaAs MQW[3], offering other optical switching and modulation devices with this material system. In addition, since the InP substrate is transparent at the wavelengths of interest, these findings are promising for modulators and optical switching devices that will be directly compatible with optical fiber communications.

Acknowledgements: We thank J. E. Henry for her help in processing the samples.

REFERENCES

[1] D. A. B. Miller, D. S. Chemla, T. C. Damen, A. C. Gossard, W. Wiegmann, T. H. Wood, and C. A. Burrus, Phys. Rev. **32**, 1043 (1985).

[2] T. H. Wood, C. A. Burrus, D. A. B. Miller, D. S. Chemla, T. C. Damen, A. C. Gossard, and W. Wiegmann, Appl. Phys. Lett. **44**, 16 (1984).

[3] D. A. B. Miller, D. S. Chemla, T. C. Damen, T. H. Wood, C. A. Burrus, A. C. Gossard, and W. Wiegmann, IEEE J. Quantum Electron **QE-21**, 1462 (1985).

[4] T. Miyazawa, S. Tarucha,Y. Ohmori, and H. Okamoto, 2nd Int. Conf. Modulated Semiconductor Struct., Kyoto, Japan, Sept. 1985.

[5] K. Wakita, Y. Kawamura, Y. Yoshikuni, H. Asahi, and S. Uehara, IEEE J. Quantum Electron. **QE-22**, 1831 (1986).

[6] J. S. Weiner, D. A. B. Miller, D. S. Chemla, T. C. Damen, C. A. Burrus, T. H. Wood, A. C. Gossard and W. Wiegmann, App. Phys. Lett. **47**, 1148 (1985).

Dynamics of Below-Gap Photoexcitation in GaAs Quantum Wells

J.E. Zucker[1], *D.S. Chemla*[1], *A. Von Lehmen*[2], *and J.P. Heritage*[2]

[1]AT&T Bell Laboratories, Crawfords Corner Road, Holmdel, NJ 07733, USA

[2]Bell Communications Research, Red Bank, NJ 07701, USA

Time-resolved transmission experiments in GaAs quantum wells show that photoexcitation in the transparent region of the heterostructure can produce a striking variety of excitonic effects.

1. Introduction

We have identified several physical processes by which below-gap photons can modify excitonic absorption in GaAs quantum wells [1-3]. These different phenomena can be distinguished by their temporal response times, which range from tens of nanoseconds to less than one picosecond. Our experiments also show that the relative contributions of these mechanisms to below-gap optical modulation of exciton absorption depend strongly on temperature as well as on the wavelength of the below-gap modulating laser. The size of the modulation indicates that a new class of absorptive and dispersive devices for optical signal processing is possible in which signals propagate in the transparent region of the crystal and experience negligible attenuation. In order to successfully interpret and calculate the magnitude of these exciting new effects,we show that it is necessary to take into account the special properties of two-dimensional excitons in semiconductor quantum wells.

2. Experiment

A differential technique [1] is used to monitor the differential changes in transmission $\Delta T/T$ of a weak probe laser induced by a below-gap pump laser. The probe is tuned near the n=1 heavy hole exciton resonance. The pump beam is tuned below the bandedge as far as 60 meV from resonance and ranges in intensity from 10^3 to 10^6 Watts/cm^2. Pump and probe pulsewidths are 6 psec with 6.7 MHz repetition rate and 10 psec cross-correlation. Quantum well samples are mounted on sapphire substrates in a cold finger and cooled to temperatures 70-300 K.

3. Free Carrier Bleaching of the Exciton via Phonon-assisted Absorption

We have found that photoexcitation of GaAs quantum wells far below the bandedge can produce a marked decrease in exciton absorption which persists for 20-30 nsec. The timescale of this phenomenon is indicative of a free carrier effect. However, in order to produce bleaching of the exciton absorption, carriers are required to be near or at the bandedge [4]. The dynamical process which links absorption of a photon in the transparent region of the crystal to a free carrier at the bandedge is the key step in this type of below-gap modulation.

Our recent measurements are the first to reveal that the process by which carriers are generated at the bandedge is phonon-assisted absorption [5]. The temperature dependence of the effect shows that the magnitude of the change in transmission induced by below-gap photoexcitation is directly proportional to the number of optical phonons present. Participation of optical phonons is even more apparent in the spectral dependence of the bleaching efficiency,shown in Fig.1. Here the below-gap absorption coefficients $\alpha(\hbar\omega_p)$,

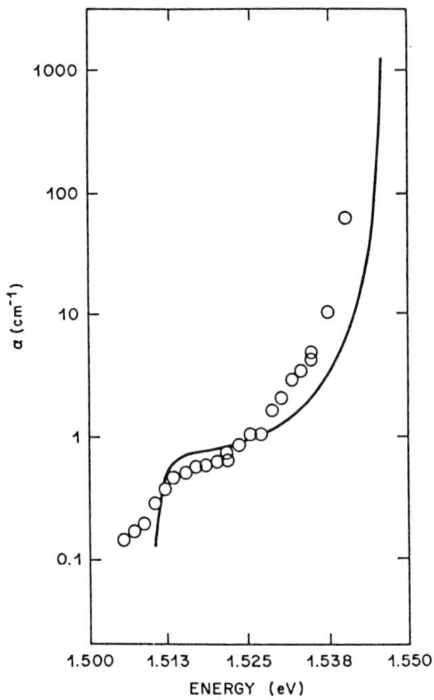

[1] Below-gap absorption coefficient near the n=1 exciton transition (1.5461 eV) as a function of pump photon energy. Pump-probe delay = 100 psec. Circles are obtained from differential transmission ΔT/T measured at 75K ; solid line calculated.

determined directly from measured values of ΔT/T, are plotted as a function of pump photon energy. The most prominent feature occurs at 1.513 eV, nearly one longitudinal optic (LO) phonon energy below the exciton resonance. The solid line represents a calculation of the one-LO phonon-assisted absorption spectrum which has been modified to include the effects of reduced dimensionality in the quantum well [5]. We find it necessary to take into account both the two-dimensional density of electronic states and the two-dimensional exciton-LO phonon Frohlich interaction. The calculated values for the absorption coefficient contain no adjustable parameters and are in good agreement with the data. The contributions of zero-phonon and two-phonon- assisted processes to the spectrum, which are not included here, are expected to further improve the agreement between theory and experiment.

4. AC Stark Shift of the Exciton Transition

At low temperature, the value of the absorption coefficient in the bandtail of the quantum well becomes small: in Fig. 1 α is between 0.1 and 10 cm^{-1}. At such low absorption , very few real carriers are created and we find that a new phenomenon appears in the differential transmission spectrum (Fig.2). The increase in probe transmission which is nearly constant for 200 psec after the pump pulse is associated with the free carrier bleaching discussed above. The new feature is the extremely fast increase in probe transmission, which rises and falls within the pump-probe correlation time. As shown in Fig.2, this ultrafast component increasingly dominates the free-carrier bleaching effect as the pump wavelength is detuned from the exciton resonance .

Further contrast between the sub-picosecond and nanosecond effects of below-gap photoexcitation are displayed in Fig.3. For a constant pump detuning of 31 meV from the exciton resonance, both fast (Fig.3b) and slow (3c) components change in magnitude and sign depending on which part of the exciton absorption feature (3a) is probed. However the lineshapes are completely different. The change in transmission shown in Fig.3c reflects a broadening and flattening of the absorption peak by free carrier bleaching, symmetric with respect to line center. On the other hand, the asymmetric lineshape shown in Fig 3b is

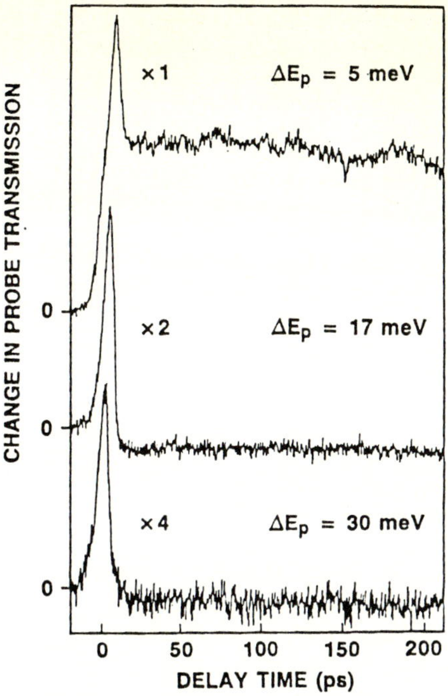

[2] Change in probe transmission as a function of the time delay between pump and probe. As the detuning ΔE_p increases the fast component due to the AC Stark shift dominates the slow component associated with free carrier bleaching of the exciton.

[3] a) Unperturbed transmission showing n=1 heavy (hh) and light (lh) hole excitons. b) and c) Percent change in probe transmission induced by 10^6Watts/cm^2 pump 31 meV below the n=1 hh transition. (b) zero pump-probe delay ; (c) delay = 100 psec. ▶

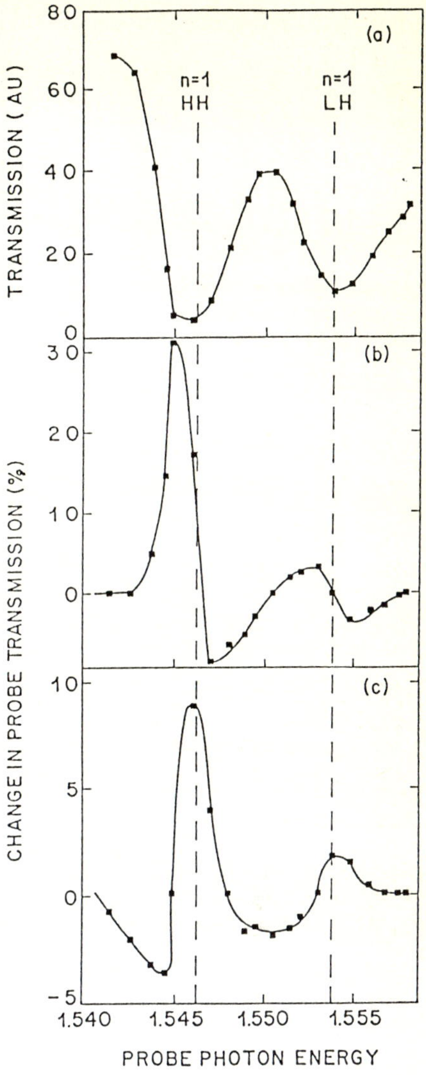

consistent with a blue shift of the entire exciton absorption line by 0.2 meV. We have interpreted this ultrafast shift in terms of an AC Stark shift of the quantum well exciton transition induced by the optical field of the below-gap laser.

The AC Stark shift is well known in atoms, where it can be viewed as a dressing of the atomic levels by the pump photons. Although it is possible to carry over the model of the atomic AC Stark shift to the case of excitons in a solid, the calculation requires several ad hoc assumptions in order to arrive at the correct magnitude of the shift. The failure of the "dressed exciton" approach is due to the nature of energy levels in a solid. Each pair of vertical transitions between conduction and valence band can be regarded as a two-level system. However these two-level systems all interact and in fact their interaction (via Coulomb forces) is responsible for formation of the exciton. Thus in the presence of a strong external field there exists, in addition to the shift of each level in the field, a change in the interaction between levels. These two effects must be calculated self-consistently and both contribute to the shift of the exciton resonance measured by a weak probe beam. Assuming that only 1s excitons participate and in the limit of low intensity, the theory predicts [6] a Stark shift

$$\delta E=\frac{2|\mu_{cv}E_p|^2|U_{1s}(r=0)|^2}{(E_{1s}-h\omega_p)N_s^{PSF}} \tag{1}$$

where E_p is the optical electric field of the pump, μ the dipole moment between conduction and valence band, $h\omega_p$ the pump photon energy, and E_{1s} the exciton transition energy in the absence of E_p. The exciton envelope function U_{1s} and the phase space filling saturation density N_s^{PSF} are defined in Ref.4. We have experimentally verified the functional form given above: we find that the shift is proportional to the intensity (E_p^2) and that it varies inversely with the detuning from resonance. Moreover the calculated magnitude of the shift $\delta E=0.15$ meV is in excellent agreement with the data shown in Fig.3 and is obtained with no adjustable parameters.

REFERENCES

[1] A. Von Lehmen, J. E. Zucker, J. P. Heritage, D. S. Chemla, A. C. Gossard, Appl. Phys. Lett. *48* 1479 (1986).

[2] A. Von Lehmen, J. E. Zucker, J. P. Heritage, D. S. Chemla, Proceedings XIV International Quantum Electronics Conferençe, PD18-1(1986)

[3] A. Von Lehmen, D. S. Chemla, J. E. Zucker,J. P. Heritage, Optics Letters *11*, 609 (1986).

[4] S. Schmitt-Rink, D. S. Chemla and D. A. B. Miller, Phys. Rev. B *32*, 6601 (1985).

[5] A.Von Lehmen, J.E.Zucker, J.P.Heritage and D.S.Chemla, submitted for publication.

[6] S. Schmitt-Rink and D. S. Chemla ,Phys.Rev.Lett. *57*, 2752 (1986).

Optical Reading of InGaAs Modulation Doped Field Effect Transistor

I. Bar-Joseph, D.S. Chemla, C. Klingshirn, D.A.B. Miller, J.M. Kuo, and T.Y. Chang

AT&T Bell Laboratories, Crawfords Corner Road, Holmdel, NJ 07733, USA

We present the first observation of absorption quenching in a semiconductor quantum well by electrical control of the carrier density. We have observed this novel effect at room temperature in a single modulation-doped InGaAs quantum well which was used as the conducting channel of a field effect transistor. The effect is extremely large and we have used it for direct optical determination of the state of the modulation doped field effect transistor (MODFET).

Modulation doping can be used to introduce carriers in quantum wells [1], producing very high mobility materials well known for their application to high speed transport [2]. Because the carriers fill the two dimensional (2D) subbands up to the Fermi energy, the absorption edge is blue-shifted with respect to the luminescence emission [3], with a shift approximately proportional to the the number of carriers. Thus electrically driven changes of the carrier density in a modulation doped quantum well will result in large changes in optical absorption. The conditions for the observation of this phenomenon correspond exactly to the situation encountered during the switching of a MODFET.

The structure we have investigated is a specially designed recessed gate InGaAs/InAlAs MODFET grown by molecular beam epitaxy (MBE) on InP:Fe. The structure is shown in Fig.1. It contains an L_z=100Å $In_{0.53}Ga_{0.47}As$ QW conducting channel for the 2D electron gas. The electrons are supplied by transfer through a 20Å $In_{0.52}Al_{0.48}As$ spacer layer from the 250Å $In_{0.52}Al_{0.48}As$ layer doped with Si$\sim 1.2\times 10^{18}$cm^{-3}. The Schottky gate pattern consists of two bonding pads of 300Å Cr

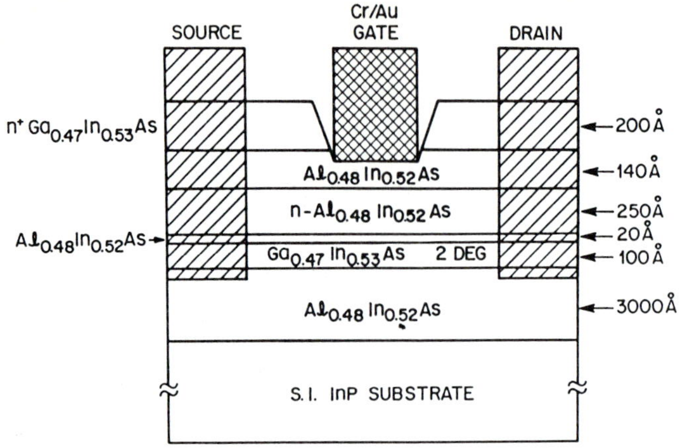

Figure 1: Drain-source current/voltage characteristics of the InGaAs/InAlAs MODFET used in our experiments. The MODFET structure is shown in the insert.

followed by 3000Å Au connected to a 1.6×100μm active region which is centered in a 5.4μm gap between the source and the drain.

The InAlAs layers as well as the InP substrate are transparent around the InGaAs gap ($\lambda \approx 1.55\mu$m) making the device compatible with lightwave communication systems.

We measured the changes in absorption in the channel by illuminating the gate through the transparent substrate, and detecting the reflection from the metal gate, thus making two passes through the channel. The change in the intensity, ΔT, of the reflected beam was measured as function of the wavelength as the gate-source voltage was modulated. A set of typical spectra for gate-source voltage modulation [(−0.6V)→(−0.2V) and (−0.6V)→(+1.2V)] are shown in Fig. 2. A large change is clearly seen at the position of the $n_z=1$ exciton peaks. The profile of the difference spectrum corresponds to a decrease in absorption without shift of the exciton resonances. The maximum amplitude is $\Delta T \sim 2\%$. At the $n_z=2$ resonance a somewhat weaker structure is also seen. The variation of absorption at this resonance has a completely

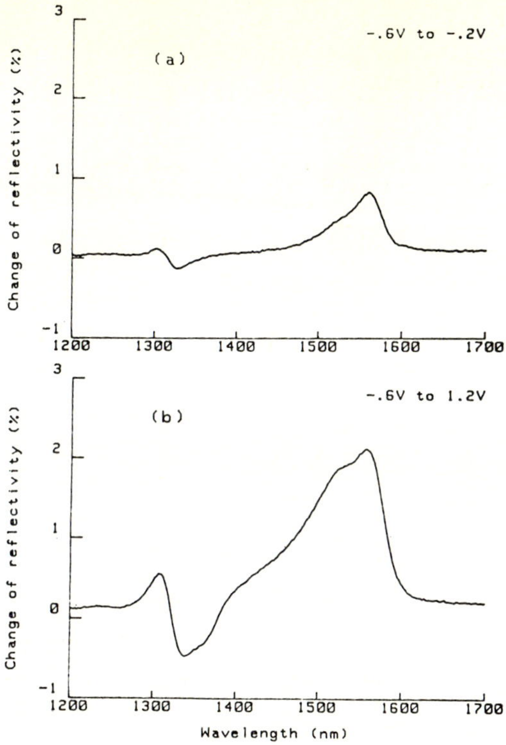

Figure 2: Differential absorption spectra showing the absorption quenching of the $n_z=1$ exciton resonances and the red shift of $n_z=2$ resonance, for two gate voltage amplitudes: a) -0.6V to -0.2V . b) -0.6V to 1.2V.

different spectral profile which corresponds to a red shift of the $n_z=2$ exciton peak due to broadening and the quantum-confined Stark effect.

Beside providing a wealth of information on the optical properties of modulation doped quantum wells, this effect has clear potential for monitoring and optically interconnecting electronic circuits based on III-V semiconductor technology. The effect is large enough to enable probing in a single QW and enables direct observations of the density of the conducting electron gas in the transistor. The contrast can be further increased in waveguides or by using more than one well.

REFERENCES

[1] H. L. Stormer, R. Dingle, A.C. Gossard, W. Wiegmann, and M. D. Sturge, J. Vac. Sci. Technol. **16**, 1517 (1979)

[2] H. L. Stormer, Surface Science, **132**, 519 (1983)

[3] A. Pinczuk, J. Shah, R.C. Miller, A.C. Gossard, W. Wiegmann, Solid State Comm. **50**, 735 (1984)

Part V

High-Speed Electronic Devices

Molecular Beam Epitaxy (MBE) for High-Speed Devices

A.Y. Cho

AT&T Bell Laboratories, 600 Mountain Avenue, Murray Hill, NJ07974, USA

Recent development in molecular beam epitaxy (MBE)[1] is to broaden the scope of the technology to cover a wider range of materials from semiconductors to metals and insulators. For III-V compounds, gas source molecular beam epitaxy (GS MBE, MO MBE or CBE) is used to grow compounds containing phosphorous[2-6]. Precisely controlled electron-gun evaporation is used for Si-MBE[7] and metal-MBE[8]. Fast loading of Hg source is used for II-VI compounds[9]. The impact of MBE does not only contribute to new discoveries in quantum physics[10,11] but also to technology for generating a whole new generation of microwave[12] and optical[13] devices. The capability of MBE to engineer the band structure by superlattices and spatially varying the composition of semiconductors to create a specific gradient at the conduction and valence band edge or to accurately and abruptly control the layer growth to atomic dimensions pushes the frontier of devices to the ultimate imagination of device physicists and engineers.

The main concerns of MBE for production are large-area uniformity, defect levels, and high through-put. MBE has made major advances in all three areas in recent years. For uniformity, variation in thickness of less than 1% over a 3-inch GaAs wafer[14] and oval defects as low as 100 cm^{-2} [Ref. 15] have been reported with commercial MBE systems. For multi-wafer deposition, a new MBE system with a 7.5-inch-diameter rotating substrate holder was introduced for the first time[16]. Seven 2-inch wafers or three 3-inch wafers can be deposited simultaneously with solid Ga and As sources. The uniformity of the thickness and carrier concentration of GaAs and $Al_xGa_{1-x}As$ layers were within $\pm 1\%$ over a diameter of 7.5 inches. Oval defects, originating from the Ga source, were reduced to a density of less than 40 cm^{-2} [Ref. 16].

High purity GaAs with Hall mobility of 163,000 cm^2/vs at 77 K with a peak value as high as 216,000 cm^2/vs at 46 K has been achieved[17]. Interface smoothness to one atomic layer is accomplished by interrupted growth or periodic high substrate temperature pulses between the heterostructures[18,19]. The precision in switching from one compound to another upon the completion of a full atomic layer (phase-locked epitaxy) may be monitored by reflection high energy electron diffraction (RHEED) intensity oscillation[20,21]. The resulting detailed understanding and control of crystal growth at the mono-atomic-layer level may be applied to architect advanced devices with local variations in band structures and transport properties.

High speed devices require size reduction in lateral and vertical dimensions. The effect of lateral variation, such as the gate-length of an FET on speed of operation, has been investigated[21]. A record propagation delay time of 5.8 ps/gate was measured for a 0.35 μm gate 19-stage ring oscillator at 77 K, with a power of 1.76 mw/gate, and a bias voltage of 0.88 V. The maximum switching speed at room temperature was 10.2 ps/gate with 1.03 mw/gate and 0.8 V bias voltage[22].

Low noise GaAs/AlGaAs modulation doped FET's with 0.25 μm gate length were operated over 40 GHz with 2.1 dB noise figure and 7dB associated gain[23]. More recently, a noise figure of 2.7 dB with an associated gain of 3.8 dB was measured at 62 GHz[24].

In the past, most of the devices were made on epitaxial layers consisting of materials having lattice constants the same as those of the substrate. This certainly limited the freedom of design of multilayered devices because of the limited pairs of compounds which have the same lattice constant. Advances in epitaxial techniques allowed us to grow thin heterostructures having different lattice constants and led to a new generation of devices which could not be realized before. For instance, a $In_{0.15}Ga_{0.85}As/Al_{0.15}Ga_{0.85}As$ pseudomorphic single quantum well modulation doped transistor was grown by MBE[25]. The higher electron saturation velocity in $In_{0.15}Ga_{0.85}As$ compared to that of GaAs provides the large saturation current and transconductance needed to drive capacitances rapidly in integrated circuits. Furthermore, this compound (with about 1% lattice mismatch to GaAs) has high electron mobility and large energy minima difference $\Delta E = E_\Gamma - E_L$ resulting in higher value of overshoot velocity. GaAs has a relatively small ΔE of 0.3 eV compared to that of InGaAs alloy of about 0.6 eV. MODFETs with 0.25 μm length gate were operated at 62 GHz with a noise figure of 2.4 dB and an associatd gain of 4.4 dB. Maximum gains achieved at 62 GHz and 94 GHz were 11.7 dB and 7 dB, respectively. The device has a maximum oscillation frequency as high as 230 GHz[25]. With InSb compounds, a further increase in oscillation frequency may be realized.

Large lattice mismatched strained $(InAs)_m(GaAs)_m$ superlattice has been grown with MBE[26]. A proposal was made for a high-speed $In_{0.52}Al_{0.48}As/In_{0.53}Ga_{0.47}As$ modulation doped FET with a $(InAs)_m$ $(GaAs)_m$ superlattice channel with $m \leq 4$ where the electrons in the active channel do not suffer any alloy scattering and thus result in an increase in mobility of about an order of magnitude at 77 K[27,28].

Monolithic integration of GaAs/AlGaAs MODFETs and N-metal-oxide-semiconductor (N-MOS) silicon circuits was reported earlier[29]. Fully monolithic integration of interconnected GaAs/AlGaAs double-heterostructure light emitting diodes (LEDs) and Si MOSFETs is demonstrated for the first time[30]. The GaAs/AlGaAs layers were grown on a Si substrate with MBE. LED modulation rates up to 27 mb/s have been achieved by applying a stream of voltage pulses to the MOSFET gate. The modulation rate is limited by the speed of the MOSFET for this preliminary demonstration. GaAs on Si substrates opened a new direction of development where Si VLSI circuits can be complemented by III-V optoelectronic components and high-speed electronic circuits.

The most significant development in telecommunications in recent years has been the evolution toward digital transmission and switching. Optical communications promises high capacity and high reliability because of its immunity to the electromagnetic radiation effects. As the capacity increases and the speed of lightwave transmission begins to exceed the processing capability of the fastest electronic circuits, we need a device technology capable of switching and processing lightwave signals in their photonic form. A recent development is to use a superlattice structure or multi-quantum well device. Bistability can be observed with 18 nW of incident power or with 30 ns switching time at 1.6 mW with a reciprocal relation between switching power and speed[31]. Electroabsorption in an InGaAs/GaAs strained-layer multiple quantum well structure has been observed. The novel quantum-confined stark effect modulator with 6.4% transmission modulation was obtained at 0.95 μm with 2 V reverse bias[32].

New avalanche multiplication phenomenon in quantum well superlattices[33], ultra sensitive photodetectors with effective mass filter[34], and novel transistors with resonant tunneling[35–38] have opened new directions in device fabrication. MBE's ability to synthesize and precisely engineer the microstructures of heterojunction thin films has led to many breakthroughs in high-speed devices.

1. A. Y. Cho: *Thin Solid Films 100,* 291 (1983).
2. M. B. Panish: *J. Electrochem. Soc. 127,* 2729 (1980).
3. A. R. Calawa: *Appl. Phys. Lett. 38,* 701 (1981).
4. M. B. Panish, H. Temkin and S. Sumski: *J. Vac. Sci. Technol. 133,* 657 (1985).
5. N. Vodjdani, A. Lemarchand and H. Paradan: *J. de Physique C5 43,* 339 (1982).
6. W. T. Tsang: *Appl. Phys. Lett. 48,* 511 (1986).
7. J. C. Bean: *J. Crystal Growth 70,* 444 (1984).
8. E. Ziegler. Y. Lepetre and I. K. Schuller: *Appl. Phys. Lett. 48,* 1354 (1986).
9. J. P. Faurie, A. Million, R. Boch and J. L. Tissot: *J. Vac. Sci. Technol. A1,* 1593 (1983).
10. For a review see H. L. Stormer: *Surf. Sci. 132,* 519 (1983).
11. L. Esaki in *The Technology and Physics of Molecular Beam Epitaxy,* Ed. E.H.C. Parker (Plenum Press 1985), pp. 143-284.
12. H. Morkoc in *The Technology and Physics of Molecular Beam Epitaxy,* Ed. E. H. C. Parker (Plenum Press 1985), pp. 195-231.
13. F. Capasso, K. Mohammed and A. Y. Cho, *IEEE J. Quantum Electronics QE 22* 1853 (1986).

14. C. Chaix; presented at Riber MBE user meeting, York, September 11, 1986.
15. H. Fronius, A. Fischer and K. Ploog: *Appl. Phys. Lett. 48,* 1392 (1986).
16. J. Saito, T. Igarashi, T. Nakamura, K. Kondo, and A. Shibatomi: Fourth International Conference on MBE, York, England, 7-10 September 1986.
17. E. C. Larkins, E. S. Hellman, D. G. Schlom, J. S. Harris, Jr., M. H. Kim and G. E. Stillman: *Appl. Phys. Lett. 49,* 391 (1986).
18. H. Sakaki, M. Tanaka, and J. Yoshino: *Jpn. J. Appl. Phys. 24,* L417 (1985).
19. J. Singh, S. Dudley and K. K. Bajoj: *J. Vac. Sci. Technol. B4,* 1986.
20. J. H. Neave, B. A. Joyce, P. J. Dobson and N. Norton: *Appl. Phys. A31,* 1 (1983); ibid *A34,* 1 (1984).
21. M. Y. Yen, T. C. Lee, P. Chen, and A. Madhukar: *J. Vac. Sci. Technol. B4* (2), 590 (1986).
22. N. J. Shah, S. S. Pei, C. W. Tu and R. C. Tiberio: *IEEE Trans. on Dev. ED-33,* 543 (1986).
23. P. C. Chao, S. C. Palmateer, P. M. Smith, U. K. Mishra, K. H. G. Duh and J. C. M. Hwang: *IEEE Electron. Dev. Lett. EDL-6,* 531 (1985).
24. P. C. Chao, P. M. Smith, K. H. G. Duh and J. C. M. Hwang: Device Research Conference, Amherst, MA, June 23-25, 1986.
25. T. Henderson, M. I. Aksun, C. K. Peng, H. Morkoc, P. C. Chao, P. M. Smith, K. H. G. Duh, and L. F. Lester: to be published in *Electron Dev. Lett.*.
26. M. C. Tamargo, R. Hull, L. H. Greene, J. R. Hayes and A. Y. Cho: *Appl. Phys. Lett. 46,* 569 (1985).
27. T. Yao: *Jpn. J. Appl. Phys. 22,* 680 (1983).
28. J. Singh: *IEEE Elec. Dev. Lett. EDL-7,* 436 (1986).
29. R. Fischer, T. Henderson, J. Klem, W. Kopp, C. K. Peng, and H. Morkoc: *Appl. Phys. Lett. 47,* 983 (1985).
30. K. K. Choi, G. W. Turner, T. H. Windhorn, and B. Y. Tsaur: to be published.
31. D. A. B. Miller, D. S. Chemla, T. C. Damen, T. H. Wood, C. A. Burrus, Jr., A. C. Gossard, and W. Wiegmann: *IEEE J. Quantum Electron QE-21,* 1462 (1985).
32. T. E. Van Eck, P. Chu, W. S. C. Chang, and H. H. Wieder: *Appl. Phys. Lett. 49,* 135 (1986).
33. F. Capasso, J. Allam, A. Y. Cho, K. Mohammed, R. J. Malik, A. L. Hutchinson and D. L. Sivco: *Apply. Phys. Lett. 48,* 1294 (1986).
34. F. Capasso, K.Mohammed, A. Y. Cho, R. Hull and A. L. Hutchinson: *Appl. Phys. Lett. 47,* 420 (1985).
35. L. L. Chang, L. Esaki and R. Tsui: *Appl. Phys. Lett. 24,* 593 (1974).
36. F. Capasso and R. A. Kiehl: *J. Appl. Phys. 58,* 1366 (1985).
37. F. Capasso, K. Mohammed and A. Y. Cho: *Appl. Phys. Lett. 48,* 478 (1986).
38. M. A. Reed, J. W. Lee and H. L. Tsai: *Appl. Phys. Lett. 49,* 158 (1986).

Multigigahertz Logic Based on InP MISFETs Exhibiting Extremely High Transconductance

A. Antreasyan, P.A. Garbinski, V.D. Mattera Jr., and H. Temkin

AT&T Bell Laboratories, 600 Mountain Avenue,
Murray Hill, NJ 07974, USA

The search for high speed electronic devices has over the years focused on the utilization of high performance semiconducting materials. Here, we describe a novel, enhancement mode metal-insulator-semiconductor field-effect transistor (MISFET) fabricated on InP. The motivation for this work is provided by two reasons. First, InP has high peak and saturation velocities [1] compared with GaAs. Secondly, optical devices on the InGaAs/InP and InGaAsP/InP material systems are well developed, therefore the development of high-speed electronic devices on InP is of utmost importance for monolithic optoelectronic integration [2]. In addition, the growth of high purity InP by vapor phase epitaxy (VPE) is a well established technique [3]. To demonstrate the suitability of these MISFET's for high-speed InP based digital circuits, we have fabricated inverters, basic building blocks of digital circuits. A propagation delay of 15 ps/stage has been measured for the inverter structure utilizing electro-optic sampling measurements, which represents, to our knowledge, the fastest FET prepared on an InP substrate.

A cross-sectional view of the structure of the InP MISFET and a load resistor associated with the inverter is shown in Fig. 1. It consists of a 100nm thick, undoped InP buffer layer, a 100nm thick undoped $Ga_{0.47}In_{0.53}As$ layer, a 400nm thick n-type $Ga_{0.47}In_{0.53}As$ layer and a 100nm thick n^+-$Ga_{0.47}In_{0.53}As$ contact layer grown by chloride VPE on an Fe-doped substrate which is intentionally misoriented 6° off the (100) towards the nearest <110> orientation. The undoped layers have a background doping level of $n \sim 5 \times 10^{14}$-$10^{15} cm^{-3}$, while the S-doped n-GaInAs and n^+-GaInAs layers have doping levels of 5×10^{17} and $10^{18} cm^{-3}$, respectively. The MISFET's are fabricated by etching mesa structures on the wafer down to the undoped InP buffer layer. Then, source-drain metallization as well as connections between individual devices are defined by lift-off. After the gate photolithography, trapezoidally shaped grooves as shown in Fig. 1, are etched until the undoped InP layer is exposed at the bottom of the groove. The definition of a self-aligned gate structure is accomplished by evaporating SiO_2 and gate metallization in a high vacuum electron beam evaporator, and consecutive lift-off procedure. Two important features of the MISFET structure need to be emphasized. First, an undoped GaInAs layer is grown between the highly n-doped GaInAs layer and the undoped InP buffer layer to prevent the back-diffusion of dopants into the InP. Hence, the gate etching can be carried out utilizing a selective etchant to yield low channel currents at zero gate bias (normally-off). Second, the substrate is kept at low temperature during the deposition of the gate dielectric. We believe this results in significant reduction of surface defects, leading to excellent device properties. The FET's utilized in this work as inverters have a gate length of 1μm, and a gate width of 50μm. In addition, discrete test FET's have been also fabricated on the same wafer

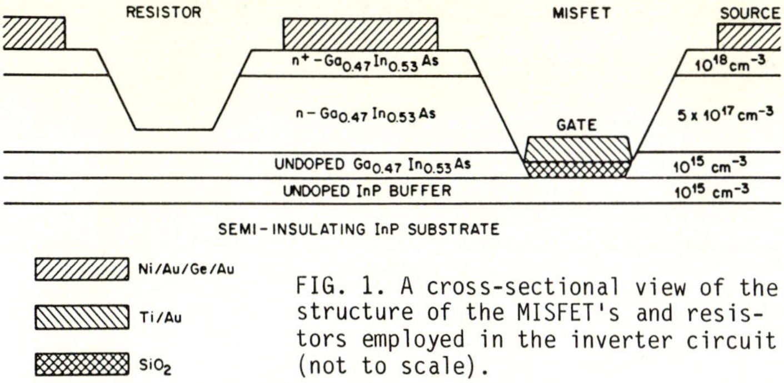

FIG. 1. A cross-sectional view of the structure of the MISFET's and resistors employed in the inverter circuit (not to scale).

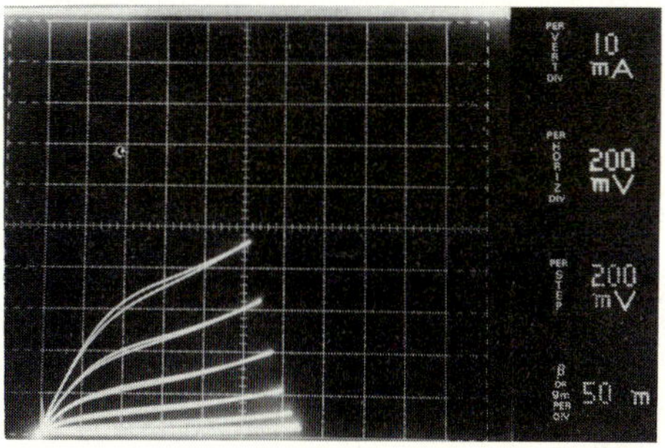

FIG. 2. Drain-source I-V characteristic of a MISFET at room temperature having a gate length and gate width of 1μm and 250μm, respectively. Vertical scale is 10mA/div, horizontal scale is 200mV/div, gate voltage is increased in steps of 200mV.

having a gate width of 250μm. The thickness of the gate insulator SiO_2 was chosen to be 500Å. A drain-source I-V characteristic is illustrated in Fig. 2 for a discrete test MISFET showing a transconductance of 80 mS corresponding to 320 mS/mm. The devices typically have threshold voltages around 0.1-0.2 volt and peak g_m values occur at about 1.5 volts. After the fabrication of the MISFET's, one additional step of photolithography is performed for etching trapezoidal grooves to fabricate the resistors (left groove on Fig. 1). In this case, however, a thin layer of n-GaInAs is preserved at the bottom of the groove to provide a resistance of about 500 ohms.

The top view of a completed inverter consisting of three MISFET's and three resistors, and its equivalent circuit are illustrated in Fig. 3 and 4, where the resistors are located next to the FET's. It should be noted that the FET on the right hand side in Fig. 3 is disconnected from the other devices and can be operated by itself as a single stage, one transistor-inverter. In practice, however, connection will be made from the output of one inverter to the input of one or more similar circuits. This is the case for the FET's on the left hand side and in the middle of Fig. 3 which form a two-stage inverter circuit. Hence, this second arrangement provides a far more realistic environment for the investigation of the inverter characteristics of a MISFET. The voltage transfer characteristics of the two-stage inverter have been plotted in Fig. 5, where two separate traces are obtained by plotting V_{OUT} vs. V_{IN}, and repeating the plot after interchanging the axis. In the course of this measurement V_{SS} was connected to ground V_{IN} was varied from 0 to 2 volts and V_{DD} was raised from 0 to 1.6 volts in steps of 0.4 volts. For V_{DD} = 1.6 volts in Fig. 5 (uppermost trace) we observe a high and low level than 4 volts and g_m's of 37mS/mm.

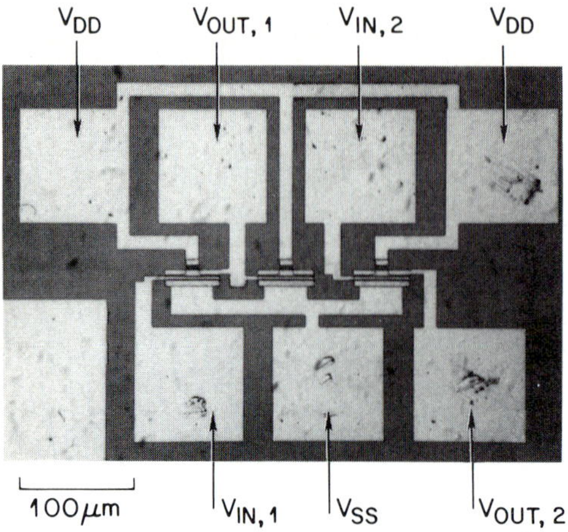

FIG. 3. Top view of the morolithic integrated inverter with three transistors and three resistors.

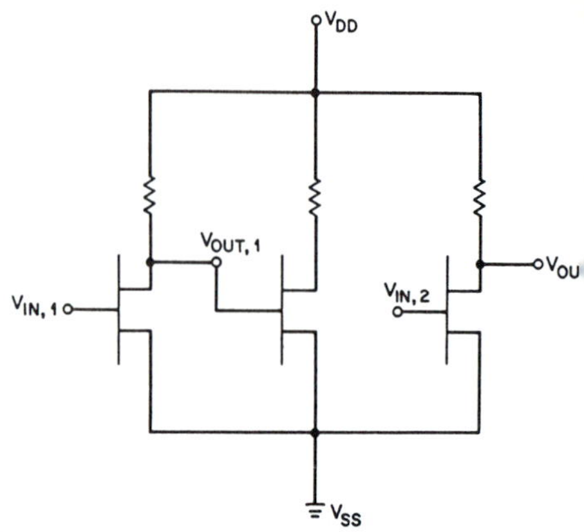

FIG. 4. Equivalent circuit for the inverter in Fig. 3.

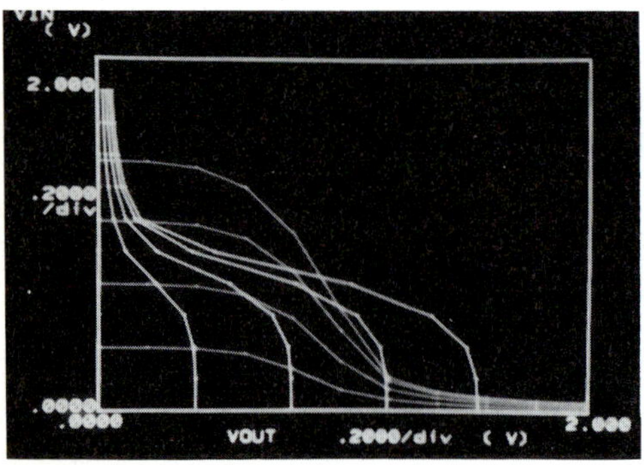

FIG. 5. A plot of voltage transfer characteristics for a two-stage inverter. Two separate traces are obtained by plotting $V_{OUT,1}$ vs $V_{IN,1}$ and repeating the plot after interchanging the axis. Horizontal and vertical axes extend linearly from 0 to 2.0 volts, while V_{DD} is a parameter varying from 0 to 1.6 volts in steps of 0.4 volt.

noise margins of 0.3 volt. We note, that the presently measured noise margins are comparable to those obtained with selectively doped heterostructure transistors [4] at similar bias level. The demonstration of InP MISFET logic operable at bias levels lower than 2 volts show a significant improvement over previous work [5], where ion-implanted InP MISFET logic was accomplished with devices operating at biases higher

An electro-optic sampling system [6] based on a 1.3μm InGaAsP laser which was capable of producing pulses of 15-20 ps duration has been utilized to evaluate the operation of the two-stage inverter circuit at high frequencies. The electro-optic sampling measurements are outlined by WIESENFELD et al. [7] in great detail. The electro-optic sampling measurements have shown that the inverters are capable of operating at a frequency of 5 GHz. The pulse propagation delay for the two-stage inverters varied in the range 15-30 ps. Furthermore, with a peak electron velocity in InP of 2.5×10^7 cm/s [1] we obtain a transit time of 4 ps for a channel length of 1μm, leading to f_t's of 40 GHz. Indeed a current gain cut-off frequency of f_t = 29 GHz corresponding to a transit time of 5.5

ps has been measured for a discrete MISFET utilizing an HP 8510 network analyzer [8]. Details of the measurement of f_t will be published elsewhere. For comparison, the shortest propagation delay of 5.8 ps/stage was measured for 0.35-μm gate GaAs/AlGaAs selectively doped heterostructure transistors [9] (SDHT) at 77K. For 1-μm gate SDHT's delays of 30 ps/stage and 20 ps/stage were measured [10] at 300 K and 77 K, respectively, which is comparable to the present result. Several MISFET structures on InP substrates have been tried for high-speed logic applications. For MISFET structures formed by n^+-implant into semi-insulating InP substrates, propagation delays of 500 ps/stage [11] and 62 ps/stage [5] were measured, while for a GaInAs MISFET structure the measured propagation delay was 300 ps/stage [12] for a 2 volt bias supply.

In conclusion, we have presented an InP MISFET structure having the highest transconductance ever measured on an FET prepared on an InP substrate. Monolithically integrated two-stage inverters have been prepared successfully utilizing these MISFET's. Electro-optic sampling measurements performed on the inverters reveal propagation delays of 15-30 ps/-stage, which is the shortest propagation delay measured on InP based high-speed circuits. Initial current gain cut-off frequency measurements revealed f_t's up to 29 GHz.

The authors would like to thank J. R. Filipe and A. Westphal for their assistance in crystal growth, and M. A. DiGiuseppe, R. H. Saul and D. V. Lang for their encouragement and support.

REFERENCES

1. S. M. Sze: Physics of Semiconductor Devices, 2nd. ed., (Wiley, New York, 1981) p. 649.

2. K. Kasahara, J. Hayashi, K. Makita, K. Taguchi, A. Suzuki, A. Nomura, and S. Matushita, Electron. Lett., 20, 314 (1984).

3. A. Antreasyan, P. A. Garbinski, V. D. Mattera, Jr., N. A. Olsson, and H. Temkin, J. Appl. Phys., 60, 1535 (1986).

4. R. H. Hendel, S. S. Pei, R. A. Kiehl, C. W. Tu, M. D. Feuer, and R. Dingle, IEEE Electron Dev. Lett., EDL-5, 406 (1984).

5. L. Messick, IEEE Trans. Electron Dev., ED-31, 763 (1984).

6. A. J. Taylor, J. M. Wiesenfeld, G. Eisenstein, R. S. Tucker, J. R. Talman, and U. Koren, Electron. Lett., 22, 61 (1986).

7. J. M. Wiesenfeld, R. S. Tucker, A. Antreasyan, C. A. Burrus, and A. J. Taylor, Second Topical Meeting on Picosecond electronics and optoelectronics, paper WA4, Incline Village, Nevada, Jan. 14-16, 1987.

8. A. Antreasyan, P. A. Garbinski, J. H. Abeles, H. Temkin, and V. D. Mattera, Jr., (unpublished).

9. N. J. Shah, S. S. Pei, C. W. Tu, and R. C. Tiberio, IEEE Transactions on Electron Dev., ED-33, 543 (1986).

10. N. J. Shah, S. S. Pei, C. W. Tu, R. H. Hendel, and R. C. Tiberio, Electron. Lett., 21, 151 (1985).

11. K. P. Pande, M. A. Fathimulla, D. Gutierrez, and L. Messick, IEEE Electron Dev. Lett., EDL-7, 407 (1986).

12. L. C. Upadhyayula, P. D. Gardner, S. G. Liu, S. Yegnarayan, IEEE Electron Dev. Lett., EDL-7, 390 (1986).

Millimeter-Wave Integrated Circuits

B.E. Spielman

Naval Research Laboratory, Electronics Technology Division, Washington, DC 20375, USA

1. BACKGROUND

The opportunities offered by monolithic millimeter-wave technology span radar, communications, electronic warfare, and missile seekers. These opportunities relate to size, weight, and cost reduction advantages over standard metal waveguide and hybrid integrated circuit implementations. However, specific technical challenges remain in realizing some important circuit building blocks.

2. CIRCUIT BUILDING BLOCKS

Circuit building blocks for monolithic millimeter-wave ICs encompass semiconductor devices and transmission line media. Table 1 lists most of the semiconductor device candidates for use in millimeter-wave source, mixer/ detector, and modulator circuit functions. Of these, the most well established candidates are: the IMPATT, TED, and MESFET for sources; the Schottky diode, Josephson junction diode, and MESFET for mixers/detectors; and the PIN and MESFET as modulator elements. As is suggested by the multitude of source candidates, current efforts are exploring new concepts

Table 1. Device candidates

SOURCES:		MIXER/DETECTOR:	MODULATORS:
• IMPATT	• PDBT	• SCHOTTKY DIODE	• PIN
• TED	• CHIRP(superlattice)	• JJ/SIS	• MESFET
• MESFET	• QUANTUM-WELL (QST)	• MESFET	
• MODFET	• DIMPATT	• MODFET	
• HJBT	• SSTWA	• CHIRP	
• CHINT	• SS-OROTRON	• QUANTUM-WELL	
• NERFET	• SS-MAGNETRON		
• OGST	• SS-GYROTRON		
• PBT	• SPACE-HARMONIC AMPLIFIER		

to achieve reasonably high output power levels (25 mW or higher) with reasonable gain and efficiency (dc to RF) for above 60 GHz applications, especially in three-terminal device configurations.

The transmission media most commonly considered for monolithic millimeter-wave ICs are standard microstrip, coplanar waveguide, and fin line or slot line. The unloaded Q value at 30 GHz for these (excluding fin line) transmission lines range from about 100 to 200. Similar values for trapped inverted microstrip, inverted microstrip, fin line, and suspended microstrip lie in the range from 400 to 700, while those for image line and metal rectangular waveguide fall in the 2500 to 3000 range. Other characteristics for such media are also important. Table 2 qualitatively summarizes some of the more significant circuit medium properties for monolithic-compatible media, hybrid IC compatible media, and metal waveguide. It is clear that application-dependent tradeoffs must be made.

It is becoming increasingly apparent that for monolithic IC designs new and more refined models are needed for semiconductor device representations and for treatment of device to circuit-medium interfaces (e.g. parasitic

Table 2. Summary of circuit medium properties

PROPERTY	WAVEGUIDE	HYBRID IC's	MONOLITHIC IC's
APPLICABLE FREQUENCY	1 GHz TO 300 GHZ	1 GHz TO APPROX 70 GHz	1 GHz TO 100 GHz+
RELATIVE COST	HIGHEST	MODERATE TO HIGH	LOWEST
SIZE	LARGEST	MEDIUM	SMALLEST
RF COMPONENT AVAILABILITY	GREATEST	EXTENSIVE	LIMITED
BANDWIDTH CAPABILITY	LESS THAN OCTAVE	GREATER THAN OCTAVE	GREATEST
CONNECTOR AVAILABILITY	COMPLETE	COMPLETE	NEEDS DEVELOPMENT
SOLID-STATE DEVICE COMPATIBILITY	POOR	FAIR/GOOD	BEST
MODELING DATA BASE	COMPREHENSIVE	MODERATE	INCOMPLETE
POWER HANDLING	BEST	MODEST-ARRAYS NEEDED	MODEST-ARRAYS NEEDED
REPEATABILITY	POOR	FAIR	POTENTIALLY EXCELLENT
LOSSES	LOW	MODERATE	HIGHEST
VOLUME PRODUCTION SUITABILITY	POOR	MODERATE	BEST

effect treatment), especially for these higher frequencies. Once such models are developed and verified, it would be desirable to have such design capability available in a mode which is widely accessible. Possibly CAD workstation equipment could incorporate a software base for such a purpose.

In the area of testing there are at least two important needs. One is to expand the commercial availability of error-correcting, automated network analyzer equipment to at least 100 GHz. Also, a need exists for on-chip wafer prober units to function accurately and reproducibly over the same frequency range.

3. RECENT/CURRENT MONOLITHIC EFFORTS

Monolithic efforts to date have focused primarily on non-transmitter related functions. This trend is observed in the listing of representative programs provided in Table 3. A notable exception in this list is the 32 GHz transceiver program. This program employs an approach where the output of a microwave source is frequency multiplied to provide a 32 GHz transmit signal. Other activities await efforts either to refine and improve existing source devices in planar format or to innovate devices for signal generation directly at the millimeter wave frequency of interest.

Table 3. GaAs millimeter-wave monolithic IC efforts

• 32 GHz TRANSCEIVER	• 110 GHZ QUASI-OPTICAL MIXER
• 75 TO 110 GHz RECEIVER TECHNOLOGY	• 215 GHz DETECTOR DIODE ARRAY
• 35 TO 38 GHz RECEIVER	• 44 GHz PIN PHASE SHIFTER
• 44 GHz RECEIVER	• K_a-BAND PIN DIODE WINDOW COMPONENTS
• 90 TO 100 GHz RECEIVER	• 30 TO 50 GHz AND 75 TO 100 GHz SWITCHES

To best illustrate the technology being pursued in the monolithic receiver area, attention is drawn to the second entry in Table 3. This program is to provide a technology base for receivers to operate instantaneously over the full 75 to 110 GHz frequency range. The focus of this development is the receiver front end configuration depicted in Fig. 1. This configuration requires: four mixers, four common IF amplifiers, four local oscillators, and an RF multiplexer. One design in GaAs pursued for the mixer is depicted in Fig. 2. This mixer has provided 7 to 9 dB conversion loss over the 75 to 110 GHz band for an IF band from 8.5 to 17.5 GHz.

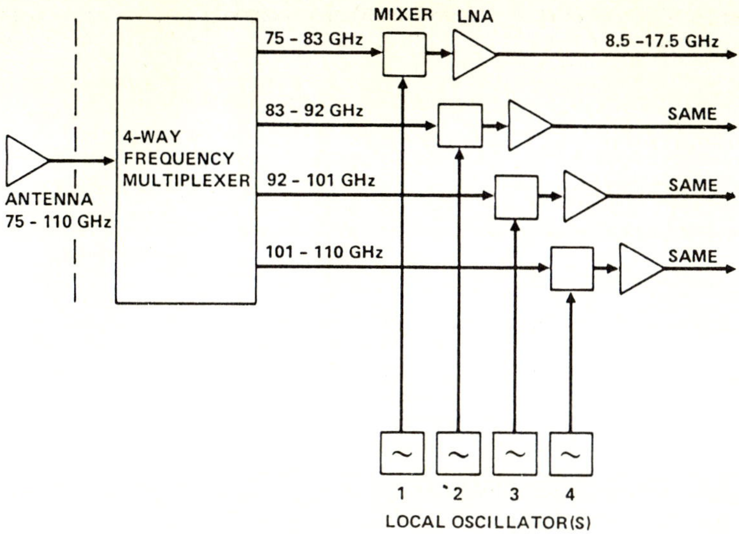

Fig. 1 Four-channel receiver front-end

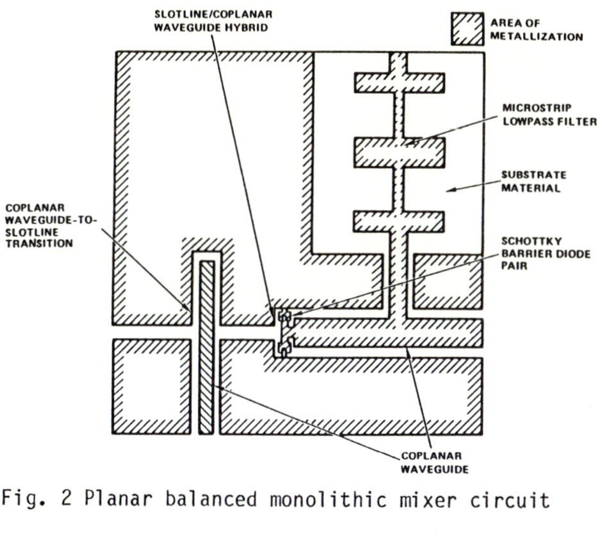

Fig. 2 Planar balanced monolithic mixer circuit

Enhanced Performance Ultrabroadband Distributed Amplifiers

G. Zdasiuk, M. Riaziat, R. LaRue, C. Yuen, and S. Bandy

Varian Research Center, 611 Hansen Way, Palo Alto, CA 94303, USA

Distributed amplifiers have been used in recent years to achieve amplification over multioctave bandwidths extending to millimeter-wave frequencies [1]. The distributed (or traveling wave) amplifier topology is especially suitable for monolithic integration using GaAs technology due to low sensitivity to component parameter variations and a relatively high active/passive component ratio. A major drawback of the approach has been the relatively low gain per stage (4-7 dB) and high current consumption. This paper describes the application of novel circuit approaches (such as cascode-connected FETs and coplanar waveguide transmission lines) and advanced devices such as high electron mobility transistors (HEMTs) to improve performance, simplify fabrication and achieve greater than 10 dB of single-stage gain over the 2-20 GHz band. This frequency range is of significance for electronic countermeasures systems and microwave measurement applications

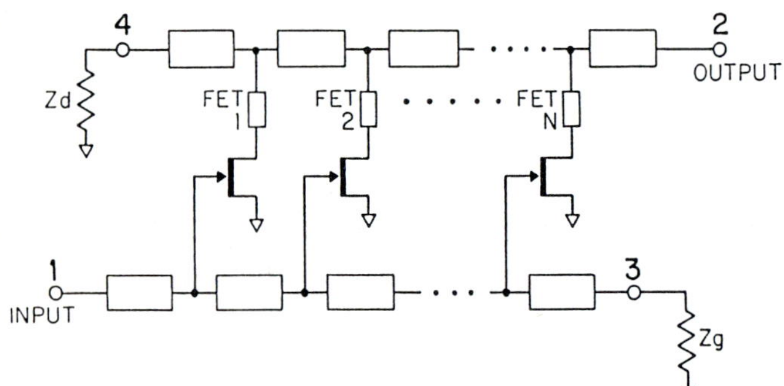

Distributed Amplifier Topology

Fig. 1 Distributed amplifier schematic.

A circuit diagram for the 2-20 GHz distributed amplifier is shown in Fig. 1. Input and output capacitances of the FET devices are incorporated into an artificial transmission line structure, which have a low pass response. A simplified gain expression, $G(\omega)$ [2], for the circuit is given by

$$G(\omega) = \frac{g_m^2 Z_o^2}{4}\left[\frac{e^{-A_g(\omega)n} - e^{-A_d(\omega)n}}{A_g(\omega) - A_d(\omega)}\right]^2 \qquad (1)$$

where g_m is the FET transconductance, Z_0 is the input and output artificial line impedance (assumed to be the same here), A_g, A_d are the gate and drain line losses (per section) respectively, and n is the number of sections. The frequency response of the amplifier is determined by the frequency behavior of the input and output line attenuation, $A_g(\omega)$, $A_d(\omega)$. The overall gain is determined by the active device g_m.

The monolithic distributed amplifier circuit fabricated with MESFET active devices on a semi-insulating GaAs substrate is shown in Fig. 2. A novel feature of this circuit is the use of coplanar waveguide (CPW) transmission lines (rather than microstrip) for forming the high impedance inductive sections of the input and output artificial transmission lines. The use of coplanar waveguide eliminates the need for through-the-substrate via holes which are difficult to fabricate. Gain is also improved, since the residual inductance of the via posts can reduce the overall amplifier gain by ~1 dB. CPW is also compatible with both high frequency contacting wafer measurements and contactless electro-optic wafer probing techniques [3].

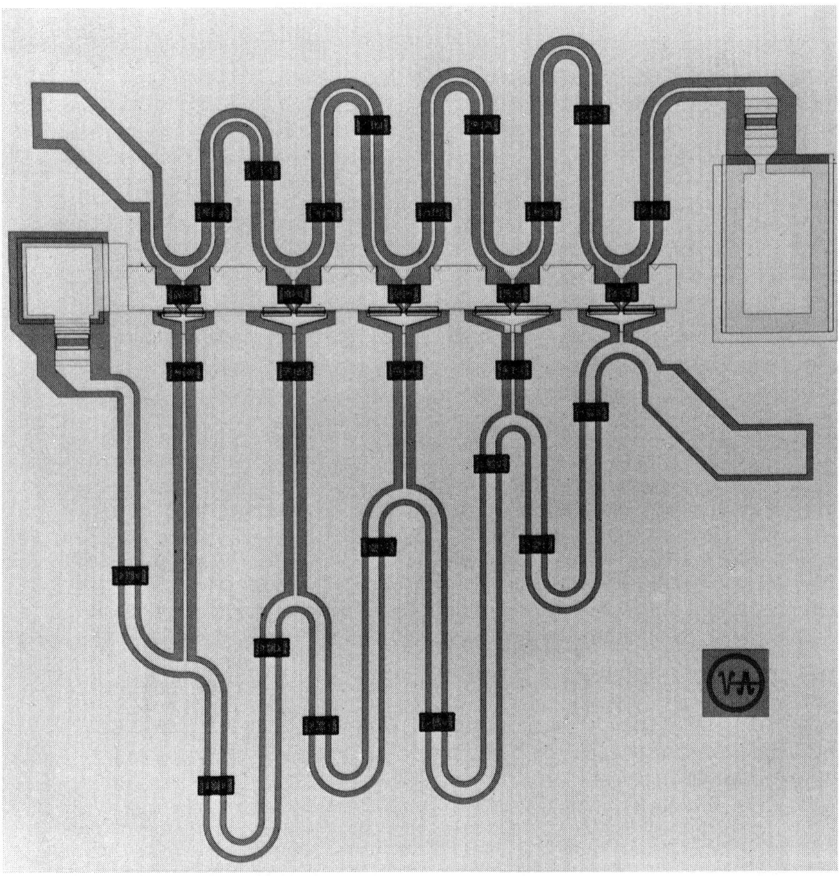

Fig. 2 2-20 GHz coplanar distributed amplifier.

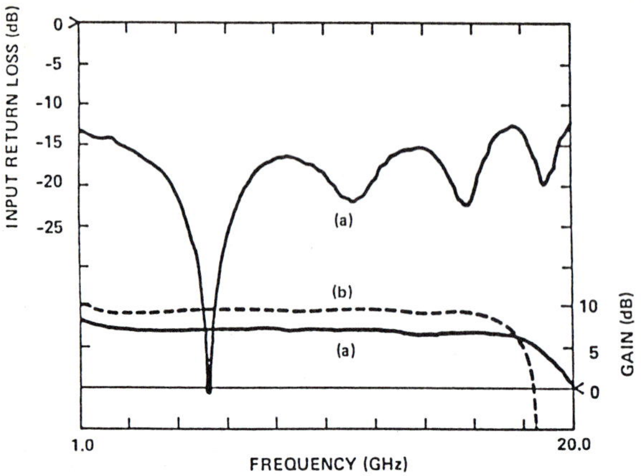

Fig. 3 (a) Coplanar waveguide distributed amplifier gain and return loss with nominal channel doping of 3-5 x $10^{17}cm^{-3}$; (b) amplifier gain with nominal doping of 5 x $10^{17}cm^{-3}$.

The active devices are GaAs MESFETs with gate lengths of approximately 0.6 μm, defined with optical lithography. The overall gate periphery is 725 μm. Active layers are epitaxially grown using metalorganic chemical vapor deposition. Figure 3 shows the gain versus frequency of the amplifier. Approximately 9.5 dB of gain is achieved from 2-17 GHz. If a wafer with slightly lower active layer doping is used, thereby lowering the gate capacitance and increasing the cutoff frequency of the transmission lines, useful gain can be obtained to above 20 GHz. Decreasing the doping also reduces the device g_m, so that the overall gain is lowered to ~7.5 dB.

By Fourier transforming the frequency domain data, the impulse response is obtained. Typically, the circuits of Fig. 2 exhibit 75 psec of pulse delay with ~100 psec of spreading.

An alternate approach to increasing distributed amplifier gain per stage is to utilize 2 cascode-connected FETs in place of single FETs as shown in Fig. 4. The output conductance of the "cascode block" of Fig. 4 is reduced by the factor $(2 + g_m r_{ds})^{-1}$ from a single FET, where r_{ds} is the output resistance of the individual FETs. This will reduce the drain line attenuation and hence increase the overall amplifier gain. Other advantages of the cascode configuration include higher output power (due to the second FET being connected as a grounded gate stage), improved isolation, and the ability to use the second gate for automatic gain control (AGC).

Figure 5 is a photomicrograph of a monolithic cascode distributed amplifier. This circuit utilized 5 "cascode sections" consisting of 2 FETs each. In this case, the FET gates were written with E-beam lithography with an average gate length of 0.3 μm. The circuit was fabricated on MOCVD-grown GaAs epitaxial material. Over 11 dB of gain was obtained over the 2-18 GHz band from the amplifier. This is currently the highest reported gain for a single-stage amplifier covering the 2-18 GHz band.

A final approach for increasing the gain and bandwidth of distributed amplifiers is to use high-transconductance (g_m) devices such as high

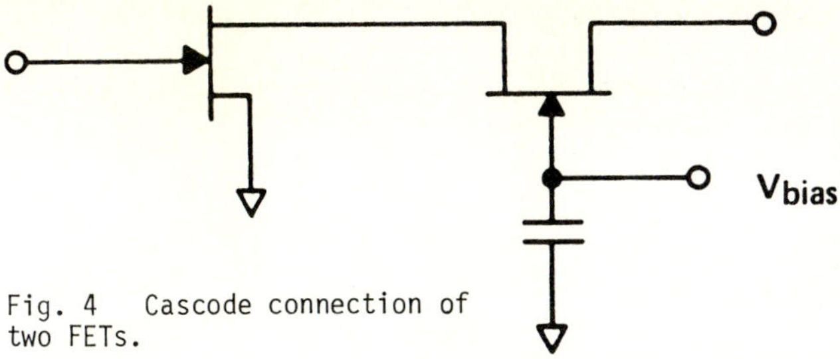

Fig. 4 Cascode connection of two FETs.

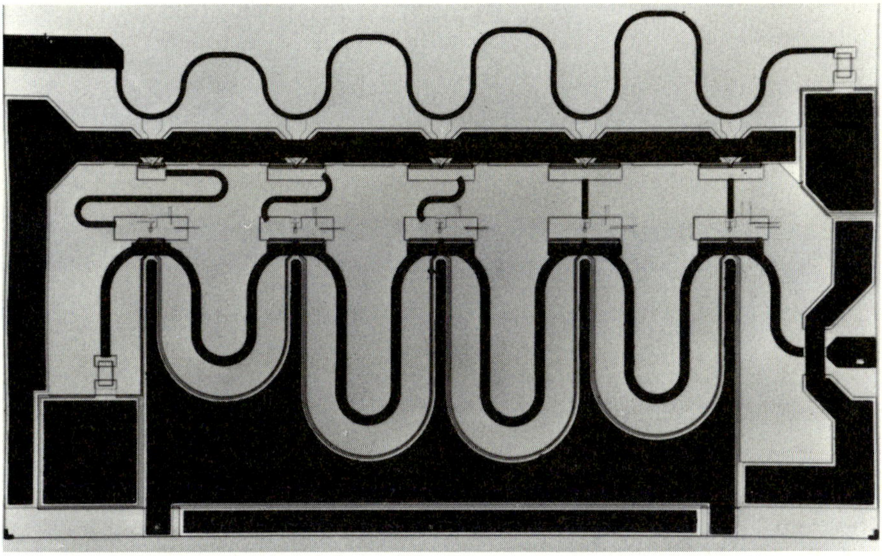

Fig. 5 2-20 GHz cascode distributed amplifier.

electron mobility transistors (HEMTs). HEMTs are especially suited to distributed amplifiers, since they have lower drain conductance than MESFETs and the gate capacitance is less dependent upon gate voltage. Other advantages include lower noise figures and higher associated gains in a 50-ohm environment and higher noise bandwidths [4].

A monolithic HEMT amplifier has been fabricated using a "conventional" topology (i.e., microstrip transmission lines, no cascode-connected devices, etc.). A photomicrograph of the circuit is shown in Fig. 6. The HEMT active layers are grown by MBE on a semi-insulating GaAs substrate. Preliminary results indicate that approximately 10 dB of gain is achieved from 2-20 GHz. These are the first reported results for a distributed amplifier fabricated with HEMT devices.

In summary, we have demonstrated enhanced performance ultrabroadband distributed amplifiers by the use of: (1) novel circuit techniques such as coplanar waveguide transmission lines and cascode-connected FETs; and (2) high-performance heterostructure active devices such as HEMTs.

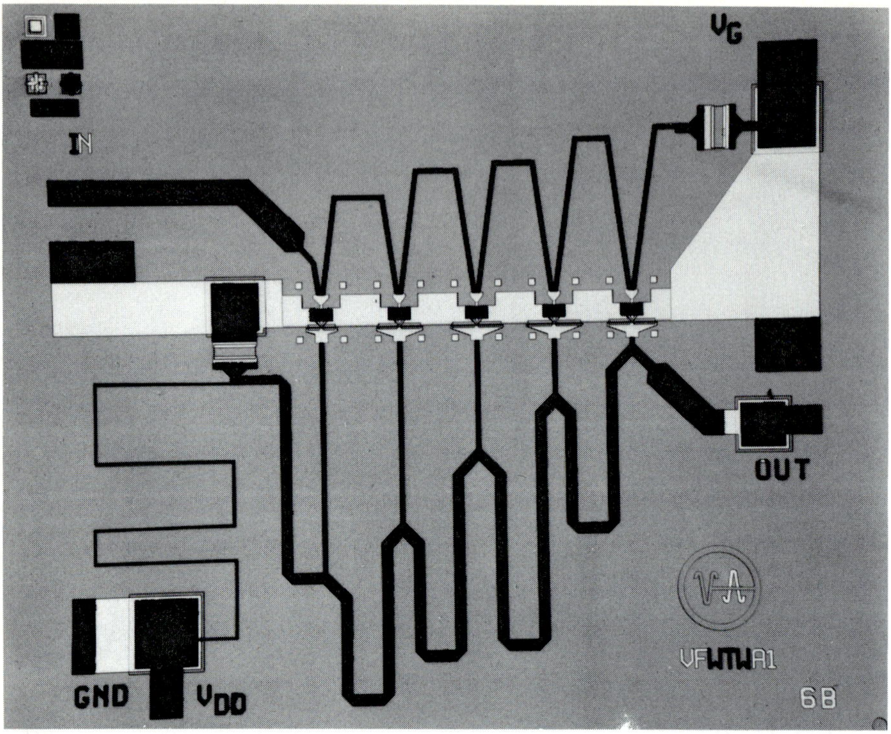

Fig. 6 HEMT distributed amplifier.

References

1. M. N. Yoder: in IEEE GaAs IC Symposium Digest (October 1984).
2. Y. Ayasli, R. L. Mozzi, J. L. Vorhaus, L. D. Reynolds and R. A. Pucel, IEEE Trans. on Microwave Theory and Techniques MTT30(7) (July 1982).
3. M. Rodwell, M. Riaziat, K. Weingarten and D. Bloom, 1986 IEEE MTT-S International Microwave Symposium Digest (June 1986).
4. S. Weinreb and M. Pospieszalski, IEEE MTT-S Symposium Digest (June 1985).

High-Performance Quarter-Micron-Gate MODFETs

J.J. Berenz

TRW Electronic Systems Group, One Space Park,
Redondo Beach, CA 90278, USA

1.0 Introduction

Modulation-doped FETs (MODFETs) provide a significant performance advantage over conventional MESFETs and other extra high frequency devices used in rf applications. They have exhibited the highest gain at 94 GHz, the lowest noise figures, and the highest power-added efficiencies reported for any solid-state devices up to 60 GHz. With the advent of new, improved materials, device structures, and further reduction of transistor gate length, it is anticipated that the maximum frequency of these devices will exceed 300 GHz.

2.0 Performance

The performance advantage of MODFETs is due to the superior transport properties of the two-dimensional electrons which form the conductive channel in these devices. Table 1 compares the material and device characteristics of different FET devices [1,2]. Quarter-micron-gate AlGaAs/GaAs MODFETs have been fabricated with cut-off frequencies as high as 80 GHz. This is a factor of two higher than comparable GaAs FETs. Noise figures as low as 0.9 dB at 18 GHz and 2.7 dB at 62 GHz have been achieved with these devices. Power-added efficiencies ranging from 42 percent at 20 GHz to 14 percent at 60 GHz have also been reported [3]. Figure 1 clearly illustrates the performance advantage which these GaAs MODFETs have over GaAs MESFETs.

Table 1. Comparison of material and device characteristics

Active Layer	Mobility [cm^2/Vs]	Velocity [cm/s]	f_T [GHz]	NF (18 GHz) [dB]
GaAs	3500	$1x10^7$	35	1.1
AlGaAs/GaAs	8000	$2x10^7$	80	0.9
AlGaAs/InGaAs	9500	$3x10^7$	80	0.9

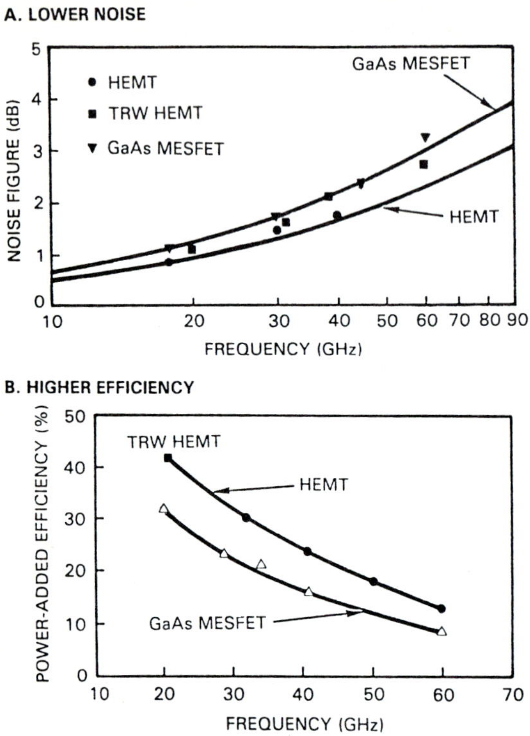

Figure 1. Comparison of GaAs MESFET and MODFET (HEMT) performance

3.0 Improvements

Significant improvement of MODFET performance has been achieved with the pseudomorphic AlGaAs/InGaAs heterostructure shown in Figure 2. A maximum available gain of 7 dB has been measured at

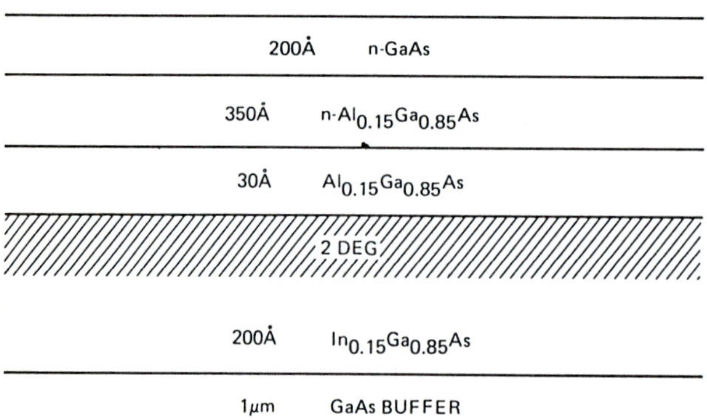

Figure 2. Pseudomorphic AlGaAs/InGaAs heterostructure

Table 2. Comparison of MODFET performance at 60 GHz

	Pseudomorphic MODFET		AlGaAs/GaAs MODFET	
Small-signal gain*	11.7	dB	9.1	dB
Extrapolated f_{max}	230	GHz	170	GHz
Min. Noise Figure	2.3	dB	2.5	dB
Associated Gain	3.9	dB	4.4	dB
Output Power Density	0.43	W/mm	0.41	W/mm
Power-added Efficiency	28	percent	14	percent

* Result measured as a single stage amplifier

94 GHz for quarter-micron gate length. This corresponds to a maximum frequency near 250 GHz. Both lower noise figure and higher power-added efficiency have been obtained with the AlGaAs/InGaAs MODFET at 60 GHz as indicated in Table 2. In comparing the power amplifier performance of MESFET and MODFET devices (Figure 3), the AlGaAs/InGaAs MODFET performance at 60 GHz is outstanding. Further improvement of MODFET performance is anticipated for optimized InGaAs heterostructures with increased indium concentration in the conductive channel [4].

Improvement of GaAs MESFET performance has been realized by reducing the gate length to 0.1-micron and optimizing the vertical channel doping profile [5]. Similar improvement in

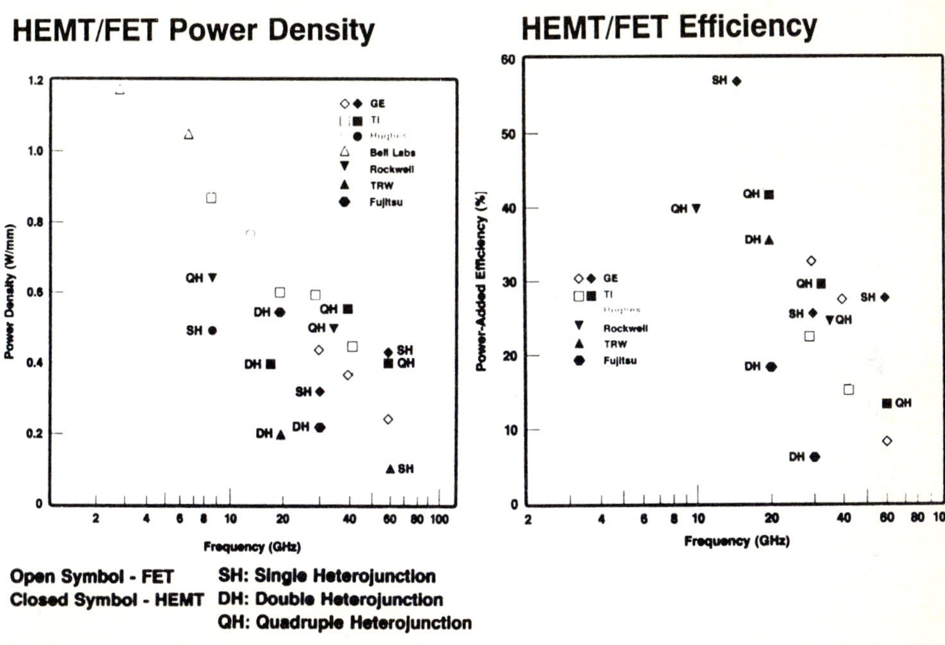

Figure 3. Comparison of power amplifier performance

MODFET performance has not been achieved to date because of "short-channel" effects encountered in operation of conventional sub-quarter-micron-gate GaAs MODFETs. The use of quantum well channels and InGaAs heterostructures has been shown to eliminate these effects due to improved carrier confinement. As a result, 0.1-micron-gate MODFET performance is expected to surpass comparable MESFET performance.

4.0 Conclusions

High performance quarter-micron-gate GaAs MODFETs have been realized with performance superior to GaAs MESFETs. Improved performance has been achieved with InGaAs MODFETs. Further improvement in performance is anticipated with optimization of these heterostructures and reduction of the transistor gate length to 0.1 micron. The maximum frequency of these devices is expected to exceed 300 GHz.

5.0 Acknowledgments

The author wishes to acknowledge the assistance of Dr. H. Morkoc of the University of Illinois and Dr. P.C. Chao of the General Electric Company in preparation of this paper.

6.0 References

1. P.C. Chao, S.C. Palmateer, P.M. Smith, U.K. Mishra, K.H.G. Duh, J.C.M. Hwang: IEEE Electron Device Letters, Vol. EDL-6, 531 (1985)
2. T. Henderson, M.I. Aksun, C.K. Peng, H. Morkoc, P.C. Chao, P.M. Smith, K.H.G. Duh, L.F. Eastman: IEEE Electron Device Letters, Vol. EDL-7, 649 (1986)
3. P. Saunier, J.W. Lee: IEEE Electron Device Letters, Vol. EDL-7, 503 (1986)
4. M.I. Aksun, C.K. Peng, A.A. Ketterson, H. Morkoc, K.R. Gleason: Proceeding of IEDM 86, p. 822
5. U.K. Mishra, R.S. Beaubien, M.J. Delaney, A.S. Brown, L.H. Hackett: Proceedings of IEDM 86, p. 829

Progress and Challenges in HEMT LSI Technology

M. Abe, T. Mimura, K. Kondo, and M. Kobayashi

Fujitsu Laboratories Ltd., Fujitsu Limited,
10-1 Morinosato-Wakamiya, Atsugi 243-01, Japan

HEMT (High Electron Mobility Transistor) is a very promising device for ultra-high-speed LSI/VLSI due to the super-mobility GaAs/AlGaAs heterojunction structure. Recent progress and technical challenges in HEMT technology for high performance VLSI are presented with the focus on material, self-alignment device fabrication, and HEMT LSI implementations.

I. Introduction

Six years have now passed since the announcement in 1980 of the High electron Mobility Transistors (HEMT)/1/, and HEMT technology has certainly opened the door to new possibilities for ultra-high-speed LSI/VLSI applications /2/. Due to the supermobility GaAs/AlGaAs heterojunction structure, the HEMT is especially attractive for low-temperature operations at liquid nitrogen temperature. A switching delay of 5.8 ps with 1.76 mW power dissipation per gate at 77K and 10.2 ps with 1.03 mW power dissipation per gate at 300 K have been achieved for 0.35μm gate devices /3/. These are the shortest switching delays achieved so far in semiconductor devices.

For LSI level complexity, HEMT technology has made it possible to develop a subnanosecond 1 kbit static RAM /4,5/, a 4 kbit static RAM /6/, a 4 x 4 bit parallel multiplier /7/, 5 bit analog-to-digital converters /8/, and 1.5 kgate gate array with an 8 x 8 bit parallel multiplier /9/, and a 16 kbit static RAM integrated with over 10^5 HEMTs /10/.

This paper first presents the performance advantages of HEMT approaches. Next we will describe a HEMT LSI technology with the focus on material, self-alignment device fabrication. Finally we will review recent progress and future technical challenges for LSI/VLSI applications.

II. Performance Advantages of HEMT Approaches

HEMT technology has new possibilities for LSI/VLSI with high speed and low-power dissipation. Because of the superior electron dynamics of HEMT channels and the unique electrical properties of HEMT structure, HEMT has a performance advantage over conventional devices. During switching, the speed of the device is limited by both low-field mobility and saturated drift velocity. Low-field mobility routinely obtained is 8 000 $cm^2/V.s$ at 300 K and 40 000 $cm^2/V.s$ at 77 K. Saturated drift velocity measured with HEMT structures at room temperature has been reported to be 1.5 to 1.9

x 10^7 cm/s. These superior transport properties in HEMT channels would result in a high average current-gain-cutoff frequency f_T value.

Going to LSI circuits with low-power dissipation per gate, logic voltage swing should be minimized. A high transconductance g_m value with a small logic voltage swing is achieved. K value of a 0.5 μm-gate HEMT at 77 K is obtained to be 900 mA/V^2 per mm gate width. This K value is about eight times higher than for conventional GaAs MESFETs.

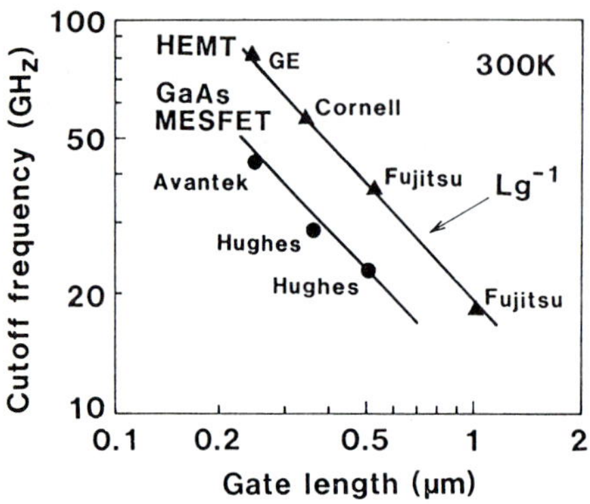

Fig.1 Current-gain-cutoff frequency versus gate length.

In Fig. 1, the current-gain-cutoff frequency f_T versus gate length summarizes for typical performance of experimental HEMTs and GaAs MESFETs reported so far /11-15/. At room temperature, the values of f_T were 38GHz /11/ and 80 GHz /13/ for HEMTs with gate length of 0.5 μm and 0.25 μm, respectively, and about twice those for GaAs MESFETs.

To evaluate the high-speed capability of HEMTs in complex logic circuits, DCFL NOR gate circuits with a 0.5 μm gate E-HEMTs were fabricated. The internal logic delays of 22 ps/gate with power dissipation of 2.8 mW/gate at 77 K and 36 ps/gate with power dissipation of 2.9 mW/gate at 300 K, were demonstrated with an average fan-out of about 2 /16/. The switching speed of HEMT is roughly three times as fast as that of GaAs MESFET.

III. Self-Alignment Device Fabrication Technology

Figure 2 is a cross-sectional view of a self-aligned gate structure of enhancement-mode (E) and depletion-mode (D) HEMTs.

The basic epilayer structure consists of a 600 nm undoped GaAs layer, a 30 nm, $Al_{0.3}Ga_{0.7}As$ layer doped to 2 x 10^{18} cm^{-3} with Si, and a 70 nm GaAs top layer successively grown on a semi-insulating substrate by MBE. A thin AlGaAs layer to act as a stopper against selective dry etching is embedded in the top GaAs layer to fabricate enhancement and depletion mode HEMTs. The key technique to achieve stable fabrication of self-aligned gate HEMTs is the selective dry etching of the GaAs/AlGaAs layer.

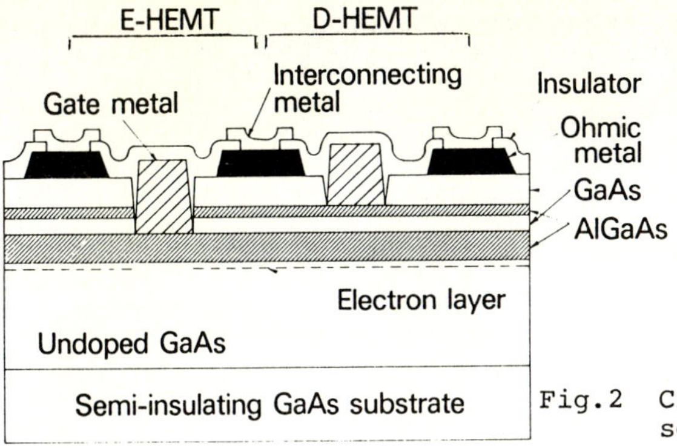

Fig.2 Cross-sectional view of a self-aligned gate E/D-HEMTs.

Etching characteristics in CCl_2F_2 + He discharges achieved a high selectivity ratio of more than 260, where the etching rate of $Al_{0.3}Ga_{0.7}As$ is as low as 2 nm/min and that of GaAs is about 520 nm/min.

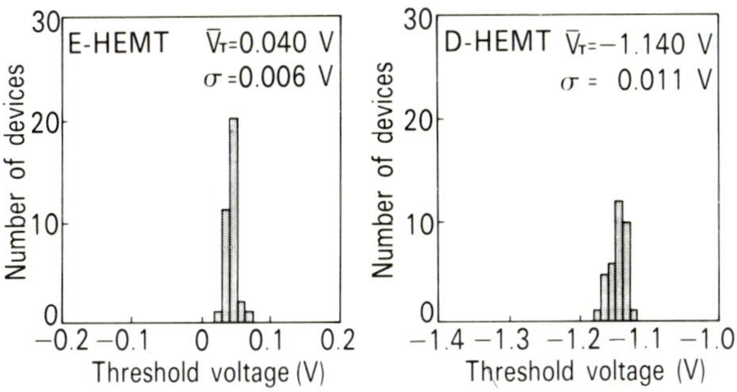

Fig.3 Histograms of threshold voltages for E/D-HEMTs.

The histograms of threshold uniformities for E- and D-HEMTs are shown in Fig.3. The standard deviations in threshold voltages are 6 mV for E-HEMTs and 11 mV for D-HEMTs over an LSI chip size of 10 mm x 10 mm area. The ratio of standard deviation 6 mV of threshold voltage to the logic voltage swing (0.8 V for DCFL) is 0.7 %, indicating excellent controllability of MBE growth and the LSI fabrication process.

The material problems in HEMT LSI fabrication have been greatly improved to achieve high quality MBE wafer with large size and high throughput./17/ Selectively doped GaAs/n-AlGaAs heterostructures were grown on semi-insulating GaAs substrates mounted on a substrate holder with a diameter 190 mm. Figure 4 shows the lateral uniformity of the thickness in n- $Al_xGa_{1-x}As$ (x=0.29). The variation of the thickness over a diameter of 180 mm was less than ± 1 %. It was also confirmed that the variations of the electron mobility and the sheet electron concentration at 77 K in the selectively doped GaAs/n-AlGaAs heterostructure was less than ± 2 % over a diameter of 180 mm.

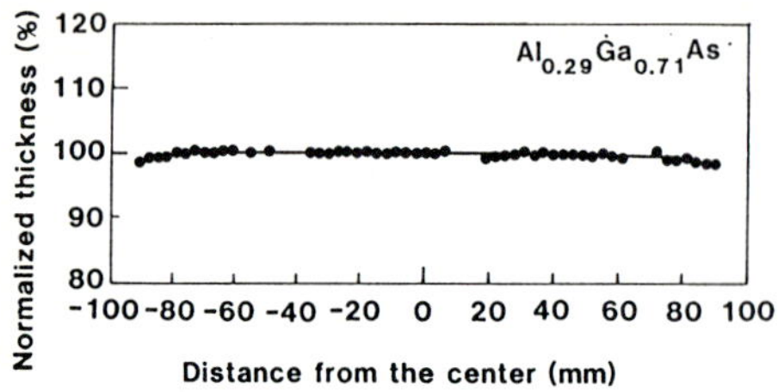

Fig.4 Thickness uniformity of n-AlGaAs layer.

The surface defect problem of MBE is a serious one for fabrication of LSI level complexity /18, 19/. We have already achieved a defect density of less than 40 cm^{-2} with a size of over 20 μm^2 by optimizing the growth conditions to make it possible to develop an LSI with complexity over a 3 kgate logic and 16 kbit static RAM circuits.

IV. HEMT LSI Implementations and Future Prospects

For logic LSI circuits, a HEMT 1.5 kgate gate array with E/D type DCFL circuits is designed and fabricated /9/.

The gate array contains 72 I/O cells and 1 520 basic cells. The basic cell measures 132 µm x 54 µm, and includes one D-HEMT and 3 E-HEMTs. The circuits used 1.2 µm gate devices. The basic cell array is arranged into 20 columns with 76 cells each, and these are 15 interconnection tracks between columns. The chip size is 5.5 mm x 5.6 mm. A microphotograph of the chip is shown in Fig.5. The gate array contains 6 656 devices. The design rule for these lines is 2 µm line width and 2 µm spacing minimum for both layers.

Fig.5 Microphotograph of HEMT 2.5 kgate gate array.

For a test vehicle an 8 x 8 bit parallel multiplier with a carry look ahead circuit and several test circuits including ring oscillators, was designed. The multiplier consists of 888 basic cells. Multiplication time of 4.9 ns at 300 K, and 3.1 ns at 77 K, was achieved, where supply voltage and total chip power consumption were 2.2 V, 5.8 W and 0.95 V, 3.2 W, respectively, including a 5 stage I/O buffer delay.

Performance of HEMT VLSI for future high-speed computers is projected and discussed, based on the results with the HEMT performance described above /2,16,20/. Chip delay time is the sum of intrinsic gate delay, logic layout delay on fan-out capability, and delay in the wiring on the chip. Here we assume that fan-out is 3, logic swing is 0.8V, wiring capacitance is 100 fF/mm, average line length is 1 mm in the chip, and heat flux for liquid cooling is 20 W/cm^2. Figure 6 shows chip delay as functions of complexity, under 0.5 µm design rule HEMT technology. At 10^4 gates, the chip delays are 70 ps at 300 K and 40 ps at 77 K. This sub-hundred-picosecond performance is sufficient for future high-speed computer requirements.

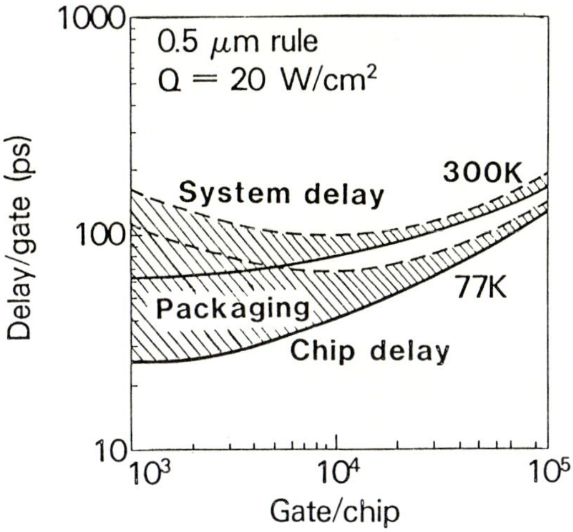

Fig.6 Delay time calculated as a function of complexity.

For memory LSI circuits, a HEMT 16K x 1 bit fully decoded static RAM has been successfully developed with the E/D-type DCFL circuit configuration. As the result of the RAM layout design, the chip size is 4.3 mm x 5.5 mm, and the RAM cell size is 23 µm x 30 µm (690 μm^2) with the switching devices having a gate length of 1.2 µm. The RAM has a total device count of 107 519. Figure 7 is a microphotograph of the 16K x 1 bit static RAM. The minimum address access time obtained was 3.4 ns, with a chip dissipation power of 1.34 W at 77 K. Figure 8 shows the address access time and power dissipation of the static RAM, compared to SiMOS, bipolar, and GaAs MESFET static RAMs. The plots show the performances of the HEMT 1 kbit, 4 kbit and 16 kbit static RAMs. The shadow area shows the projected performance assuming device technologies between 1 and 0.5 µm. By using 0.5 µm gate device technology, subnanosecond address access time can be projected for the 64 kbit static RAM.

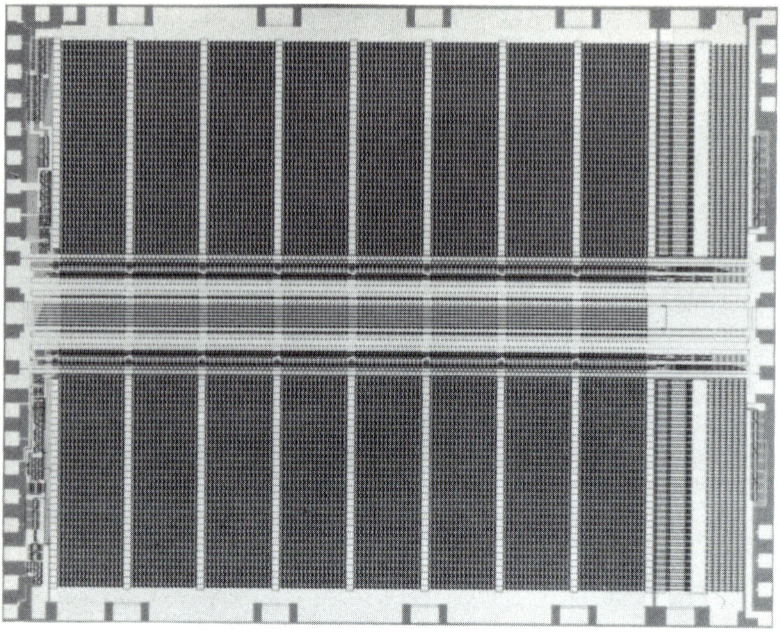

Fig.7 Microphotograph of HEMT 16 kbit static RAM.

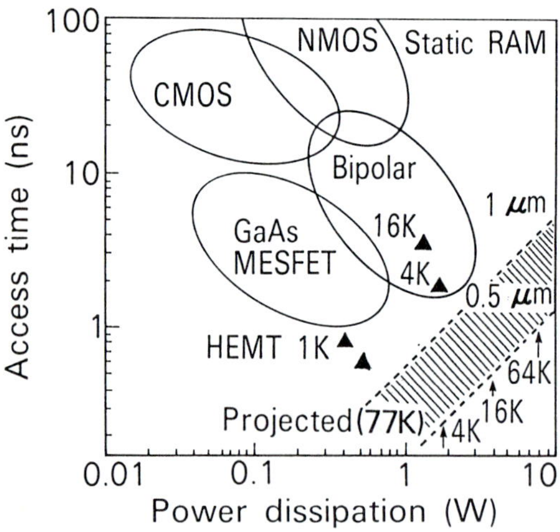

Fig.8 Address access time and power dissipation of the static RAM.

V. Summary

The current status and recent advances in HEMT technology for high-performance VLSI were reviewed with a focus on material, self-alignment device fabrication, and HEMT LSI implementations.

HEMTs are very promising devices for VLSI, especially operating at liquid nitrogen temperature , because of their ultrahigh speed and low power dissipation. The performance described here will achieve speeds required for future large-scale computers.

Acknowledgment

The authors wish to thank Dr. T. Misugi,and Dr. M. Fukuta for their encouragement and support. The present research effort is the National Research and Development Program on the "Scientific Computing System," conducted under a program set by the Agency of Industrial Science and Technology, Ministry of International Trade and Industry, Japan.

References

1. T. Mimura, S.Hiyamizu, T. Fujii, and K. Nanbu, Japan. J. Appl. Phys., vol. 19, pp. L225-L227, 1980.
2. M.Abe, T. Mimura, N. Yokoyama, and H. Ishikawa, IEEE Trans. Electron Devices, vol. ED-29, pp. 1088-1093, 1982.
3. N. J. Shah, S.-S. Pei, C. W. Tu, and R. C. Tiberio, IEEE Trans. Electron Devices, vol. ED-33, pp. 543-547, 1986.
4. K. Nishiuchi, N. Kobayashi, S. Kuroda, S. Notomi, T. Mimura, M. Abe, and M.Kobayashi, ISSCC Dig. Tech. Papers, vol. XXVII, pp.48-49, 314, 1984.
5. N. H. Sheng,H. T. Wang, S. J. Lee, C. P. Lee, G. J. Sullivan, and D. L. Miller, IEEE GaAs IC Symp. Tech. Dig. pp.97-100, 1986.
6. S. Kuroda, T. Mimura, M.Suzuki, N. Kobayashi, K. Nishiuchi, A. Shibatomi, and M. Abe, IEEE GaAs IC Symp. Tech. Dig. pp. 125-128, 1984.
7. A. R. Shlier S.-S. Pei, N. J. Shah, C. W. Tu, and G. E. Mahoney, IEEE GaAs IC Symp. Tech. Dig. pp.91-93, 1985.
8 C.P. Lee, N. H. Sheng, H. F. Lewis, H. T. Wang, D. L. Miller and J. Donovan, IEDM Tech. Dig., pp.324-327, 1985.
9. Y. Watanabe, K. Kajii, K. Nishiuchi, M. Suzuki, I. Hanyu, M. Kosugi, K. Odani, A. Shibatomi, T. Mimura, M. Abe, and M. Kobayashi, ISSCC Dig. Tech. Papers, vol. XXIX, pp.80-81, 314, 1986.
10. M. Abe, T. Mimura, S. Notomi, K. Odani, K. Kondo, and M. Kobayashi, 10th Int. Vacuum Congress (IVC-10), EM-Mo5, 1986.
11. K. Joshin, T. Mimura, M. Niori, M. Yamashita, K. Kosemura, and J. Saito, IEEE MTT-S Dig. pp.563-565, 1983.
12. L. H. Caminitz, P. J. Tasker, H. Lee, D. V. D. Merwe, and L. F. Eastman, IEDM Tech. Dig. pp.360-363, 1984.
13. P. C. Chao, S. C. Palmateer, P. M. Smith, U. K. Mishra, K. H. G. Duh, and J. C. M. Hwang, IEEE Electron Device Letters, vol. EDL-6, pp.531-533, 1985.
14. M. Feng, H. Kanber, V. K. Eu, E. Watkins, and L. R. Hackett, Appl. Phys. Letters, vol. 44, pp.231-233, 1984.
15. W. Chye and C. Huang, IEEE Trans. Electron Device Letters, EDL-3, pp.401-403, 1982.
16. M. Abe, T. Mimura, K. Nishiuchi, A. Shibatomi, and M. Kobayashi, IEEE GaAs IC Symp. Tech. Dig. pp.158-161, 1983.
17. J. Saito, T. Igarashi, T. Nakamura, K. Kondo, and A. Shibatomi, Proc. the 4th Int. Conf. on Molecular Beam Epitaxy, D-12, 1986.
18. T. Mimura, M. Abe, and M. Kobayashi, Fujitsu Sci. Tech. J., vol. 21, pp. 370-379, 1985.

19. K. Nanbu, J. Saito, T. Ishikawa, K. Kondo, and A. Shibatomi, J. Electrochemical Society, vol. 133, pp. 601-604, 1986.
20. M. Abe, T. Mimura, K. Nishiuchi, and N. Yokoyama, VLSI Electronics; Microstructure Science, vol. 11: GaAs Microelectronics, N. G. Einspruch and W. R. Wisseman, Eds. New York: Academic, pp. 333-365, 1985.

Picosecond Switching in Josephson Tunnel Junctions

R. Sobolewski[1*], *D.R. Dykaar*[1,2], *T.Y. Hsiang*[1], *and G.A. Mourou*[2]

[1]Department of Electrical Engineering, The University of Rochester, Rochester, NY 14627, USA

[2]Laboratory for Laser Energetics, The University of Rochester, 250 East River Road, Rochester, NY 14623, USA

We report experimental and numerical studies of the dynamics of the Josephson tunnel junction driven by picosecond current pulses. Both single pulse switching, and the response to the train of pulses are discussed. The observed behavior is explained in terms of the "critical pulse charge" concept.

In this work we have followed our earlier attempts, DYKAAR *et al.* [1], to apply the electro-optic sampler to superconducting electronics, and developed a fully cryogenic electro-optic sampler by integrating and immersing in superfluid helium both the photoconductive switch and the birefringent lithium tantalate sensor. The ultimate temporal response of the system [2] as determined by measuring the transient onset of photoconductivity in a GaAs switch was less than 400 fs. The cryogenic part was arranged in two configurations presented in Fig. 1, and used to probe the switching of a 30 x 30 μm^2 Josephson tunnel junction. The input pulse duration was about 8 ps.

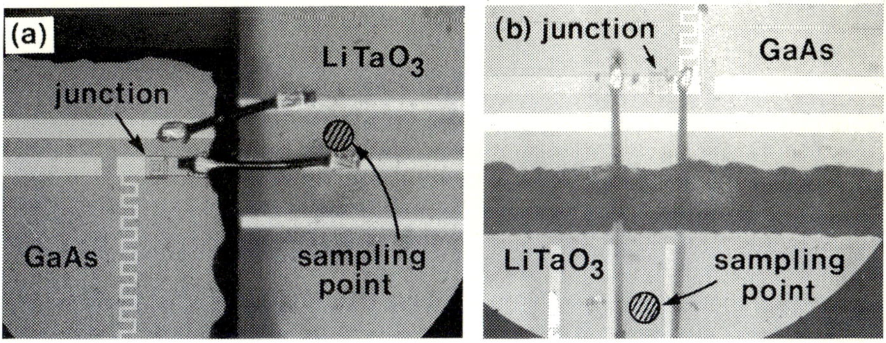

Figure 1. Micrographs of the samples used in this work. (a) – transmission line configuration. (b) – voltmeter configuration.

The coplanar transmission line configuration (Fig. 1a) was arranged in the same way as a typical transfer function measurement in two-port microwave network. For a fixed input signal, we measured the signal waveform with a resistor substituted for the junction and compared this with the waveform obtained with the junction included in the transmission line. We attribute the

*Permanent address: Institute of Physics, Polish Academy of Sciences, PL-02668 Warszawa, Poland.

difference in the waveforms, observed in Fig. 2, to the different propagating conditions, caused by the junction switching from the zero-voltage to the finite-voltage state.

The voltmeter or "side sampling" configuration (Fig. 1b) was designed to measure directly the response of the junction. In this case the input pulse propagated in the main transmission line on GaAs, while the junction signal was transmitted into the $LiTaO_3$ coplanar waveguide. This scheme enabled us to measure the switching voltage across the junction. However, the major difficulty of these measurements lay in discriminating between the junction response of less than three millivolts, and a much larger signal radiated by the photoconductive switch. In order to achieve the desired sensitivity a new, improved data acquisition system is currently being developed in our laboratory.

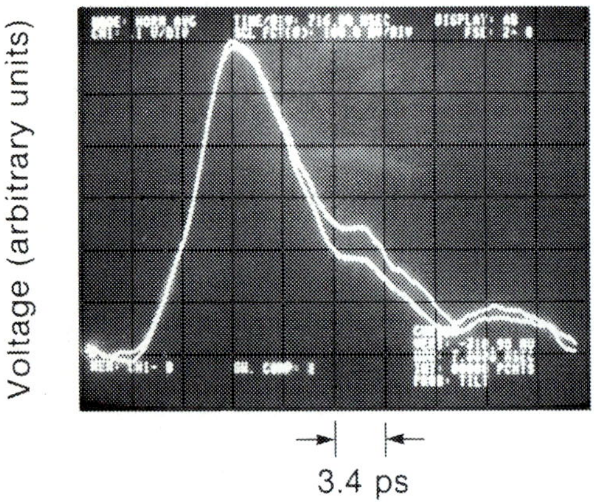

Figure 2. Junction voltage waveforms. Upper curve corresponds to the "no junction" condition in the transmission line, and the lower curve is the unbiased junction response. The temperature was 1.8 K.

The results presented in Fig. 2 clearly indicate that the signal experienced reflection during the early stage of propagation. It therefore appears that the junction started to respond when the input pulse reached its maximum, and the full voltage across the junction was developed at the time corresponding to the input pulse duration. No significant delay was present in junction response. Based on our measurements the response time of the junction was at most 12 ps, but it should be noted here that virtually all the dispersion evident in Fig. 2 was incurred after the pulse had already passed through the junction. The distance from the switch to the junction was only 70 μm, while the sampling point (see Fig. 1a) was several hundred micrometers, most of which was in the much more dispersive $LiTaO_3$ sampling crystal.

As was discused by DYKAAR *et al.* [3] the observed response is characteristic for junctions in the pulse regime, where the input pulse width is of order of ω_0^{-1} (ω_0 is the plasma frequency). In this case the junction critical current I_0 is no longer a single threshold parameter, and should be replaced by a new critical parameter, namely the critical "pulse charge" defined as the time integral of the threshold current pulse.

A typical simulated junction response for a picosecond input pulse with amplitude equal to 1.2 I_0 is presented in Fig. 3. The simulations are based on

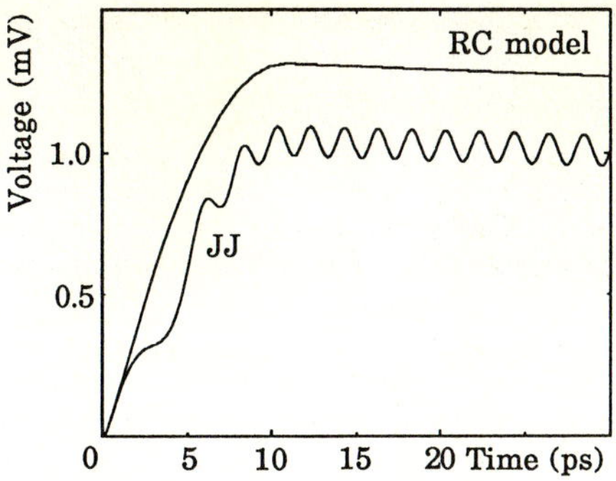

Figure 3. The simulated Josephson junction response (JJ curve) to a stepped input pulse of amplitude 1.2 I_0 and duration $t_p \approx 8$ ps. The full solution of the RCSJ model is compared to the prediction of the charging capacitance model (RC model).

the resistively and capacitively shunted junction (RCSJ) model [4]. One notices that despite the fact that the input pulse amplitude is larger than I_0, the junction output voltage peak reaches only about half of the value corresponding to the gap voltage. The pulse exhibits a very short rise, followed by a substantial, exponential decay. The turn-on delay time is negligible. The physical mechanism can be easily understood in terms of a simple model that approximates the junction response as the charging of the junction capacitance [3]. The major factor in the junction response time is how fast the input pulse charge, Q, is delivered to the junction capacitance, C_J. The switching voltage is just $Q(t_p)/C_J$, and the corresponding rise time is roughly the input pulse width itself. On the other hand the junction decay is totally dominated by the properties of the dielectric layer and the associated RC time constant. These features agree very well with the experimental observations shown in Fig. 2.

Finally we note that the above situation is characteristic of very short pulses, which change significantly during the junction switching time. For longer, rectangular-type input pulses the delivered charge can be approximated by constant current times time, and our approach leads immediately to the static limit discussed by McDONALD *et al.* [4].

Another interesting phenomenon was observed in these experiments: the junction current-voltage (I-V) characteristics showed radical changes for different values of pulse train amplitude. We demonstrated that, under proper conditions, the periodic kicks gave rise to chaotic oscillations in the junction. The experimental I-V curves for three different excitation pulse amplitudes (switch voltages, V_s) are shown in Fig. 4. As can be seen in Fig. 4a, for small input pulses only I_0 on the positive branch of the I-V curve was reduced (here "positive" is defined as the direction of the current pulse). The negative branch was practically unaffected. Upon increasing the input pulse amplitude to V_s equal to 1.2 V, complete compensation of the positive branch of I_0 was reached. This threshold value provided a calibration to convert the experimental switch voltage into a corresponding simulated current pulse. For current pulse amplitudes above this threshold (Figs. 4b and 4c) "noisy" or "oscillatory," and as was proved by SOBOLEWSKI *et al.* [5] - chaotic, behavior set in on the negative branch, while the positive branch displayed the typical quasiparticle tunneling curve. When the input pulse was too small to compensate fully the negative branch of I_0 (Fig. 4b), we observed a novel "reentrance" process in which, with the increase of the dc bias, the junction switched from the chaotic regime back to the stable, zero-voltage state.

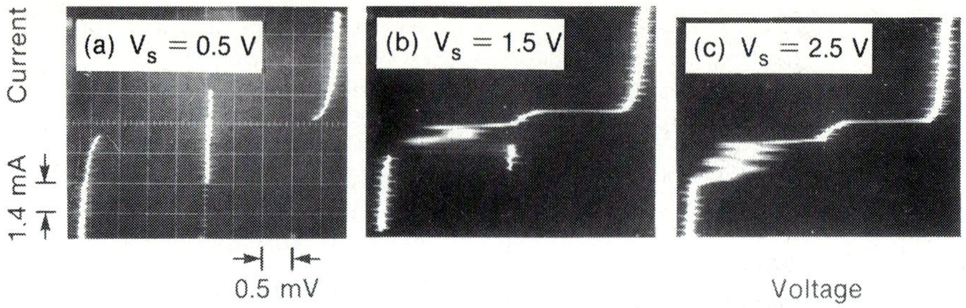

Figure 4. Experimental I–V characteristics. Due to the corrected, two-wire measurements, the gap voltages displayed appear to be slightly reduced. The small noise present on the gap-voltage sections of the quasiparticle I-V curves is due to the 60-amps laser power supply noise picked up by the oscilloscope amplifier outside of the cryostat. All I-V measurements were preceded by a 1 MHz low-pass filter. The temperature was 1.8 K.

The numerical studies served to confirm most of the observed features, and helped to establish that our system exhibited chaotic motion. The proof was obtained numerically, by studying the "noisy" portion of the I-V curve on the Poincaré section, and the route to chaos was established as the intermittent chaotic behavior through an inverse tangent bifurcation.

Acknowledgment

This work was supported by the National Science Foundation (DMR-85-06689). In addition, the laser facility was partially supported by the Laser Fusion Feasibility Project at the Laboratory for Laser Energetics. One of us (R. S.) would like to thank C. Hamilton and other members of the Electromagnetic Technology Division for their hospitality and help during his visit at the National Bureau of Standards, (Boulder, Colorado), where the Josephson junctions used in these experiments were fabricated.

References

1. D. R. Dykaar, T. Y. Hsiang and G. A. Mourou, in **Picosecond Electronics and Optoelectronics,** *ed. by* G. A. Mourou, D. M. Bloom and C. -H. Lee, Springer Ser. in Electrophysics Vol. 21 (Springer, Berlin, Heidelberg 1985) p. 249; IEEE Trans. Magn. **MAG-21**, 230 (1985).
2. D. R. Dykaar, R. Sobolewski, J. F. Whitaker, T. Y. Hsiang, G. A. Mourou, M. A. Holis, B. J. Clifton, K. B. Nichols, C. O. Bozler and R. A. Murphy, in **Ultrafast Phenomena V**, *ed. by* G.R. Fleming and A.E. Siegman, Springer Ser. in Chemical Physics Vol. 46 (Springer, Berlin, Heidelberg 1986) p. 103.
3. D. R. Dykaar, R. Sobolewski, T. Y. Hsiang and G. A. Mourou, IEEE Trans. Magn. – in print.
4. D. G. McDonald, R. L. Peterson, C. A. Hamilton, R. E. Harris and R. L. Kautz, IEEE Trans. Electron Device, **ED–27**, 1945 (1980), and references therein.
5. R. Sobolewski, D. R. Dykaar, T. Y. Hsiang, C. Vanneste, and C. C. Chi, Phys. Rev. Lett. – submitted for publication.

Modulation Efficiency Limited High Frequency Performance of the MODFET

M.C. Foisy, J.C. Huang, P.J. Tasker, and L.F.Eastman*

School of Electrical Engineering and National Research and Resource Facility of Submicron Structures, Cornell University, Phillips Hall, Ithaca, NY14853, USA

*Department of Applied Physics, Cornell University

1. Introduction

Epilayer design for the modulation doped field-effect transistor (MODFET) is frequently performed using the depletion approximation [1,2]. The work of authors such as GRAY and LUNDSTROM [3], STERN and DAS SARMA [4], and PONSE et al. [5] have provided tools by which electron distributions and capacitances can be calculated without this assumption. Their results demonstrate that for two-dimensional electron gas (2DEG) sheet densities significantly below the saturation value, free and bound electrons are present in the electron supplying layer. Because these electrons must be modulated with the 2DEG electrons, the gate capacitance is increased while the transconductance is decreased [6]. The unity current gain frequency, f_T, is thus decreased.

In this work we take the above analysis a step further by introducing a figure of merit, the modulation efficiency (ME), which gives an indication of the charge control-limited f_T as a function of 2DEG sheet density. In this paper, the effect of each epilayer design parameter on the ME is studied via a numerical charge control model which incorporates Fermi-Dirac statistics and Al composition-dependent donor energies [7]. It is found that at room temperature, unlike 77K where the depletion approximation is more valid, ME is sensitive to epilayer design even for low sheet densities.

2. Modulation Efficiency

We define the modulation efficiency as the ratio of 2DEG capacitance to total capacitance for the 1-D system extending from the gate to the buffer/substrate interface:

$$ME = C_{2DEG}/C_{TOT}. \tag{1}$$

A classical model is used to calculate electron densities and capacitances. It is found to give excellent agreement with quantum-mechanical calculations by VINTER [8] as predicted by YOSHIDA [9]. Agreement with experimentally measured capacitance was also obtained [10]. Although two-dimensional effects such as the variation in channel potential and real space transfer are ignored by such an analysis, the ME calculated by this procedure represents a best case result. If the electrons entering the charge control region under the gate (i.e. those electrons at the source end of the gate) cannot be modulated efficiently or are present in the supply layer in excess number, then the situation can only be worsened as the carriers are heated. The correlation between f_T and ME is best illustrated by using the short channel approximation, neglecting carrier heating, and assuming electron velocity is independent of sheet density:

$$f_T = \frac{g_m}{2\pi C_G} = \frac{v_{eff}}{2\pi} \frac{dQ_{2DEG}/dV_G}{dQ_{TOT}/dV_G} = \frac{v_{eff}}{2\pi} ME. \qquad (2)$$

Obviously, these simplistic assumptions will not hold under all device operating conditions. However, since ME will always retain its degrading effect on f_T, optimization of ME is a minimum design objective.

3. Results

To illustrate the use of modulation efficiency in evaluating epilayer design, an depletion-mode MODFET is simulated. Figure 1 shows the effect of varying the aluminum mole fraction. The doped AlGaAs thickness is varied so as to maintain a constant threshold voltage; all other parameters are held constant. Note that the curve representing 77K operation is almost box-like. In the limit of zero temperature (an impossibility due to carrier heating) the depletion approximation is accurate and the ME vs. 2DEG sheet density curve would have an exact box form. In contrast, the design curves for 300K operation show significant modulation efficiency degradation for 2DEG sheet densities well below the maximum attainable density. The decrease in f_T for higher sheet densities and hence higher drain currents predicted by this model is in qualitative agreement with experimental data [12]. It should be noted that two-dimensional effects not included in this model are responsible for decreases in f_T at lower currents. [13]

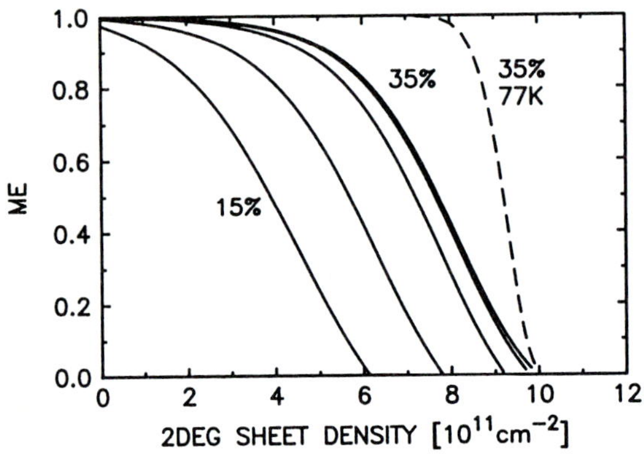

Fig. 1. Effect of electron supplying layer aluminum concentration on modulation efficiency. 300K ME vs. 2DEG sheet density is plotted for 15, 20, 25, 30, and 35% Al (solid lines). For comparison, 77K ME is also plotted for 35% Al (dashed line)

Comparison of material parameters with Fig. 1 shows that ME degradation increases as the difference between donor energy and the GaAs conduction band energy at the interface $(\Delta E_C)-E_D$, as measured in thermal volts, is decreased. This quantity is a measure of the ability to fill the 2-D channel without populating the donor levels in the electron supplying layer. Since in the AlGaAs/GaAs system it saturates near 30% Al composition [7,11], further improvement in ME cannot be obtained by increasing the aluminum content. Fig. 1 also illustrates the ME consequences of avoiding low-temperature operation or reducing the Al concentration in an attempt to avoid light sensitivity.

The effect of spacer layer thickness on ME is investigated using the 30% Al MODFET simulated in Fig. 1 as the starting point. (Note, all the epilayers reported in Fig. 1 have no spacer.) Spacer thicknesses between 0 and 100Å are modeled. To maintain the same threshold voltage, the doped AlGaAs thickness is kept constant. The results plotted in Fig. 2 suggest that variations in the designed spacer may be responsible for the wide range of f_T values reported in the literature for submicron gate MODFETs. For example, if the device is biased for a 6×10^{11} cm^{-2} 2DEG sheet density, changing from a 25Å spacer to a 50Å spacer will reduce ME by more than a factor of two.

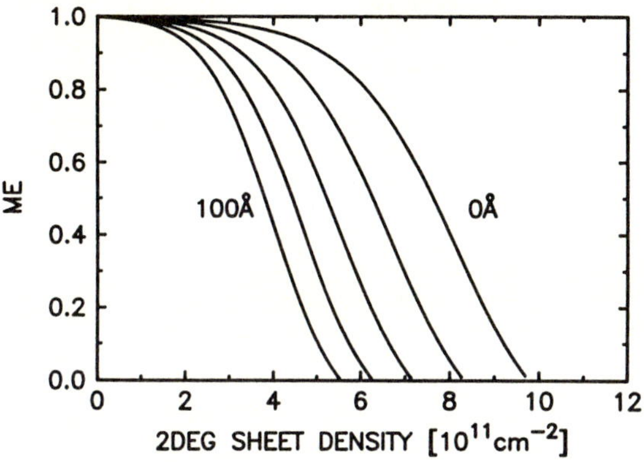

Fig.2. Effect of spacer layer thickness on modulation efficiency. 300K ME vs. 2DEG sheet density is plotted for spacer thicknesses of 0, 25, 50, 75, and 100 Å

4. Conclusion

We have shown that inefficient charge control is a primary degrading mechanism in high-frequency MODFET operation. Numerical simulation of the ME is an important first step in the optimization of an epilayer design.

Acknowledgement

We are grateful to L.D. Nguyen and D.C. Radulescu for valuable technical discussions. This work was supported by McDonnell Douglas and the Joint Services Electronics Program.

1. K. Lee, M.S. Shur, T.J. Drummond, H. Morkoç: IEEE Trans. ED-30, 207-212 (1983)
2. K. Lee, M.S. Shur, T.J. Drummond, H. Morkoç: IEEE Trans. ED-31, 29-35 (1984)
3. J.L. Gray, M.S. Lundstrom: IEEE Trans. ED-32, 2102-2109 (1985)
4. F. Stern, S. Das Sarma: Phys. Rev. B 30, 840-848 (1984)
5. F. Ponse, W.T. Masselink, and H. Morkoç: IEEE Trans. ED-32, 1017-1023 (1985)
6. M. Moloney, F. Ponse, H. Morkoç: IEEE Trans. ED-32, 1675-1684 (1985)
7. N. Chand, T. Henderson, J. Klem, W.T. Masselink, R. Fischer, Y. Chang, H. Morkoc: Phys. Rev. B 30, 4481-4492 (1984)
8. B. Vinter: Appl. Phys. Lett. 44, 307-309 (1984)
9. T. Yoshida: IEEE Trans. ED-33, 154-156 (1986)
10. J.C. Huang, M.C. Foisy, P.J. Tasker, S.R. Seidman, G.W. Wicks, L.F. Eastman: (to be submitted for publication)
11. S. Adachi: J. Appl. Phys. 58, R1-R29 (1985)
12. L.H. Camnitz: Ph.D. Thesis, Cornell University, Ithaca, NY (1986)
13. G. Wang, W.H. Ku: IEEE Trans. ED-33, 657-663 (1986)

Development of 18 GHz GaAs Static Frequency Dividers and Their Evaluation by Electrooptic Sampling

J.F. Jensen[1], *K.J. Weingarten*[2], *and D.M. Bloom*[2]

[1]Hughes Research Laboratories, 3011 Malibu Canyon Road, Malibu, CA 90265, USA

[2]Stanford University, Edward L. Ginzton Laboratory, Stanford, CA 94305, USA

This paper describes high-speed GaAs static divider circuits operating up to 18.05 GHz at room temperature, and the use of electrooptic testing for the evaluation of their performance. The dividers were implemented in both buffered FET logic (BFL) and capacitively enhanced logic (CEL), and were fabricated with 0.2-μm gatelength MESFETs, molecular beam epitaxy (MBE)-grown active channels, and airbridge interconnects.

1. Introduction

Digital integrated circuits (ICs) operating at clock rates above 10 GHz will have a significant impact on the performance of digital systems in the next decade. We have successfully employed advanced design and fabrication techniques to raise the operational frequency of room-temperature GaAs MESFET divider circuits to 18 GHz, which exceeds speeds achieved by any other technology at any temperature.

Because of the limited bandwidth of conventional electronic instrumentation, measurement of digital performance above 10 GHz is generally a very formidable task. In this work, however, we have verified large signal digital operation and characterized signal timing on internal circuit nodes using electrooptic testing techniques. Internal gate delay measurements from electrooptic sampling agree very well with gate delays estimated from the measurements of the maximum flip-flop toggle frequency.

2. Flip-Flop Design and Fabrication

The divider circuits are implemented with a biphase clock master-slave flip-flop that has a maximum toggle rate of 1/2 τg, where τg is the loaded AND/NOR gate delay [1]. Two logic families are used to implement the divider circuits. The first logic family is buffered-FET logic (BFL) (Fig. 1) [2]. BFL generally produces the fastest logic gates at high fanouts, but it also dissipates the most power per gate. The BFL gate employs only DFETs, so level-shifting is required to make input and output logic levels compatible. This extra circuitry increases both gate delay and power dissipation. However, the level-shifter stage also provides buffering to the following logic gates and, therefore, has excellent fanout and interconnect drive capability.

An alternative to BFL in depletion-mode technology is unbuffered capacitively enhanced logic (CEL) (Fig. 2) [3]. CEL uses diode level-shifting to establish the dc operating points of the circuit; however, a

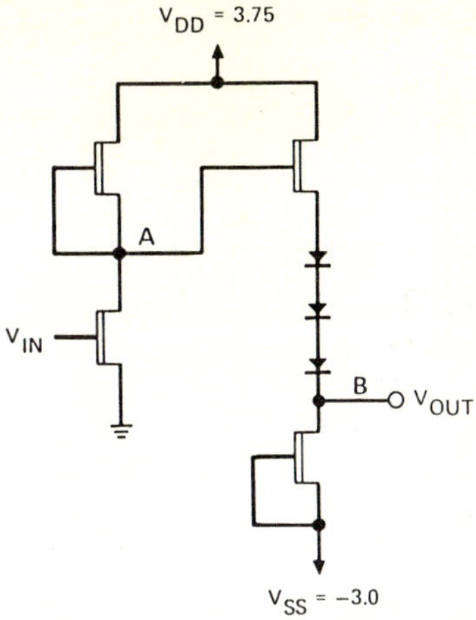

Fig. 1. Buffered FET logic gate

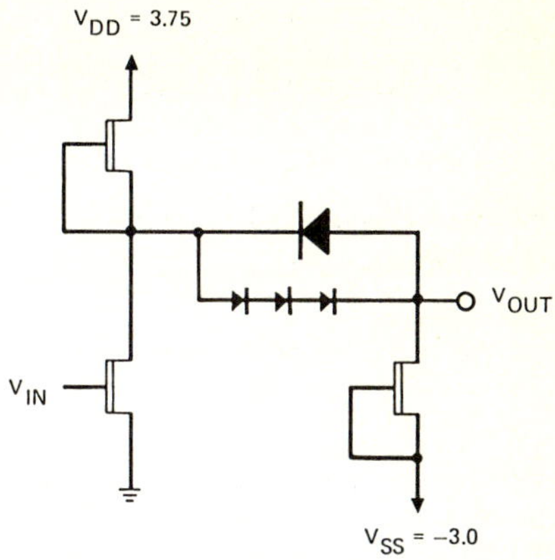

Fig. 2. Capacitively enhanced logic gate

reverse biased diode, which functions as a capacitor, is placed in parallel with the level-shift diodes to capacitively couple the logic gates. During fast logic-state transitions, all the switching currents pass through the capacitor. Since the speed of the logic gate does not depend on the operating current in the level-shift diodes, this current can be minimized to achieve very low dc power dissipation. The fanout capability of the CEL gate is not as good as BFL because the output is not buffered, but at low fanout, CEL is potentially faster than BFL.

To maximize the speed of GaAs ICs, several issues are addressed: (1) the device/circuit design rules are reduced to provide subquarter-micrometer gatelengths, (2) shorter gatelengths require thin active channels (<700 A) grown by MBE, (3) source resistance is minimized by using n^+ contact layers grown on top of the channel, and (4) airbridge interconnects are used to reduce parasitic capacitances. Fig. 3 is an SEM photomicrograph of a portion of a completed IC with 0.2 μm gates and airbridges. More complete fabrication details are given in Ref. [1].

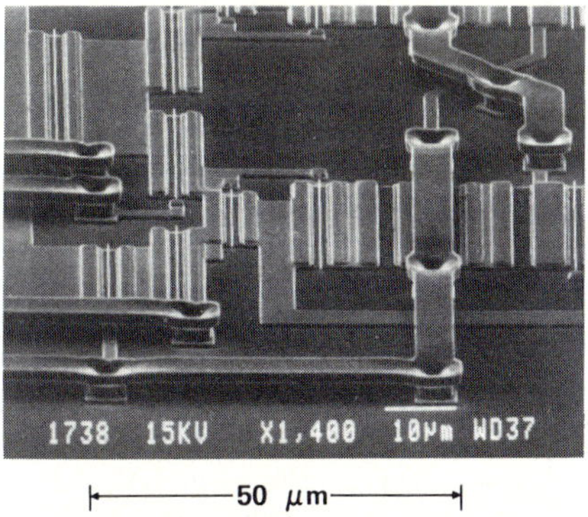

Fig. 3. SEM photomicrograph of a completed divider circuit with 0.2 μm gates and airbridge

3. Test Results

The dividers were tested at Stanford University on-wafer using the electrooptic sampling test station [4]. The clock inputs were generated using a microwave frequency synthesizer set to an exact multiple of the laser pulse repetition rate, plus some small frequency offset. To improve the signal-to-noise ratio of the electrooptic detected signal, clock inputs were also pulse modulated at 10 MHz for synchronous detection. The clock signals were then amplified with a microwave amplifier to obtain a 2 V logic swing into 50 Ω. A 180° phase balun was used to generate the biphase signals, and bias tees were used to dc-offset the clock inputs to their optimum values.

Figure 4 shows the electrooptic sampled waveforms from a CEL divider circuit measured at a clock frequency of 12.1 GHz. The top waveform is the clock signal measured internal to the divider circuit at the input of the master latch. The middle waveform is the Q-bar output of the slave, which is fed back to the D input of the master. The bottom waveform is the Q output of the master latch. The measured waveforms verify the correct operation of the divider. The data at the D input arrive before the rising edge of the clock. After the rising edge of the clock, the master output changes state. The critical delay that determines the toggle rate of the flip-flop is the time between the 50% point of the rising edge of the clock, to the 50% point of the rising or falling edge (whichever is longer) of the output waveform. As measured from the waveform, the delay is 20 ps to the rising edge, and 28 ps to the falling edge. In this case, the delay from the clock to the falling edge sets the toggle rate. The maximum toggle frequency of this divider was measured to be 17.8 GHz at 170 mW power dissipation, which corresponds to a calculated gate delay of 28 ps. The electrooptic sampled waveform therefore confirms correct operation at 17.8 GHz.

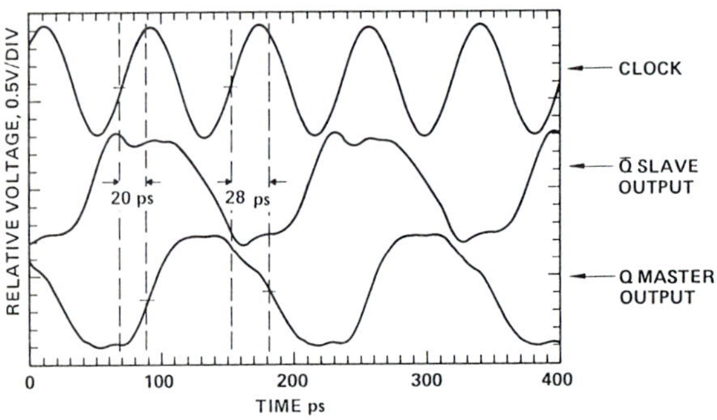

Fig. 4. Electrooptic waveform from a CEL divider circuit measured at 12.1 GHz

Electrooptic measurements are also used to determine the internal delays in the BFL divider circuits. These internal delays give us insight into improving divider circuit performance. Figure 5 is the measured electrooptic waveform of the clock, the high impedance node of the master output BFL gate (node A, Fig. 1), and the level-shifted source follower output of the same gate (node B, Fig. 1). The delay from the clock edge to the high impedance node is 13.6 ps. The delay from the high impedance node to the output is 12.1 ps. Comparing these delays to values reported previously for longer FET gatelengths [4], we find that the delay between the clock input and the high

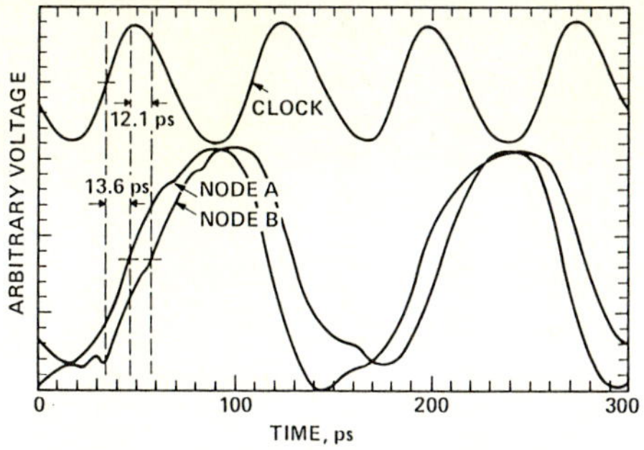

Fig. 5. Electrooptic waveform measured at 13.46 GHz of the clock, high impedance point (node A), and source follower level shift output (node B) of a BFL divider

impedance node scaled with gatelength; however, the delay through the source follower and level-shift diodes remained about the same. The latter is dominated by the RC delay of the resistance of the level-shift diodes and the gate capacitance of the following stage. The speed of this divider circuit can be improved by reducing the diode resistance. The maximum measured toggle frequency of the BFL divider was 18.05 GHz, with 658 mW power dissipation. This frequency corresponds to a calculated gate delay of 27.7 ps, which correlates well with the total measured gate delay of 25.6 ps.

4. Summary

Electrooptic sampling techniques have been used to evaluate the performance of ultrahigh-speed frequency dividers. Electrooptic testing is a powerful technique to both confirm operation of high-speed digital circuits on-wafer, and to provide insight into maximizing their performance.

5. References

1. J. Jensen, L.G. Salmon, D.S. Deakin, and M.J. Delaney, IEDM Tech. Digest, Dec. 1986, pp. 476-479.

2. A. Barna and C.A. Liechti, IEEE J. Solid-State Circuits SC-14, 708-715 (1979).

3. P.J.T. Mellor and A.W. Livingstone, Electron. Lett. 16, 749-750 (1980).

4. K.J. Weingarten et al., "Electrooptic sampling of gallium arsenide integrated circuits," Topical Meeting on Ultrafast Phenomena, 1986.

Characteristics of Shielded Microstrip Lines on GaAs-Si at Millimeter-Wave Frequencies

M.I. Aksun and H. Morkoc

University of Illinois at Urbana-Champaign, Coordinated Science Laboratory, 1101 W. Springfield Avenue, Urbana, IL 61801, USA

Recent advances in the growth of GaAs on Si substrates have necessitated the consideration of this composite structure as a substrate material for microwave and millimeter-wave monolithic integrated circuits. It is, therefore, important to investigate the dielectric properties at frequencies of operation for assessing the viability of this technology. We report the results of a calculation of the dielectric losses of this composite material when used as a transmission line medium and a comparison with other possible substrates, such as GaAs and Si alone, in the frequency range of 10-100 GHz. Results show that the growth of high resistivity GaAs on Si substrates with moderate resistivity reduces the dielectric loss depending upon the thickness of the GaAs overlay material.

1. INTRODUCTION

The advantages of GaAs growth on Si substrates, such as using the advanced processing technology for Si, taking advantage of the speed of GaAs and the possibility of monolithic integration GaAs and Si devices, have stimulated a great deal of interest [1,2]. GaAs based devices on Si substrates, such as metal semiconductor field effect transistors (MESFETs) and heterojunction bipolar transistors (HBTs) have already been demonstrated, with performances nearly identical to those grown on GaAs substrates [3,4].

As mentioned above, one of the advantages of GaAs/Si as a substrate medium is that it opens up the possibility of monolithically integrating GaAs based and Si based devices [5]. Although Si has been studied extensively as the substrate material for microwave monolithic integration because of its high thermal conductivity, well-established technology, mechanical strength and availability in larger diameters [6-8], it lacks millimeter-wave three terminal devices which may now be possible with GaAs/Si. Dielectric losses in GaAs/Si, however, must be small in order for this composite technology to be practical. We have thus undertaken a theoretical study of the shielded microstrip lines on GaAs/Si up to 100 GHz for various parameters in Si and GaAs. The dielectric losses in GaAs, Si, and GaAs/Si have been investigated and compared.

2. THEORY

The characteristics of a shielded microstrip line with a double layer substrate were investigated by using spectral domain analysis [9,10]. Since the hybrid field components consist of the superposition of TE and TM fields, usually magnetic and electric scalar potentials ψ^h, ψ^e are defined to find the total field in the structure. Hence the following total field components can be obtained:

$$E_{zi} = j \frac{k_i^2 - \beta^2}{\beta} \psi_i^e (x,y) e^{-j\beta z}, \tag{1}$$

$$H_{zi} = j \frac{k_i^2 - \beta^2}{\beta} \psi_i^h (x,y) e^{-j\beta z}, \tag{2}$$

$$E_{ti} = \nabla_t \psi_i^e (x,y) e^{-j\beta z} - \left(\frac{\omega \mu}{\beta} \right) \hat{a}_z \times \nabla_t \psi_i^h (x,y) e^{-j\beta z}, \tag{3}$$

$$H_{ti} = \left(\frac{\omega \epsilon_i}{\beta} \right) \hat{a}_z \times \nabla_t \psi_i^e (x,y) e^{-j\beta z} + \nabla_t \psi_i^h (x,y) e^{-j\beta z} \ , \tag{4}$$

where the subscript i designates the regions inside the shield and the superscripts (e) and (h) stand for the TE and TM fields, respectively. After transforming the scalar potential functions into the Fourier domain, the appropriate boundary conditions are applied. Using Galarkin's method in the Fourier domain and after some mathematical manipulations, one can get the following algebraic equations:

$$\sum_{m=1}^{M} a_m K_{im}^{11}(\beta) + \sum_{m=1}^{N} b_m K_{im}^{12}(\beta) = 0 \qquad i=1, \cdots ,N, \tag{5a}$$

$$\sum_{m=1}^{M} a_m K_{im}^{21}(\beta) + \sum_{m=1}^{N} b_m K_{im}^{22}(\beta) = 0 \qquad i=1, \cdots ,M \ . \tag{5b}$$

Solving these two algebraic equations simultaneously yields the propagation constant for a specified frequency. The details of this analysis can be found in Refs. [9-11].

After having found the propagation constant and the field components in each layer, (6) was used to calculate the dielectric loss,

$$\alpha_d = \frac{\omega \epsilon \tan\delta \int_{sdiel} |E|^2 ds}{2\mathrm{Re} \int_s E \times H^* \cdot \hat{a}_z ds} \ . \tag{6}$$

This dielectric loss calculation assumes that the loss tangents of the dielectric materials being used are sufficiently small.

3. RESULTS AND DISCUSSION

Data were obtained for three different substrates, namely GaAs, Si, and GaAs on Si. Dielectric loss and effective dielectric constant vs. frequency of Si and GaAs substrates are shown in Figs. 1 and 2. The same parameters of GaAs/Si substrate were plotted for two different resistivities of the GaAs overlay material, each at three different GaAs overlayer thickness as shown in Figs. 3 and 4.

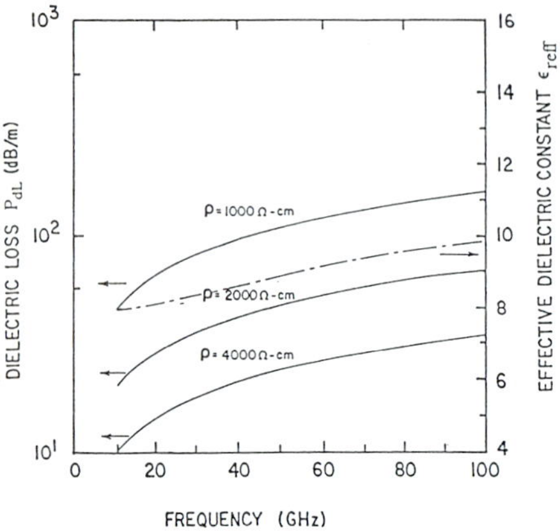

Fig. 1. Dielectric loss and effective dielectric constant vs. frequency for Si substrate, $\epsilon_r = 11.8$

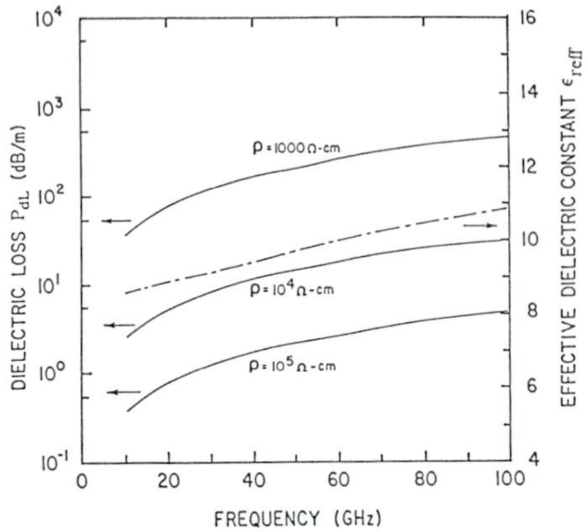

Fig. 2. Dielectric loss and effective dielectric constant vs. frequency for GaAs substrate, $\epsilon_r = 12.91$

Because the resistivities of available GaAs substrates (typical 10^5–10^7 Ω–cm) are much higher than that of available Si substrates (up to 10,000 Ω–cm), the dielectric loss in GaAs is quite small, as can be seen from Figs. 1 and 2. However, GaAs suffers from poor mechanical strength, immature processing technology, and the unavailability of large diameter substrates. Therefore, GaAs and Si have been studied for different kinds of applications which can utilize the advantages and tolerate the disadvantages of the substrate, such as GaAs substrates for low noise communication systems [12], and Si substrates for systems which require high power and higher millimeter-wave frequency operation [13]. Si-based millimeter-wave high power IMPATT diodes, GaAs-based high frequency MESFETs, modulation doped field effect transistors, and several millimeter-wave passive circuit elements [14] have already been designed, and may be monolithically integrated on GaAs/Si substrates. The dielectric loss of this composite substrate is lower than that of Si substrates alone provided that the resistivity of the GaAs overlayer is higher than the resistivity of the Si substrate, shown in Fig. 4.

From Figs. 3 and 4 one can conclude that a) the resistivity of the substrate material will not be degraded considerably if the resistivity of

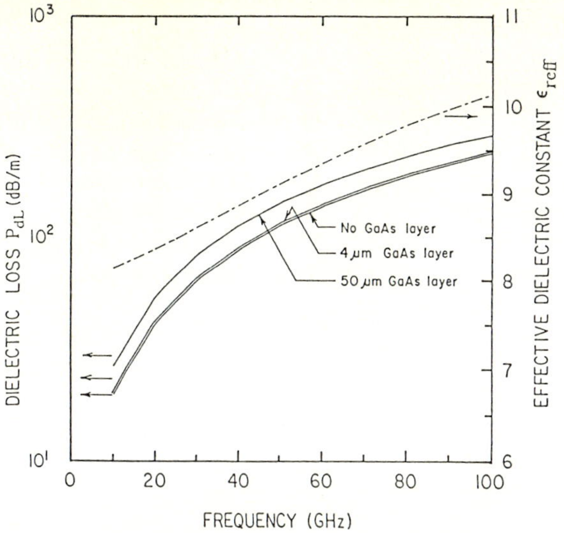

Fig. 3. Dielectric loss and effective dielectric constant vs. frequency for $\rho_{si} = 2000\ \Omega-\text{cm}$, $\rho_{GaAs} = 1000\ \Omega-\text{cm}$

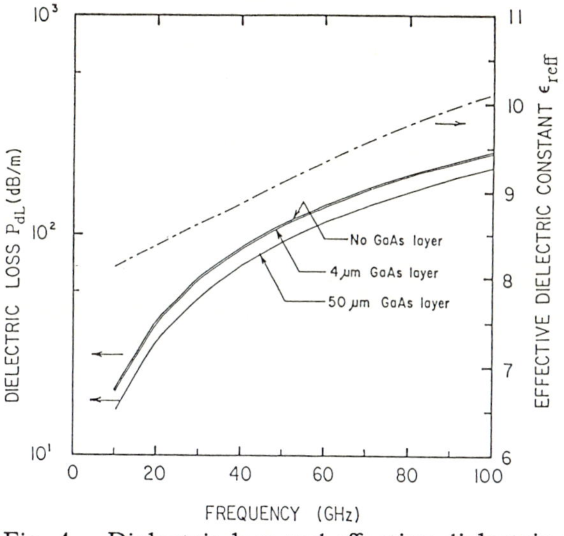

Fig. 4. Dielectric loss and effective dielectric constant vs. frequency for $\rho_{si} = 2000\ \Omega-\text{cm}$, $\rho_{GaAs} = 4000\ \Omega-\text{cm}$

the GaAs overlay material is less than that of the Si substrate provided that the thickness of the overlay material is relatively small, b) depending upon the GaAs overlayer thickness, the equivalent resistivity of the entire structure can be higher than that of the Si substrate when high resistivity GaAs overlay material is used.

4. CONCLUSION

The dielectric loss in GaAs/Si material has been investigated as substrate material for 50 Ω shielded microstrip line and it has been found that GaAs overlay material modifies the resistivity of the entire structure depending upon overlay thickness and resistivity relative to the Si material. Additionally, by using a high resistivity GaAs overlayer on Si, the dielectric loss can be reduced. Since some microwave and millimeter-wave passive components, active components, and sources have already been established monolithically on either GaAs substrates or Si substrates, they might be combined on GaAs/Si substrates to yield a monolithically integrated system. Therefore, it may be concluded that this combination offers exciting possibilities for monolithic microwave and millimeter-wave integration.

1. G.M. Metze, H.K. Choi, and B.-Y. Tsaur, *Appl. Phys. Lett.*, **45,** 1107 (1984)
2. H. Morkoç, N. Otsuka, and H. Zabel, *Science*, in press.
3. R. Fischer, J. Klem, C.K. Peng, J.S. Gedymin, and H. Morkoç, *IEEE Electron Dev. Lett.*, **EDL-7,** 112 (1986)
4. R. Fischer, N. Chand, W. Kopp, C.K. Peng, H. Morkoç, K.R. Gleason, and D. Scheitlin, *IEEE Trans. Electron Dev.*, **ED-33,** 206 (1986)
5. R. Fischer, T. Henderson, J. Klem, W. Kopp, C.K. Peng, H. Morkoç, J. Detry, and S.C. Blackstone, *Appl. Phys. Lett.*, **47,** 983 (1985)
6. A. Rosen, M. Caulton, P. Stabile, A.M. Bombar, W.M. Janton, C.P. Wu, J.F. Corboy, and C.W. Magee, *RCA Review*, **42,** 663 (1981)
7. H. Sobol and M. Caulton, *Advances in Microwaves*, **8,** 11, (1974)
8. K.M. Strohm, J. Buechler, P. Russer, and E. Kasper, *presented at IEEE 1986 Microwave and Millimeter-Wave Monolithic Circuits* Symposium, June 4-5, 1986.
9. T. Itoh and R. Mittra, *IEEE Trans. Microwave Theory and Techniques*, **MTT-21,** 496 (1973)
10. T. Itoh and R. Mittra, *IEEE Trans. Microwave Theory and Techniques*, **MTT-22,** 896 (1974)
11. J.B. Davies and D.M. Syahkal, *IEEE Trans. Microwave Theory and Techniques*, **MTT-25,** 143 (1977)
12. K.B. Bhasin and D.J. Connolly, *IEEE Trans. Microwave Theory and Techniques*, **MTT-34,** 994 (1986).
13. Glenn R. Thoren, *RCA Review*, **45,** 557 (1984)
14. S.C. Binari, R.E. Neidert, G. Kelner, and J.B. Boss, *RCA Review*, **45,** 579 (1984)

Part VI

Fast Lasers and Detectors

Characteristics, Packaging and Physics of Ultra High Speed Diode Lasers and Detectors

J.E. Bowers and C.A. Burrus

AT&T Bell Laboratories, Crawfords Corner Road, Holmdel, NJ 07733, USA

Recently high speed, high power semiconductor lasers with impulse responses of 13 ps, peak powers of tens of milliwatts, and modulation capabilities to 40 GHz have been demonstrated. These small, efficient sources are useful for fiber optic communications, microwave signal transmision for instrument, satellite and military applications, and they can serve as convenient benchtop sources for physics experiments and optical device characterization.

High speed photoconductive detectors and Schottky photodiodes with impulse responses under 5 ps have been demonstrated. More recently, high speed PIN detectors have been made for use at 1.1 to 1.6 μm and 1.8 to 2.5 μm wavelengths, regions important for optical communications using silica fibers and fluoride glass fibers, respectively. InGaAs detectors have an impluse response under 10 ps and are reasonably efficient (35 percent), making them useful in optical communications, microwave signal transmission, and the measurement of very low level (0.1 fJ) repetitive signals with ps time resolution.

Semiconductor lasers can be modulated at rates many orders of magnitude faster than other lasers because the cavity length (10^{-4}m) is so short, and the photon lifetime in the cavity is correspondingly short (~1 ps). Consequently, the relaxation oscillation frequency

$$f_r = \frac{1}{2\pi\sqrt{\tau_{ph}\,\tau_{elec}}}$$

can be several tens of GHz for a photon lifetime (τ_{ph}) of 1 ps and an electron lifetime under lasing conditions $(\tau_{elec}^{-1} \approx g'S)$ of 100 ps. This has resulted in modulation at rates to 8 Gb/s and 3 dB bandwidths as high as 31 GHz in cooled lasers.[1] Because of these large modulation bandwidths, mode locking at rates up to 20 GHz has been demonstrated. The pulse widths of active and passive mode locked semiconductor lasers are longer than have been achieved with other lasers. This is not because the gain bandwidth is narrow; in fact it is 500-1000 Å in quaternary lasers. Rather, it is largely due to the strong gain-phase coupling in the active layer waveguide. This phase distortion limits the minimum pulse width.

Quaternary semiconductor lasers with output powers of 200 mW have been demonstrated.[2] This power translates to a stimulated electron lifetime of 60 ps and a calculated relaxation oscillation frequency of 20 GHz, although the actual relaxation oscillation frequency may be less due to gain compression. This effect of gain compression has been inferred from damping of the intensity modulation response and from measurements of transient chirping in lasers. The same value $(\epsilon = 1.8 \times 10^{-23} m^3$ for 1.3 μm InGaAsP for a differential gain dependence on photon density of $g' = g_0/(1 + \epsilon s))$ is inferred from both measurements, and recent subpicosecond measurements of carrier dynamics in GaAs have confirmed it.[4] However, further measurements are needed, particularly on InGaAsP material. This gain compression results in an ultimate maximum relaxation oscillation frequency, given by[3]

$$f_r = \frac{g_0}{2\pi \sqrt{2}\,(\epsilon + g_0 \tau_p)} .$$

For single longitudinal mode GaAs and InGaAsP lasers, this limit is on the order of 20 GHz.

Achieving such bandwidths requires optimization of the laser structure to provide a high photon density with low parasitics, namely a low series resistance and a low parasitic capacitance due to bonding pads and current confinement. Several structures have been demonstrated with a capacitance of 5 pF or less: constricted mesa lasers,[3] vapor phase regrown (VPR) lasers[5] and mass transport lasers[6]. It is obviously also important to package the device with low parasitic connections. Coaxial K connector (over-moded at 46 GHz) packages have been used for modulation to 40 GHz.[1]

GaAs Schottky diodes[7] and InGaAs p-i-n detectors[8] with bandwidths of 100 and 58 GHz, respectively, have been demonstrated. The main limitations to the speed are (1) the transit time across the depleted region, (2) the capacitance of the diode, (3) diffusion of carriers out of absorbing but undepleted regions and (4) charge trapping at heterojunction interfaces. The capacitance can be reduced by making detectors with small area (which, even for the highest speed detectors, is still comparable in size to the output of a single mode fiber). Diffusion current can be reduced by the use of high doping outside the depleted region or larger bandgap cladding layers (to make the undepleted region transparent). Charge trapping can be reduced by grading the heterojunctions. The transit time limitation forces the depleted layer to be thin, which causes a reduced quantum efficiency. The bandwidth efficiency product of detectors in this limit ($\alpha L \ll 1$) is

$$f_{3dB}\eta = 0.44\ \alpha\ v_s$$

where α is the absorption coefficient and v_s is the saturated carrier velocity. This limit is theoretically 40 GHz for InGaAs detectors at 1.3 µm. The measured bandwidth efficiency products for the GaAs Schottky diode and InGaAs pin described above are 5 and 27 GHz.

The bandwidth efficiency limit occurs because the light absorption and current collection directions are parallel. The use of edge absorbing or waveguide detectors, where the light propagation direction and carrier transit direction are perpendicular, allows simultaneous short transit times and long absorption lengths. Such waveguide devices have been demonstrated[9] in InGaAsP that have high speed and high efficiency, even at zero bias, since the absorbing layer is fully depleted at zero bias.

The packaging of high speed detectors often limits the response time. Parasitic capacitance is more detrimental to detectors than to semiconductor lasers, because the impedance of a laser is only a few ohms while the effective load seen by the detector is typically 50Ω -- and can be even higher in microwave waveguide circuits.

1. J. E. Bowers, Electron. Lett. 21 1195 (1985).

2. T. Asane and T. Okumara, J. Quantum Electron. QE-12, 619 (1985).

3. J. E. Bowers, B. R. Hemenway, A. H. Gnauck, and D. P. Wilt, J. Quantum Electron. QE-22, 833 (1986).

4. B. L. Johnson and A. Mooradian, CLEO, paper THH3, 1986.

5. R. Olshansky, V. Lanziscra, C. B. Su, W. Powazinik, and R. B. Lauer, Appl. Phys. Lett. 49 128 (1986).

6. D. Z. Tsang and Z. L. Liou, J. Lightwave Tech., LT-5 300 (1987).

7. S. Y. Wang and D. M. Bloom, Electron. Lett. 19, 554 (1983).

8. J. E. Bowers, C. A. Burrus, and F. Mitschke, Electron. Lett. 22, 633 (1986).

9. J. E. Bowers and C. A. Burrus, Electron. Lett. 22 905 (1986).

Energy Variations in Optical Pulses from Gain-Switched AlGaAs Diode Lasers

*R.T. Hawkins, J.H. Goll, and N.C. Anheier**

Tektronix, Inc., P.O. Box 500, Beaverton, OR97077, USA

One method for producing optical pulses as short as <20 ps FWHM with a diode laser is by gain switching. For such a laser, gain switching is readily achieved by applying an electrical (forward bias) pulse excitation to the diode. The average duration of the resulting optical pulses depends on the detailed shape of the electrical drive pulse and the dynamic response of the laser diode (in its package) to that drive pulse. Streak camera measurements[1-2] have demonstrated that significant temporal substructure is present on the optical pulse, with subpulses of duration below 10 ps. This substructure changes dramatically from pulse to pulse. Previously reported work on gain switching of laser diodes has been almost exclusively confined to minimizing the time-averaged optical pulse duration, spectral width, and time-bandwidth product. An important additional consideration for ultrafast electro-optical sampling applications is the energy contained in the optical pulse, which should ideally be large and constant to minimize both shot noise and averaging time in the optical detection system. Little work has been done heretofore to quantify the variation in energy of diode laser optical pulses.[3] We present here the results of a series of measurements of this energy variance in pulses produced by gain switching commercially available double heterostructure AlGaAs laser diodes.

The measurement apparatus is shown schematically in Fig. 1. Both Mitsubishi ML4102 and Hitachi HLP 1400 diode lasers were used in this study. The gain-switched laser output was split to allow observation of the temporal structure with a fast photodetector (Ortel PD050-OM) and electronic sampling system while simultaneously measuring pulse energy with a silicon charge-coupled device (CCD). Monitoring the pulse temporal structure was necessary to ensure that only a single relaxation oscillation was generated by the laser under test. The optical pulses were focussed onto the Tektronix four-phase, 96 element linear array CCD to maximize the illumination of a single well (to minimize the effects of shot noise). The CCD was always operated well below saturation. Individual CCD waveforms were digitized to 10-bit accuracy (no internal averaging) with a digitizing oscilloscope under computer control. The CCD reset pulse train was used to provide a synchronous trigger for the pulse generator and the oscilloscopes. This triggering scheme minimized the timing jitter in the measurement system. It also ensured that charge generated by one optical pulse could be clocked out of the illuminated region of the CCD before the arrival of the next pulse. The optical pulse is triggered when no charge is being transferred within the CCD. Noise due to CCD dark current was found to be less than the digitization noise of the digitizing oscilloscope. To examine the dependence of variation in laser diode optical pulse energy on variation in drive pulse energy, a system was added, as shown, to digitize the energy of each electrical drive pulse.

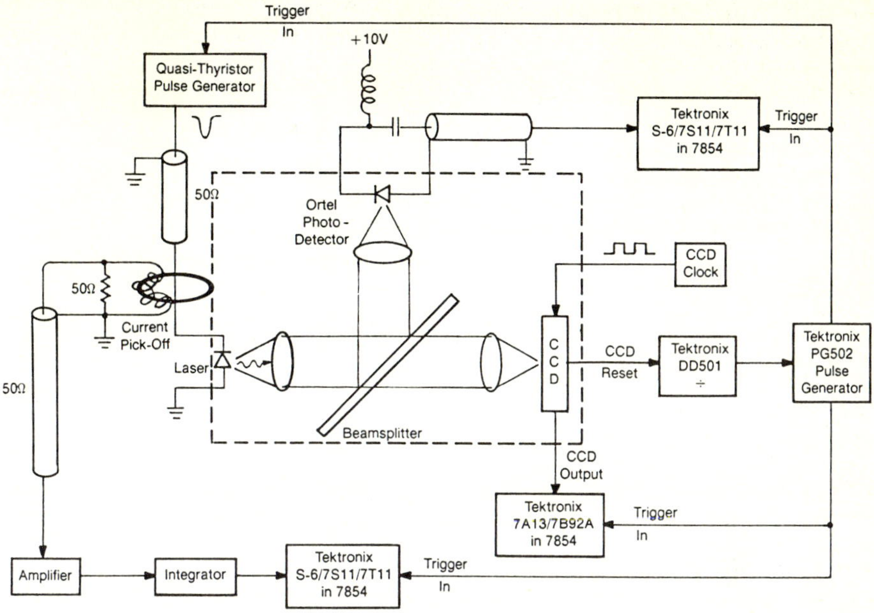

Fig. 1. Schematic of measurement apparatus

Typical optical pulse energy variation is seen in Fig. 2(a), which is a multiple exposure of the output waveform from the illuminated portion of the CCD for nine separate optical pulse energies. The voltage corresponding to the charge in each well appears as a negative-going square pulse of ~20 μs duration. (Other voltage levels corresponding to CCD charge transfers are not relevant to this discussion and are not seen above the top of the figure.) The sixth well in the figure is the target well for the optical pulses. The camera shutter speed was chosen so that there are

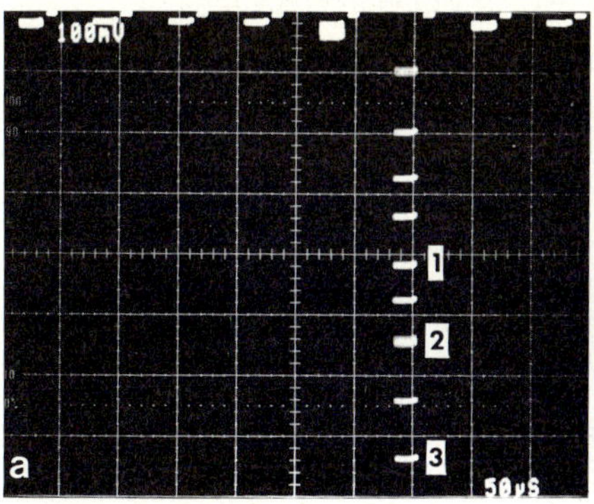

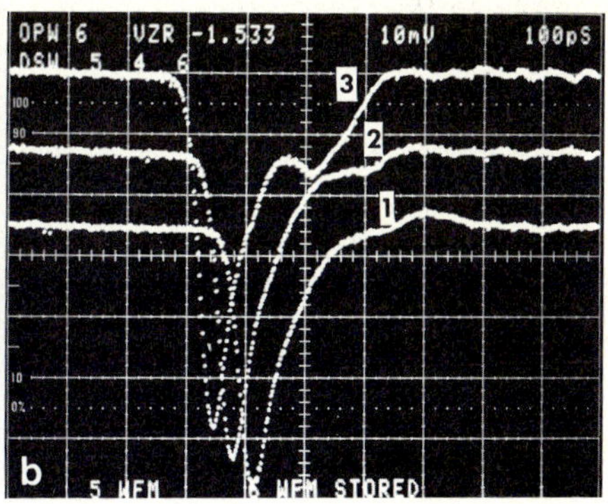

Fig. 2(a). Output from illuminated portion of CCD for nine optical pulse energies — more negative voltages correspond to greater optical energies

Fig. 2(b). Photodetector + sampling system response to three of the optical pulses from (a)

~1000 CCD output waveforms for each exposure. Figure 2(b) shows the high-speed photodetector output that corresponds to the energies labeled '1'-'3' in Fig. 2(a). Note the progression from a single optical pulse ('1') to two optical pulses ('3') in Fig. 2(b). The width of the trace in Fig. 2(a) labeled '2', which corresponds to operation at or slightly above threshold for the second relaxation oscillation, is significantly greater than the width of any other trace. This shows a local maximum in the variation of the optical pulse energy. Furthermore, at an optical output energy just below that associated with the initial appearance of the second pulse, a local minimum in the optical energy variation is seen in the width of the trace between those labeled '1' and '2' in Fig. 2(a). This energy variation behavior was observed for all the diode lasers tested.

The observed minimum in optical pulse energy variations is well defined. The fast photodetector measurements show that this minimum energy fluctuation point occurs just below the energy at which a second pulse is detectable. Therefore, we define the optical pulse energy at this minimum variation point as the maximum single pulse energy, and refer to its variation as the optical pulse energy variation.

In general, lowest energy variation was found to correlate with minimum pulse duration, as well as with maximum pulse energy (since maximum pulse energy corresponds to minimum duration). Total variation in pulse energy lay in the 2-4% range for all of the lasers tested, with rms energy variation below 1% for pulses with duration < 60 ps FWHM. No correlation was observed between variation ($\ll$1%) in the energy of the electronic drive pulse to the laser and variation in optical pulse energy. Based on the observation of a local maximum in energy fluctuations near the threshold for a second relaxation oscillation, we suggest that these enhanced fluctuations in total optical pulse energy may result from threshold amplification of effects due to competition between radiative and non-radiative relaxation processes in the diode laser active region.

References

* Summer Intern, 1985. Present address: Student, Department of Physics and Astronomy, Western Washington University, Bellingham, WA 98225

1. R. A. Elliot, H. DeXiu, R. K. DeFreez, J. M. Hunt, and P. G. Rickman, Appl. Phys. Lett. 42, 1012 (1983)

2. D. Bimberg, K. Ketterer, H. E. Schöll, and H. P. Vollmer, Electron. Lett. 15, 640 (1984)

3. F. Mengel, C. Lin, and N. Gade, In Picosecond Electronics and Optoelectronics, ed. by G. A. Mourou, D. M. Bloom, and C.-H. Lee (Springer-Verlag, Berlin, 1985), p. 163

Frequency Chirping in Pulse Modulated Gain and Index Guided Single Quantum Well Lasers

A. Larsson[1], *P. Andersson*[1], *and A. Yariv*[2]

[1]Department of Electrical Measurements, Chalmers University of Technology, S-41296 Göteborg, Sweden

[2]Department of Applied Physics, California Institute of Technology, Pasadena, CA 91125, USA

The dynamic frequency shift, or frequency chirp, in pulse modulated semiconductor lasers has recently received great attention since it limits the bandwidth-distance product in high bit rate-dispersive optical communication systems emploing dynamic single mode lasers [1]. In the search for a laser structure that exhibits a small frequency chirp, the single quantum well (SQW) structure has been suggested [2-4]. In this paper, we report on an experimental investigation of the frequency chirp in picosecond pulses from pulse modulated GaAs/AlGaAs lasers with SQW active layers and the chirp dependence on the lateral guiding mechanism.

SQW lasers offers a number of potential advantages from a frequency chirping point of view. During excitation with short current pulses, the carrier density in the active layer undergoes relaxation oscillations. This results in a shift in the lasing wavelength since the refractive index and hence the spectral position of the cavity modes depends on the carrier concentration. It is important, however, to realize that the modal index, and its dependence on the carrier density, that the guided optical mode experiences when propagating along the waveguide, can be considerably different from the refractive index of the active layer. Both theoretical [5] and experimental [4] investigations show that the change in modal index with carrier density varies linearly with the optical confinement factor. This suggests that SQW lasers may exhibit improved frequency stability under direct current modulation. It has also been shown [6-7] that the time-bandwidth-product (TBP), which we use as a measure of the frequency chirp, of a linearly chirped Gaussian pulse is given by

$$\mathrm{TBP} = \Delta t \cdot \Delta f = 0.44\,\sqrt{1 + \alpha^2_{\mathrm{eff}}} \tag{1}$$

where Δt and Δf is the pulse intensity FWHM and the power spectrum FWHM of a single longitudinal mode, respectively. α_{eff} is the effective linewidth enhancement factor and is defined as the ratio between the change in the real part of the modal refractive index and the change in the imaginary part of the modal refractive index when the carrier density in the active layer is changed. Equation (1) is also a good approximation for more realistic pulse shapes when the pulse is strongly chirped, i.e. for a laser with a large α_{eff} [8]. Theoretical calculations on SQW lasers predict a small intrinsic α of the active layer [9], α_a, which might be reflected in α_{eff} and hence in the chirp behavior of these lasers.

The lasers used in the experiment are graded index waveguide separate confinement heterostructure single quantum well (GRIN-SCH SQW) GaAs/AlGaAs lasers. Both gain-guided and weakly index-guided lasers were investigated. The index-guided laser is a ridge waveguide laser fabricated by us from a low threshold current density (300 A/cm^2 for a 500 µm long cavity) MBE-grown wafer with an active layer thickness of 60 Å (Fig. 1a). Evidence for index

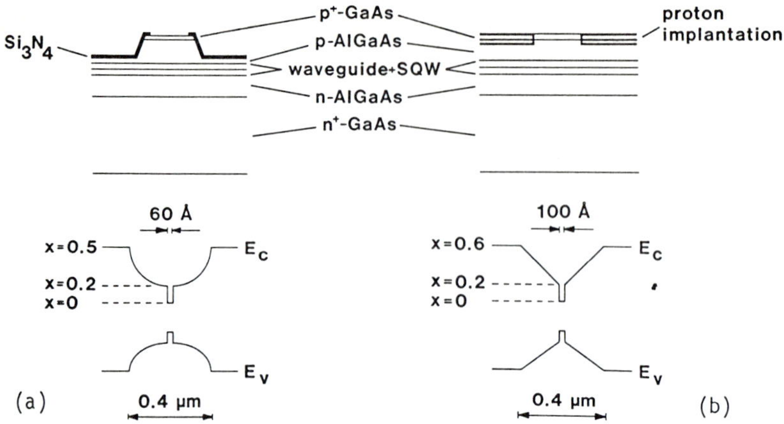

Fig. 1. GRIN-SCH SQW laser structures a) Index guided, b) Gain guided

guiding is provided by the single longitudinal mode spectrum with a sidemode suppression of 14 dB obtained at certain bias levels during CW operation. The gain-guided laser is a commercial GRIN-SCH SQW laser (Thomson CSF, SE-1000) with a narrow stripe defined by shallow proton implantation and an active layer thickness of 100 Å (Fig. 1b). Both lasers have a 0.4 µm thick waveguide region and similar Al-mode fractions.

The TBP:s are measured using an interferometric technique [10]. The experimental setup is shown in Fig. 2. The modulation consists of short electrical pulses with a repetition frequency of 1 GHz, superimposed on a bias current at or above threshold. The fringe pattern from the interferometer is detected by a fast avalanche photodiode. This gives the modulus of the field autocorrelation function (ACF) of the radiation. After adding the phase of the

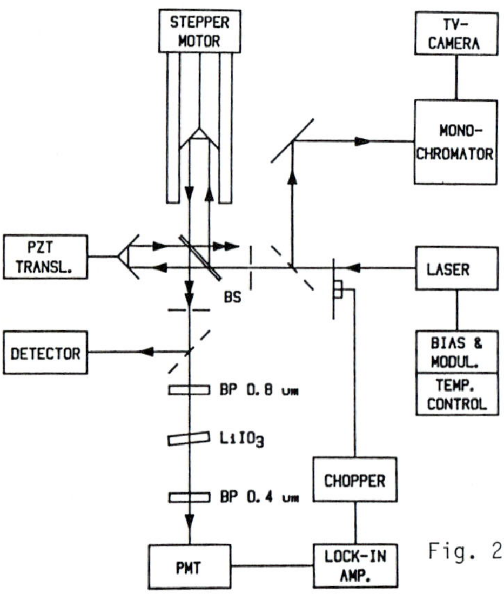

Fig. 2. Experimental setup for temporal coherence measurements

ACF, obtained from computer simulations, the frequency power spectrum can be calculated by Fourier transformation. The output from the interferometer can also be detected by a standard setup for second harmonic generation. In this way, the intensity ACF is obtained, from which the pulse width is deduced using a Gaussian pulse approximation. Finally, the TBP is obtained by multiplying the power spectrum and the pulsewidth FWHM. As an example we show in Fig. 3a a measured modulus of the ACF for a pulsed ridge waveguide GRIN-SCH SQW laser. The corresponding calculated power spectrum is shown in Fig. 3b.

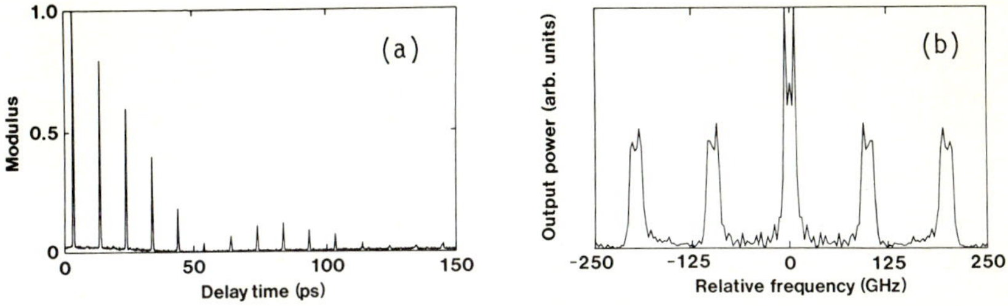

Fig. 3. a) Measured modulus of the ACF for a pulsed ridge waveguide SQW laser
b) Power spectrum calculated from Fig. 3a

In Fig. 4 we present results from the TBP measurements for different bias and modulation conditions. Results from measurements on other lasers are also included for comparison. Here we notice a significant difference in the chirping behavior between gain and index-guided SQW lasers. We also notice a comparably small chirp in the index-guided SQW laser. The estimated effective α parameter, using (1) for the SQW lasers are 4.2 and 16 for the index-guided and the gain-guided, respectively. The latter is the highest value yet observed for semiconductor lasers. A notable feature of the gain-guided SQW laser is that for high modulation amplitudes, the total chirp exceeds the longitudinal mode spacing. Consequently, the chirp bandwidth can only be assigned a minimum value equal to the mode spacing as indicated in Fig. 4a. We also note that the TBP is approximately constant over a wide range of modulation conditions, in agreement with computer simulations [8].

Our results support the theoretically predicted laser structure dependence of the linewidth enhancement factor [11]. Here it was suggested that in index guided lasers the effective α, α_{eff}, is close to the intrinsic α of the active layer, α_a. In gain guided lasers, however, α_{eff} can be substantially larger than α_a. The waveguiding parallel to the junction plane in gain-guided lasers is due to the combined effect of carrier-induced gain guiding and index antiguiding. Therefore the modal index is very sensitive to perturbations in the carrier density. This results in a large frequency chirp. The advantages of using ridge waveguide structures with respect to the frequency chirp has been noted earlier [3]. These lasers exhibit a good spatial overlap of the fundamental optical mode and the carrier density profile along the junction plane which results in a small frequency chirp. The fact that the measured TBP:s for our ridge waveguide SQW laser is similar to that of the CSP laser is also worth noting (Fig. 4b). This is not surprising when the similarity in laser structures is considered. The lateral mode guiding is in both cases provided by evanescent field coupling. Any effects that can be attributed to the use of SQW active layers (small α_a) were not observed in our measurements. These results point out the importance of optimizing the lateral waveguiding to reduce the frequency chirp in semiconductor lasers.

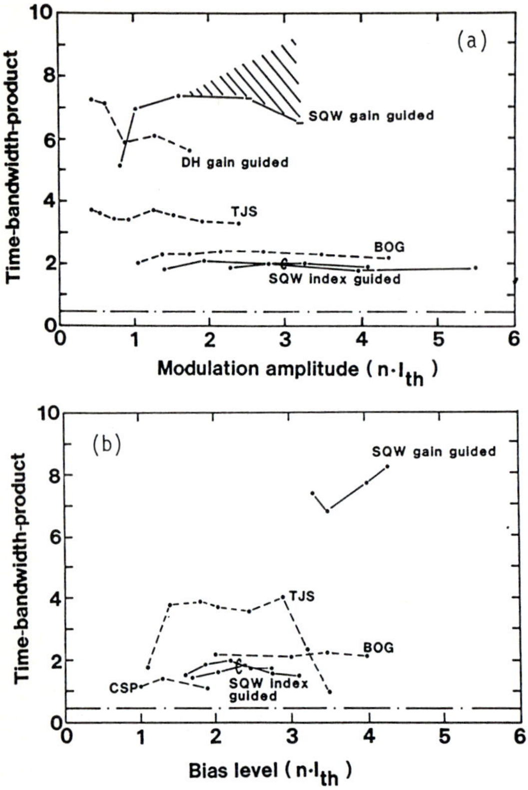

Fig.4. Results from TBP measurements as a function of a) modulation amplitude and b) bias level. The transform limited value of the TBP for a Gaussian pulse (0.44) is also indicated.

In conclusion, we have investigated the dynamic frequency shift in pulse-modulated gain and index-guided SQW lasers. A strong influence of the guiding mechanism and a small frequency chirp in the weakly index-guided lasers was found. This suggests that the lateral guiding mechanism is of primary importance for designing lasers with a small frequency chirp. The effect of modifying the active layer by using quantum wells (small intrinsic α) becomes apparent only when the guiding mechanism is optimized.

1. R.A. Linke, Electron. Lett. 20, 472 (1984)
2. N.A. Olsson, N.K. Dutta, and K.Y. Liou, Electron. Lett. 20, 121 (1984)
3. N.K. Dutta, N.A. Olsson, L.A. Koszi, P. Besami, R.B. Wilson, and R.J. Nelson, J. Appl. Phys. 56, 2167 (1984)
4. N.K. Dutta, N.A. Olsson, and W.T. Tsang, Appl. Phys. Lett. 45, 836 (1984)
5. J.S. Manning and R. Olshansky, Electronc. Lett. 17, 507 (1981)
6. T.L. Koch and J.E. Bowers, Electron. Lett. 20, 1039 (1984)
7. F. Koyama and Y. Suematsu, IEEE J. Quantum Electron. QE-21, 292 (1985)
8. P. Andersson and T. Andersson, IEEE J. Lightwave Techn. LT-4, 795 (1986)
9. Y. Arakawa, K. Vahala, and A. Yariv, Appl. Phys. Lett. 45, 950 (1984)
10. P. Andersson, T. Andersson, S. Lundqvist, and S.T. Eng, IEEE J. Lightwave Techn. LT-2, 146 (1984)
11. K. Furuya, Electron. Lett. 21, 200 (1985)

InP/GaInAs/InP PIN Photodiode with FWHM<18 ps

S.Y. Wang, K.W. Carey, and B.H. Kolner

Hewlett-Packard Laboratories, 1651 Page Mill Road, Palo Alto, CA 94304, USA

Introduction:

High-speed optical detectors in the 1.3-1.55 μm wavelength range are required for a new generation of microwave-optical instruments to characterize high data rate systems, microwave transmission via lightwave, and other applications requiring true replication of the modulation signal. Front-side illuminated photodiodes have an advantage over back-side illuminated devices in simplicity of packaging. In addition, the transit time of the photocarriers is determined by electrons rather than by holes as for backside illuminated devices fabricated on n-type substrates. The speed of response of the double heterostructure photodiode is wavelength independent since photocarriers can only be generated in the i-layer which is fully depleted. For single heterostructure backside illuminated photodiodes with p-GaInAs, the speed response degrades as the wavelength increases since photocarriers can be generated in the low field p-layer resulting in carrier diffussion . For high-speed operation of the device, the i-layer thickness is selected such that the device is transit time limited and is of the order of one absorption length. However one potential obstacle exists that could degrade the speed of a front-side illuminated device, and that is the trapping of holes at the discontinuity in the valance band between the InP and GaInAs. The effect of hole trapping was not evident in the results of Susa et al[1] which utilized a separate absorption and multiplication (SAM) structure for a front-side illuminated avalanche photodiode. This suggests that the quality of interface could play a dominant role in the transit time at heterojunction interfaces.

Structure:

The photodiode is a $P-i-N$ mesa structure whose cross-section is shown in Figure 1. On an N^+ (100) S-doped InP substrate, an undoped GaInAs lattice matched layer 1.5 μm thick and then a Zn doped InP window layer 1.0 μm thick are grown by organometallic vapor phase epitaxy (OMVPE). The photosensitive area is circular in geometry with a diameter of 25 μm in contact with a 25 μm bond pad. The p-contact metallization is Ti-Au on the front surface and the n-contact is NiCr-Ge-Au/Ti-Au on the back surface. SiO_x was used as an antireflection film. The mesa edges were left unpassivated. The operating voltages for our devices were less than 4 volts reverse bias. Dark current density was $6 \cdot 10^{-6}$ A cm^{-2} at -4 V.

Growth technique:

The epitaxial layers were grown using OMVPE at atmospheric pressure in a horizontal reactor. Trimethylgallium (TMGa) and trimethylindium (TMIn) were the sources for

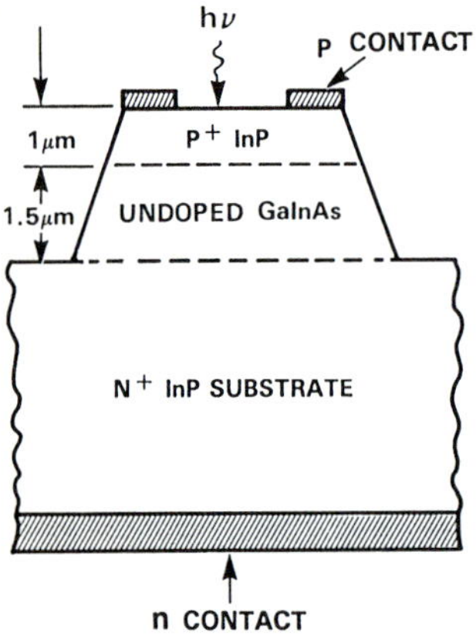

Figure 1:
Layer structure of the InP/GaInAs/InP front-side illuminated photodiode; the absorbing layer GaInAs is 1.5 μm thick, and the window layer P-InP is 1 μm thick. 1000 Å SiO_x was used as antireflection. Photosensitive area is 25 μm in diameter in contact with a 25 μm diameter bondpad.

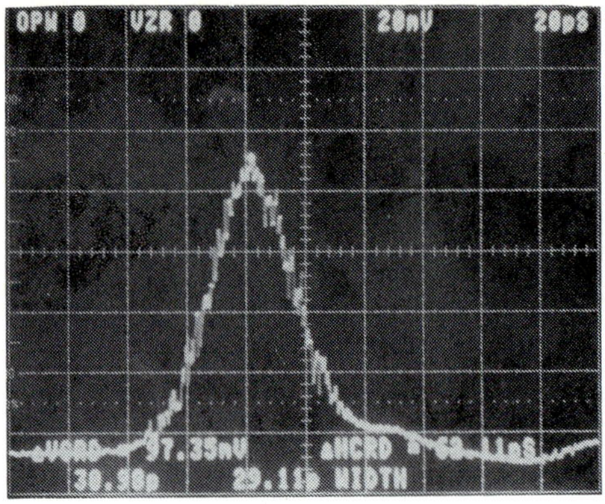

Figure 2:
Transient time response of the photodiode to 1.3 μm diode laser pulse showing a full width at half maximum of 29 ps and a fall time of 31 ps. The impulse response of the sampling head is 23ps. Using a Gaussian approximation a combined photodiode and laser pulse width of 18 ps is calculated.

the group III elements, 10% AsH3 in hydrogen was used as the As source and 10% PH3 in hydrogen as the P source. The p-type dopant source was diethylzinc (DEZn) and the carrier gas was hydrogen. The growth rates for GaInAs and InP were 1.5 μm/h and 0.8 μm/h, respectively. Other details of the growth of atmospheric pressure OMVPE $Ga_{0.47}In_{0.53}As$ can be found in publictions by Carey [2] and Kuo et al[3]. The carrier concentration in P-InP is approximately $1 \cdot 10^{18}cm^{-3}$, and the n-GaInAs was nominally undoped with residual carrier concentrations of $< 4 \cdot 10^{15}cm^{-3}$. Using high-resolution transmission electron microscopy the InP(substrate)/GaInAs transition is determined to be about two monolayers (5.8 angstroms) and the transition from GaInAs/InP is found to be about 5 monolayers (< 15 angstroms)

High-Speed Measurement Technique and Results:

A pulsed diode laser was used to excite the photodiode whose transient response was measured on a sampling oscilloscope.

A 1.3 μm diode laser (Hitachi HLP5400) driven by a step recovery diode at 100 MHz was used as the optical impulse source. The laser pulses were passed through a low pass filter to eliminate internally generated second harmonic radiation and were attenuated to less than 15 microwatts average power before illuminating the photodiode. The photodiode output was connected to a Tektronix S-4 sampling head via a short microwave cable (Gortex) for the time domain analysis. The measured temporal response is shown in Figure 2 where the pulse width at full width half maximum is 29 ps. The impulse

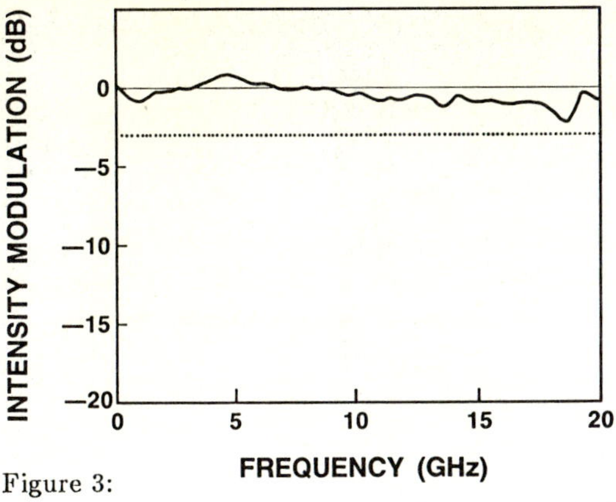

Figure 3:
Combined frequency response of the GaAs optical modulator and the photodiode. The vertical axis is optical modulation intensity and the horizontal axis is 0-20 GHz. The dotted line is -3 dB optical.

response of the S-4 sampling head is 23 ps ; using Gaussian approximation a combined photodiode and laser pulse width of 17.7 ps is calculated. The photodiode response is therefore $<$ 17.7 ps or correspondingly $>$ 18 GHz.
Recently[4], we have developed a high bandwidth travelling-wave electro- optic GaAs polarization modulator. In testing the combined modulator- photodiode frequency response at 1.3μm wavelength an optical -3 dB bandwidth in excess of 20 GHz or an electrical -3 dB bandwidth of appro ximately 20 GHz is demonstrated as shown in Figure 3. The photodiode bandwidth is greater than 20 GHz.

Discussion:

We have developed front-side illuminated double heterostructure P-i-N mesa InP/GaInAs/InP photodiodes with bandwidth in excess of 18 GHz or a FHWM of less than 18 ps at 1.3 μm wavelength. The device has an external quantum efficiency of greater than 85% and a dark current density of $6 \cdot 10^{-6}$ A cm^{-2} at the operating bias of -4 volts. The bandwidth predicted for our device with the given geometries is 20 GHz taking into account only the RC time and transit time which agrees well with our observation. The measured fall time of the photodiode is 31 ps, the slight hump at the trailing edge Figure 2 may be due to ringing in the laser, reflection, carrier diffusion due to dopant redistribution. Hole trapping at the interface has not been observed to significantly degrade the speed of our device[5].
From our recent combined GaAs modulator-InP/GaInAs photodiode frequency response measurements, the photodiode -3 dB bandwidth is at least 20 GHz.

[1] N. Susa, H. Nakagome, O. Mikami, H. Ando and H. Kanbe, "New InGaAs/InP Avalanche Photodiode Structure for the 1-1.6 um Wavelength Region", IEEE J. Quantum Electron.,vol. QE-16(8),pp.864-870,1980

[2] K.W. Carey, "Organometallic vapor phase epitaxial growth and characterization of high purity GaInAs on InP", Appl.Phys. Lett., vol.46(1),pp.89-91,1985

[3] C.P. Kuo, J.S. Yuan, R.M. Cohen, J. Dunn and G.B. Stringfellow, "Organometallic vapor phase epitaxial growth of high purity GaInAs using Trimethylindium", ibid,vol.44,pp.550-552,1984

[4] S.Y.Wang, S.H. Lin and Y.M. Houng,"GaAs travelling-wave electro-optic waveguide modulator with bandwidth >20 GHz at 1.3μm", Sixth International Conference on Integrated Optics and Optical Fiber Communication, paper WK3,Reno, Nevada Jan. 1987

[5] S.Y. Wang and K.W. Carey,"Front-side illuminated InP/GaInAs/InP p-i-n photodiode with FWHM < 26ps",43rd Annual Device Research Conference, paper IIIA-8, Boulder, Colorado, June 1985.

Fast GaAs Photoconductive Detectors with High Sensitivity Integrated in Coplanar Systems onto GaAs Substrates

H. Schumacher, U. Salz, and H. Beneking

Institute of Semiconductor Electronics Aachen, University of Technology, D-5100 Aachen, Fed. Rep. of Germany

Introduction

Since its introduction in 1975 /1,2/, high-speed sampling of voltage transients on transmission lines using photoconductive switches has been of great interest. Besides superior equivalent bandwidth, it offers a dynamic range considerably larger than conventional sampling oscilloscopes, and complete isolation between the transmission line under test and the sampling circuit. Despite these advantages, it has never found widespread use outside the research area, mainly due to the costly ps dye lasers required. A sampling system using a semiconductor laser would resolve this problem, provided it has a temporal resolution equal or superior to conventional sampling oscilloscopes (≤25 ps). Besides the generation of ps pulses with semiconductor lasers /3/, which will not be discussed here, a photoconductive gate with a carrier lifetime ≤10 ps is necessary, yet sensitive enough to efficiently sample when controlled by pJ pulses from a laser diode.

The sensitivity of a photoconductive sampling gate, i.e., the change of conductance δG for a given optical pulse energy E is inversely proportional to the square of the illuminated gap length:

$$\delta G = (1-R)\cdot q\cdot(\mu_n + \mu_p) \cdot \frac{E}{\ell^2 \cdot h\nu} \qquad (1)$$

where R is the surface reflectivity, q the elementary charge, $\mu_{n,p}$ the mobilities and $h\nu$ the photon energy. Lowering the gap width to enhance sensitivity, however, faces some limitations:

— the laser cannot be focussed to efficiently couple to a μm gap;

— the parasitic parallel capacitance increases.

To maintain a conveniently sized active area (20μm × 20μm) with enhanced sensitivity an interdigital layout (Fig. 1) was chosen. The

Permanent Address: Bell Communications Research 331 Newman Springs Rd., Red Bank, N.J. 07701, USA

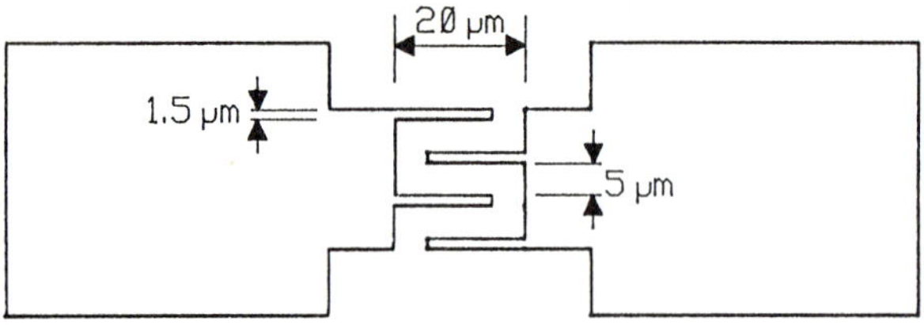

Fig. 1 Lay-out of the interdigital photoconductive sampling gate.

sensitivity of the interdigitated gate is increased twelvefold over a straight gap of the same active area, including the loss due to the finite contact finger width. The parasitic capacitance parallel to the gate was calculated to be $C_p = 7$ fF, which can be considered sufficiently low for a sampling aperture < 5 ps.

Both Cr-doped and nominally undoped semiinsulating GaAs substrates were examined. The nominally undoped material, however, was preferred due to its superior homogeneity. Ni-AuGe-Ni layers were evaporated and annealed at 465°C to form the Ohmic contacts. Pre-implantation carrier lifetime was determined to be $\tau = 231$ ps using an optoelectronic pulse technique.

To reduce the carrier lifetime, proton-bombardment was used to obtain the necessary deep damage profile (≈ 3 µm) with the 350 keV maximum implantation energy available. Near-homogeneous damage profile is ensured by a triple implantation of H^+:

$$
\begin{aligned}
&2\times 10^{12}\ cm^{-2} \text{ at } 20 \text{ keV,}\\
&4\times 10^{12}\ cm^{-2} \text{ at } 110 \text{ keV,}\\
&5\times 10^{12}\ cm^{-2} \text{ at } 280 \text{ keV.}
\end{aligned}
\qquad (2)
$$

The damage density obtained is $>10^{16}$ cm^{-3} upto 3 µm depth. Figure 2 shows the effect of varying implantation density on the dark I-V-characteristic of the sampling gate. Implant #2 lowers the dark conductance and improves the linearity. Increasing the damage concentration further will increase the dark conductance again. We also observed an increased carrier lifetime which is not yet well understood. With proton bombardment, an optimum implantation density seems to exist, both with respect to dark conductance (which influences the sampling system's noise floor) and carrier lifetime (determining the sampling aperture).

A coplanar waveguide was chosen for the integrated sampling system, offering increased design flexibility over microstrip lines and lower dispersion below the first TE-mode cut-off /4/. For the high-speed part of the system between the sampling gate and the pulse

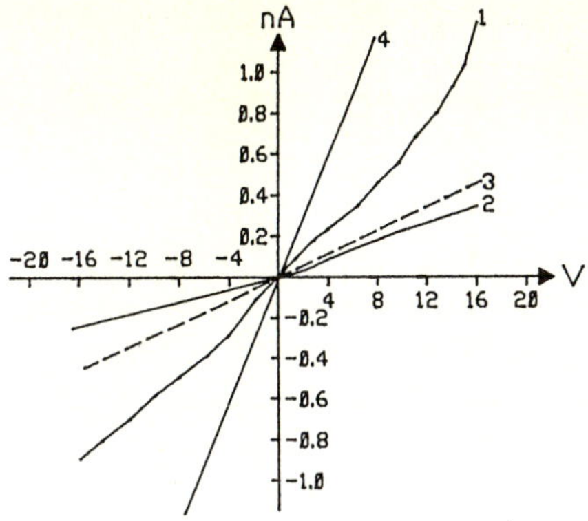

Fig. 2 Influence of the implantation density on the dark 1-V-characteristics: 1: not implanted; 2: implanted as given by (2); 3: dose × 5; 4: dose × 25.

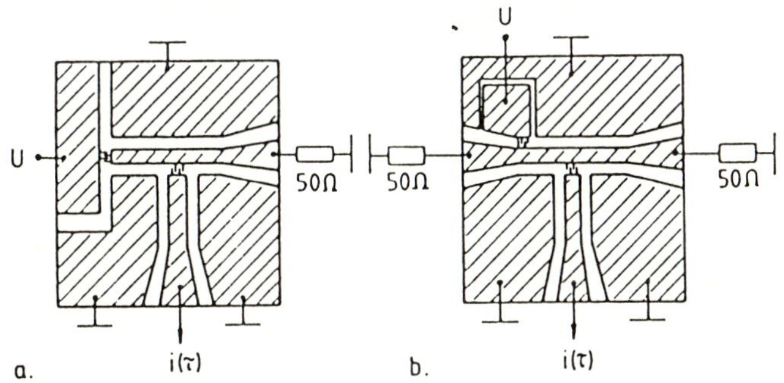

Fig. 3 Two different optoelectronic sampling heads.

injector (see Fig. 3), a waveguide with 60 μm center conductor width and 41 μm slot offers improved waveguide dispersion ($\epsilon_{r.eff}$ vs. frequency) over a microstrip line on the 380 μm thick substrate. As Ohmic transmission line loss is also a source of dispersion, the coplanar metalization is galvanically reinforced to 2.5 μm thickness.

Figure 3 schematically shows the two different sampling circuits examined. Both circuits use a gate prebiased at a fixed voltage as a signal injector. When illuminated by a laser pulse, this gate will generate a short voltage transient on the transmission line. This voltage will bias the other photoconductor, the sampling gate, which will be opened by a second laser pulse. The mean current through this gate will hence depend on the delay τ between the two light pulses, and $i(\tau)$ will represent the cross-correlation function of the two gate responses. The circuit (a) has the injector at the end of the transmission line, modeling the characterization e.g., of a high-speed photodiode. Circuit (b) injects the voltage transient in a "T"-configuration. As shown by Smith et al., /5/, this configuration is useful to characterize electrical twoports like FETs in the time domain.

In the experiment, the sampling gates were illuminated by a passively mode-locked dye laser (FWHM<0.1ps at 620 nm). Both beams travelled through a rotating glass block acting as the variable delay line. A lock-in technique was not necessary due to the high gate sensitivity. Therefore, direct observation of the correlation function on an oscilloscope was possible. The measurement results, finally, demonstrate a width of the correlation function of 7.2 ps, see Fig. 4. Assuming a single-sided exponential waveform for the electrical pulse on the transmission line, this leads to a sampling aperture

$$T = 5.2 \text{ ps.}$$

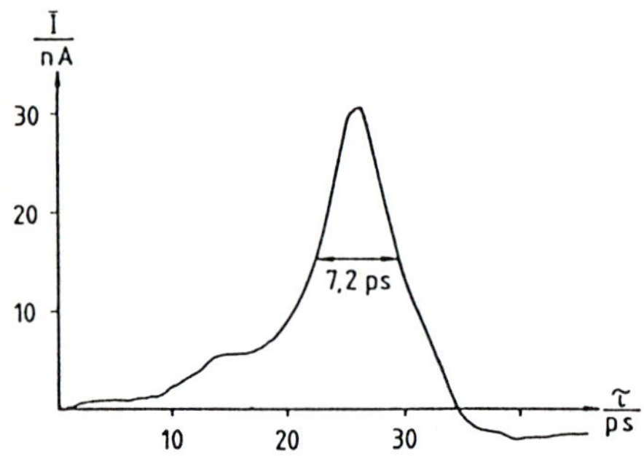

Fig. 4   Cross-correlation function obtained from the circuit in Fig. 3(a).

The same sampling gates have also been used to demonstrate a semiconductor laser-driven sampling oscilloscope with 56 ps sampling aperture and electronic delay that will be covered in a different publication. In conclusion, we have demonstrated an optoelectronic sampling gate on GaAs substrate with ps resolution and improved sensitivity.

This work was funded by the German Research Council.

Literature

[1] Auston, D. H., Picosecond optoelectronic switching and gating in silicon; Appl. Phys. Lett. 26 (1975), pp. 101.

[2] Lawton, R. A., and Andrews, J. R., Pulsed-laser application to sampling oscilloscope; Electron. Lett. 11 (1975), pp. 138.

[3] Schöll, E., Bimberg, D., Schumacher, H., and Landsberg, P. T., Kinetics of picosecond pulse generation in semiconductor lasers with bimolecular recombination at high current injection; IEEE Journ. Quantum Electron. QE−20 (1984), pp. 394.

[4] Hasnain, G., Dienes, A., and Whinnery, J. R., Dispersion of picosecond pulses in coplanar transmission lines; IEEE Trans. Microwave Theory Tech. MTT−34 (1986), pp. 738.

[5] Smith, P. R., Auston, D. H., and Augustyniak, W. M., Picosecond photoconductivity in radiation-damaged silicon-on-sapphire films; Appl. Phys. Lett. 38 (1981), pp. 47.

MESFET Compatible IMSM Detectors

D.L. Rogers

IBM T.J. Watson Research Center, P.O. Box 218,
Yorktown Heights, NY 10598, USA

Introduction

In order to be useful in high speed, gigabit/s, fiber optic communication receivers the photo-detector must not only be capable of very fast response but also must be capable of providing a close proximity, low capacitance connection to the input preamplifier in order to optimize electronic performance and minimize noise coupled from adjacent circuits.

The monolithic integration of the photodetector with it's companion amplification circuits on the same chip can solve many of these problems. The interdigitated Metal-Semiconductor-Metal (IMSM) detector[1] is a attractive candidate for use in such integration due to its, low capacitance, potential compatibility with GaAs circuit processes, and its potential high performance [2]. Indeed previous workers have demonstrated that high performance detectors of this type can be fabricated on high purity epitaxial buffer layers along with MESFET preamplifiers [3] but only with the use of relatively low temperature, non-self-aligned MESFET processes. On the other hand, detectors fabricated directly on semi-insulating substrates using a high temperature, self-aligned process have suffered from an anomalous, low frequency gain [4]. This gain increases the noise level in the detector and would require subsequent equalization for use in any data transmission system. We have found experimentally that this low frequency gain can be reduced by using shallow implant under the active area of the detector, thereby allowing the successful fabrication of low noise integrated photoreceivers using a high-temperature self-aligned process [5]. In this paper we present calculations supporting what we believe is the mechanism responsible for the low frequency gain and demonstrate how this gain is in fact reduced by means of the shallow implant.

Theory

Most mechanisms offered to explain the gain observed in IMSM detectors are based on a photo-induced injection of electrons or holes at the contacts [6]. Such injection implies, however, a significant bending of the bands and has led to the proposal of several complex mechanisms for accumulating sufficient charge in the region between the contacts. We have found, however, that such large band bending can be generated simply due to the charging of a low density of traps by the photo-generated carriers.

In order to calculate the change in potential between the contacts due to trapped charge, the geometry of the detector in the region between the contacts was approximated by a gap between two semi-infinite conducting planes. The potential for the uncharged substrate and the Green's function for this geometry were then calculated using conformal mapping. This mapping was used to define a coordinate system in which the current density and continuity equations could be simply solved using a Runge-Kutta numerical integration yielding the electron and hole carrier densities. It was then assumed that the trap levels attain equilibrium with these carrier densities and the resultant trap charging calculated using a simple rate equation. Published values for the capture cross sections and detrapping times for the EL2 trap were used.

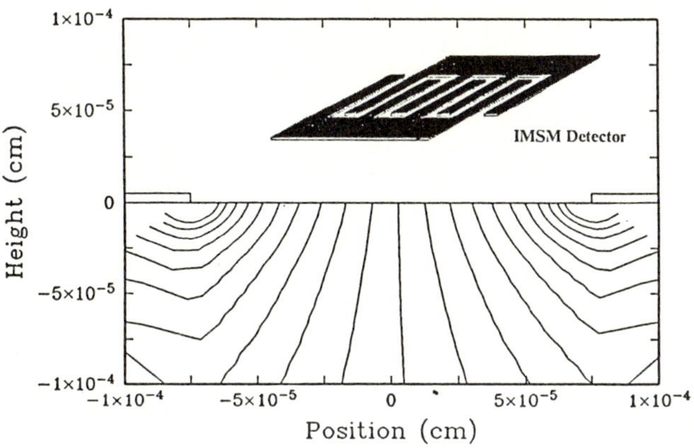

1. Equipotential plot of charge density between adjacent fingers.

Figure 1 shows an equipotential plot of the induced charge density for an incident radiation flux of 0.005 W/cm^2, and a trap density of $1x10^{15}$ cm^{-3} which are conditions in which gain has been observed in these detectors. Shown in the inset is a schematic of an IMSM detector. The equipotential plot is for the region between two adjacent fingers in the detector. Each equipotential line corresponds to a change in charge density of $1x10^{-6}$ c/cm^3 with the charge being zero at the center. Using this charge density the potential was calculated by a numerical integration using the Green's function. Figure 2 shows this potential calculated along a line parallel to the surface of the semiconductor, perpendicular to the fingers. Notice that the tunneling length at 0.1 eV below the top of the barrier is reduced by a factor of 2 due to trap charging. Since the tunneling current depends exponentially on the tunneling length we feel this supports the conjecture that trap charging alone can give rise to significant modulation of the tunneling barrier.

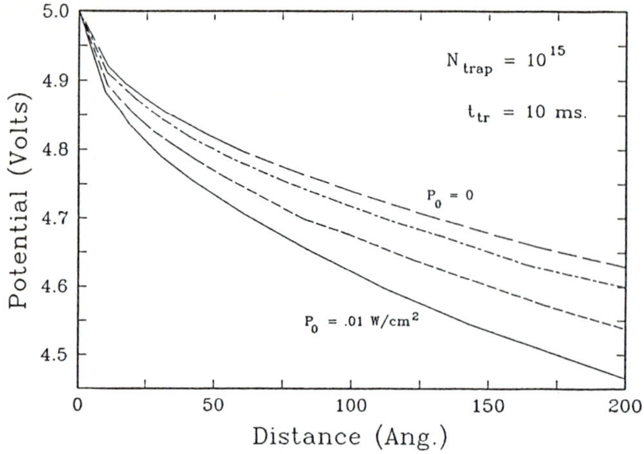

2. Potential near electrode edge for various illuminations.

Note that the degree of substrate charging and band bending illustrated in these calculations would not be possible in the more conventional type of detector in which the junctions are parallel to the surface. The large amount of tunneling in the IMSM detector is due to both the large electric fields near the electrode edges and the fact that the crowding of field lines at these points increases the carrier density and thus the trap charge density in the region where it has the most effect on the band bending.

Although the bulk trap density can be kept quite low by using high purity substrates or epitaxial layers, the carriers still suffer from the possibility of being trapped at the surface where, especially in GaAs, there is a large density of surface states. Whether the photo-induced carriers will be trapped there depends on the electric field profile in the direction perpendicular to the surface. One way to prevent the attraction of the photo-generated carriers to the surface is to create a thin doped layer, by implanting for instance, at the surface of the semiconductor with the doping density and thickness chosen such that the semiconductor remains depleted. Figure 3 shows the effect of such a layer on the band structure in a direction perpendicular to the surface. Clearly the photo-generated electrons, although attracted to the surface deep in the substrate, will be repelled from the surface by the electric field in the doped layer. On the other hand the photo-generated holes generated deep in the substrate are repelled from the surface except those in the thin doped region. Assuming that the doped layer is much thinner than the absorption depth of the radiation almost all of the holes will be repelled from the surface.

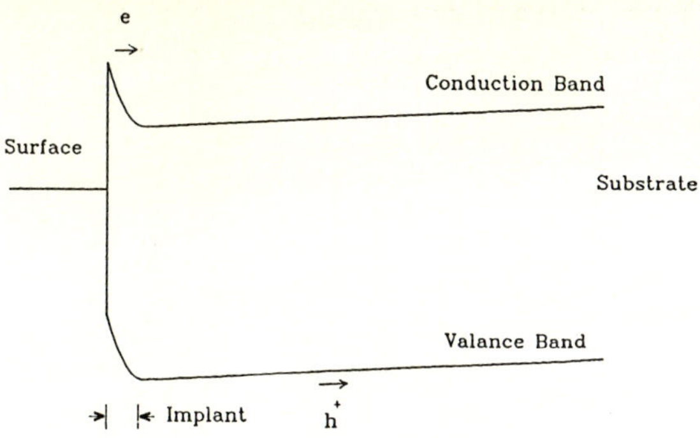

3. Band structure due to surface implant.

Fabrication and Design

Photodetectors were fabricated on semi-insulating GaAs substrates using the same process that we use to fabricate refractory gate, enhance/deplete MESFET circuits[7]. Thin doped regions were prepared under some detectors by implantation. These detectors were first implanted with 100 keV SiF ions with a dose of $1.3x10^{12}$ cm^{-3} and all the detectors were annealed at 850^oC for 20 minutes. The surface was then sputter etch cleaned and approximately 2000 Angstroms of WSi sub 0.11 was deposited and delineated using a subtractive reactive ion etch. Subsequent to these steps the detectors were further subjected to a 800^oC anneal for 10.minutes to activate the contact implants used in the IC process. Detectors were fabricated adjacent to each other on the substrate both with and without the implant. The detectors described below have 100 μm long fingers with 1 μm finger widths and 2.25 μm spaces between fingers. The total active area was $1x10^{-4}cm^{-2}$.

Results

The photodiodes were characterized using a probe station equipped with a fiber optic probe. The dark current for the detectors was measured at 14 locations on a 2 inch wafer. Dark currents of 88 nA with a standard deviation of 40 nA. were measured for the un-implanted detectors and 11 nA with a standard deviation of 7 nA for the implanted detectors. Figure 4 shows IV curves under illumination for typical implanted and un-implanted detectors. The optical power incident on the detector is 140 μW and the horizontal line labeled I_0 in this figure represents the ideal photocurrent coresponding to an internal quantum efficiency of one, taking into account the light masked by the metal fingers and the reflection coeficient at the GaAs-air interface. The

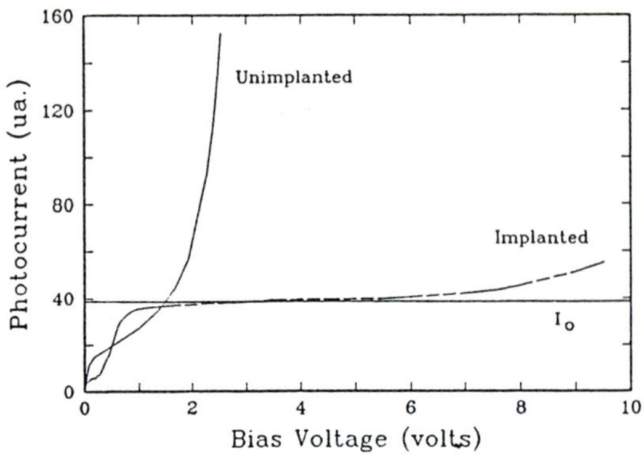

4. Illuminated IV curves.

implanted detectors demonstrate an internal quantum efficiency very close to one and only begin to show low frequency gain at bias voltages above 7 V. On the other hand the un-implanted detector shows large amounts of gain at biases as low as 1.5 V.

Conclusion

We have shown that the gain previously observed in the IMSM detector can be accounted for in terms of a simple trap induced tunneling mechanism and have demonstrated how a shallow implanted region can be used to greatly reduce these effects.

Acknowledgements

The author would like to thank J.H. Magerlein, D.J. Webb, A Callegari, T. Fryxell, H.C. Guthrie, Peter D. Hoh, J.W. Mitchell, A. Pomerene, S. Scontras and G. Spiers for the fabrication of the chips containing the detectors.

References

1. T. Sugeta, T. Urisu, S. Sakata and Y. Mizushima, "Metal-Semiconductor Metal Photodetector for High-Speed Optoelectronic Circuits", Jap. Jour. Appl. Phys., V19, Suppl 19-1, 1980, p459-464.

2. W. Roth,H. Schumacher,J. Kluge,H. Geelen, and H. Beneking,"The DSI Diode - A Fast Large-Area Optoelectronic Detector", IEEE Trans. Elect. Dev.,v ED-32,n6,1985,p1034.

3. M. Makiuchi, H. Hamaguchi, T. Kumai, M. Ito, O. Wada, "A Monolithic Four-Channel Photoreceiver Integrated on a GaAs Substrate Using Metal-Semiconductor-Metal Photodiodes and FET's", Elect. Dev. Lett., v. EDL-6, n12, 1985, pp634-635.

4. W.S. Lee, G.R. Adams, J. Mun, J. Smith, "Monolithic GaAs Photoreceiver for High-Speed Signal Processing Applications", Elect. Letts., v. 22, no. 3, 1986, pp147-148.

5. D.L. Rogers, "Monolithic Integration of a 3 GHz Detector/Preamplifier using a Refractory Gate, Ion Implanted MESFET Process", Elect. Dev. Lett., v. EDL-7, #11, p600.

6. M. Ito and O. Wada, "Low Dark Current GaAs Metal-Semiconductor-Metal (MSM) Photodiodes Using WSi Contacts", J. Quan. Elect., v. QE-22, no. 7, July 1986.

7. D. J. Webb, J.H. Magerlein, A. Callegari, J.D. Feder, T. Fryxell, H.C. Guthrie, Peter D. Hoh, J.W. Mitchell, A.T.S. Pomerene, S. Scontras, Guy D. Spiers and J.H. Greiner, "Characterisation of GaAs Self-Aligned Refractory-Gate MESFET Integrated Circuits", Jour. Appl. Phys., to be published.

Picosecond Optical Autocorrelation Experiments on Fast Photodetectors

T.F. Carruthers and J.F. Weller

Naval Research Laboratory, Washington, DC 20375, USA

1. Introduction

Typically the response time of a very high speed photodetector is measured by correlating its response to an optical pulse with that of another fast photodetector [1]. The resulting signal is ideally a convolution of the time responses of the two devices, but the signal may be distorted by dispersion in the transmission line joining the two devices [2], by reflections of the original transient from discontinuities, or by signals continually radiated by the propagating electrical transient [3].

We describe an alternate correlation technique which exploits the nonlinear response of a single photodetector to pairs of optical pulses and which is therefore to a large degree not subject to distortion from the sources described above; the technique has been used to measure response times as short as 3 ps in a high-speed photodiode [4]. Since bandwidth-preserving connections between pairs of devices are not necessary, the response times of many devices which cannot be so linked become measurable. Examples of measurements are presented for a planar photoconductor and a Schottky-barrier photodiode.

2. Experiment

In our experiments, the arrangement of which is sketched in Fig. 1, a beam of 1.5-ps optical pulses at a wavelength of 600 nm is split in two, and one beam is delayed a time τ with respect to the other. The two beams are chopped at different frequencies f_1 and f_2 and are focused to overlapping spots on the surface of the device being investigated.

A tandem arrangement of lock-in amplifiers detects the frequency components of the photodetector's output at the sum and difference of the chopping frequencies; in this respect the measurement resembles a low-frequency mixing experiment. Components at $f_1 \pm f_2$ are present only if the photodetector's response is nonlinear with respect to the incident optical energies. The magnitude of the nonlinear response will depend upon the arrival time difference τ between the two trains of pulses, ranging from a maximum mutual effect for simultaneously arriving pulses to no interaction for pulses widely separated in time.

In the simplest experimental cases such a measurement yields directly the autocorrelation function of the temporal response of the device. If the output $I(t)$ of a photodetector is assumed to depend upon the instantaneous number of photocarriers $N(t)$ generated by each pulse, the detected signal $Q(\tau)$ at $f_1 \pm f_2$ has been shown [4] to be, to lowest order,

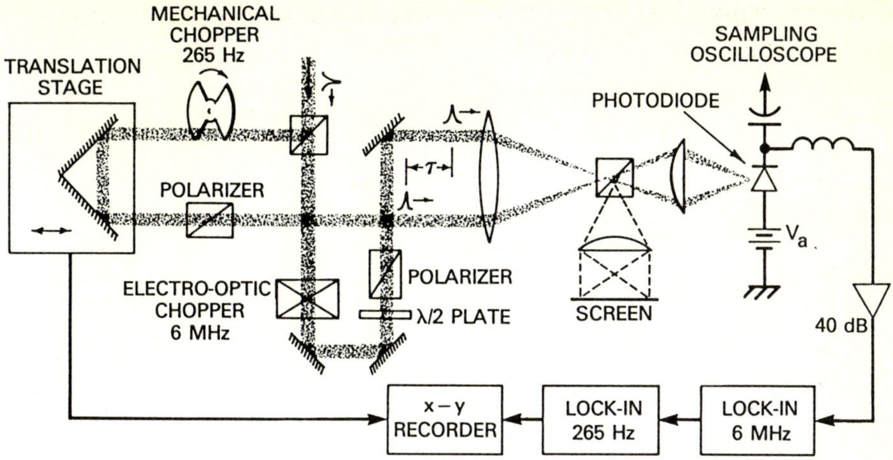

Figure 1. A sketch of the experimental arrangement

$$Q(\tau) \simeq \frac{d^2I/dN^2}{(dI/dN)^2}\int_{-\infty}^{\infty} dt\ I(t)\ I(t+\tau) \tag{1}$$

where dI/dN and d^2I/dN^2 are coefficients in the Taylor expansion of I(N). (We have assumed infinitesimally short optical pulses). The signal detected at one of the chopping frequencies--say, f_1--is equal to the sum of $Q(\tau)$ and the τ-independent signal which would be present if beam 2 were blocked.

3. Schottky-Barrier Photodiode

The response of a nominally 20 GHz GaAs Schottky-barrier photodiode [5] is presented at one of the chopping frequencies (6 MHz) in Fig. 2a and at the sum and difference of the chopping frequencies in Fig. 2b. The two measurements provide the same information, but the relative noise amplitude is much less in the frequency mixing experiment.

The peak centered about $\tau=0$ corresponds to a mutually attenuating effect of the two beams upon one another which we attribute to a decreased quantum efficiency due to lower ambient electric fields in the device when photocarriers are present [4]; its full width at half-maximum (FWHM) of 11 ps corresponds closely to the expected transit time of photocarriers through the ~1 μm thick active region of the device. At larger delay times the sign of the nonlinear signal eventually changes and the pulses begin to enhance the sensitivity of the device to one another. The large excursion across zero implies a deviation from the simple correlation model described in (1). The bias and power dependence of the sensitivity enhancement, especially at forward bias, show that the phenomenon is most likely due to the storage of holes near the interface between the n^- epitaxial active region and the n^+ substrate [6].

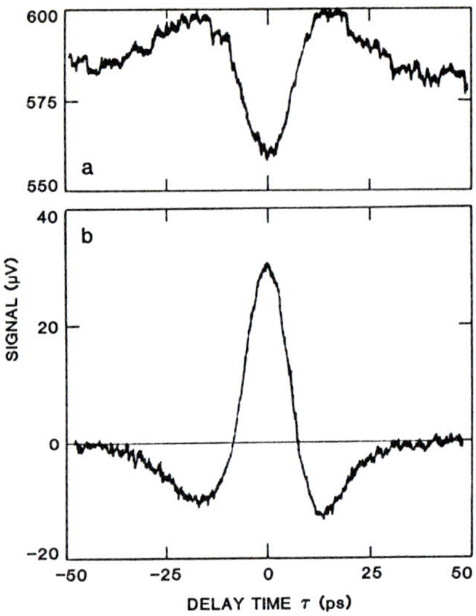

Figure 2. The signal detected at (a) one of the chopping frequencies and at (b) the sum and difference of the chopping frequencies by a 20 GHz GaAs photodiode. Incident optical energy was 1 pJ per pulse

4. Photoconductor

The responses of planar photoconductors consisting of gaps in microstrip transmission lines deposited upon semiconductor substrates have been extensively investigated. Ion bombardment damage can reduce the free carrier lifetime to less than a picosecond, as careful correlation measurements have demonstrated [3]. When the two gaps of a correlation circuit can be radiatively coupled, however, reflections of and radiation by the original transient mask and broaden the correlation signal.

Figure 3 shows the results of a nonlinear correlation measurement upon an implantation-damaged InP photodetector known to have a subpicosecond free carrier lifetime [3]. An inset shows the photocurrent as measured with a 25-ps sampling oscilloscope; a prominent reflection off the sample boundary is visible 130 ps after the initial transient. The nonlinear correlation signal is centered about $\tau = 0$ and, although it is superimposed upon a nearly time-independent background signal, it shows no evidence of reflected transients or of a slow radiatively-coupled signal.

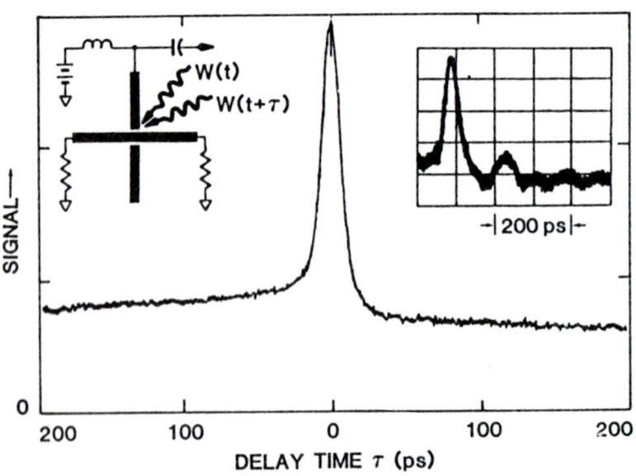

Figure 3. A nonlinear correlation measurement on a subpicosecond InP correlation circuit. The illumination scheme is illustrated in the left inset; the pulse response as monitored with a sampling oscilloscope is shown in the right

The full-width at half-maximum of 8 ps is greater than either the optical pulse duration or the semiconductor's photocarrier lifetime.

The response time is, however, very close to the expected circuit relaxation time. The photocarriers generated by one pulse cause a transient decrease in the electric field within the gap; what is likely being observed is the change in the collection efficiency of the device for the carriers produced by one pulse due to the field transient generated by the other. The lack of spurious signals due to reflections is very simply explained by observing that the transient is both generated in and detected by a single gap with an applied dc bias voltage. Signals due to the relatively small bias field changes caused by reflected or radiated transients should be very small in comparison with the primary signal. An offset signal is present in all planar photoconductive devices we have investigated. We attribute this signal to very slow carrier accumulation processes whose rates may correspond more closely to the chopping frequencies than to device response speeds. Deep traps provide one such process; a steady offset is observed to dominate the response of FETs which are known to trap photogenerated charge in interfacial regions.

5. Conclusions

The measurement technique described here makes possible response time measurements on devices which cannot be combined with other high-speed devices in a cross-correlation circuit. The reliability of the technique in any given application turns upon an understanding of the origin of the nonlinear response of the device being tested. The devices studied here are on loan to us from Hewlett-Packard and AT&T Bell Laboratories. The authors are grateful for extensive discussions with D. K. Donald and P. M. Downey.

6. References

1. D. H. Auston: In Picosecond Optoelectronic Devices, ed. by Chi H. Lee (Academic Press, Orlando, 1984), p. 73.
2. I. N. Duling III, C.-C. Chi, W. J. Gallagher, D. Grischkowsky, N. J. Halas, M. B. Ketchen, and A. W. Kleinsasser: In Ultrafast Phenomena V, Springer Ser. Chem. Phys., Vol. 46 (Springer, Berlin, 1986), p. 110.
3. J. R. Karin, P. M. Downey, and R. J. Martin: IEEE J. Quantum Electron. QE-22, 677 (1986).
4. T. F. Carruthers and J. F. Weller: Appl. Phys. Lett. 48, 460 (1986); T. F. Carruthers and J. F. Weller: In Ultrafast Phenomena V, Springer Ser. Chem. Phys., Vol. 46 (Springer, Berlin, 1986), p. 131.
5. S. Y. Wang, D. M. Bloom, and D. M. Collins: Appl. Phys. Lett. 42, 190 (1983).
6. D. L. Scharfetter: Solid-State Electron. 8, 299 (1965).

Detection of 10 μm Infrared Radiation via Intersubband Absorption in Doped GaAs Quantum Wells

B.F. Levine, K.K. Choi, C.G. Bethea, J. Walker, and R.J. Malik

AT&T Bell Laboratories, 600 Mountain Avenue,
Murray Hill, NJ 07974, USA

Abstract

A new type of long wavelength detector is demonstrated. It is based on intersubband absorption in a superlattice of doped $GaAs/Al_xGa_{1-x}As$ quantum wells. The photoexcited electron, which is promoted to the first excited state of the quantum well, rapidly tunnels out and produces a photocurrent. The measured quantum efficiency is $\eta = 6\%$ at a wavelength of $\lambda = 10.8$ μm.

For the detection of 10 μm infrared radiation, II-VI compounds particularly $Hg_xCd_{1-x}Te$ have been extensively studied and excellent detectors fabricated. However, if sensitive detectors could be made from $Al_xGa_{1-x}As$ [1,2] this would allow advantageous use of their highly developed growth and processing technologies, as compared with II-VI compounds. [1,2] Furthermore, device parameters (e.g. bandgap, operating temperature, bandwidth and speed) can be tailored in ways that are difficult to do with either II-VI's or extrinsic Si detectors. We report here a successful demonstration of a novel high-speed infrared detector based on intersubband absorption [3,4] and sequential resonant tunneling in doped $GaAs/Al_xGa_{1-x}As$ quantum well superlattices. We achieved a responsivity of 0.52 A/W at $\lambda = 10.8$ μm, an advantageous narrow bandwidth response $\Delta\lambda/\lambda \simeq 10\%$, and estimate the speed to be ≈ 30 psec. From our experiments we have determined that the mean free path of the photogenerated hot electrons through the superlattice is 2500 Å.

The superlattice consisted of 98 Å $Al_{0.24}Ga_{0.76}As$ barriers and 65 Å GaAs wells (doped 1.4×10^{18} cm^{-3}). Detectors were fabricated by etching 50 μm mesas and back illuminating them through a 45° angle polished on the substrate. The responsivity vs bias voltage is shown in Fig. 1 for $\lambda = 10.8$ μm. The solid line is theory based on absorption of the infrared light by the intersubband resonance, followed by the rapid tunneling of the photoexcited electron out of the well (shown in Fig. 2).

The hot electron then travels a mean free path L (determined to be L = 2500 Å) thereby producing a photocurrent before being recaptured by one of the wells.

In order to further understand the physics of the operation of the STAIR detector (*S*uperlattice *T*unneling and *A*bsorption by *I*ntersubband *R*esonance detector), we can express the contribution to the photocurrent, $I_p(x)$, due to an electron photogenerated at a distance x from the anode as $I_p(x) = (e/h\nu)P(x)\alpha p L(x)$, where $P(x) = Pe^{-\alpha x}$ is the optical power at x, α is the

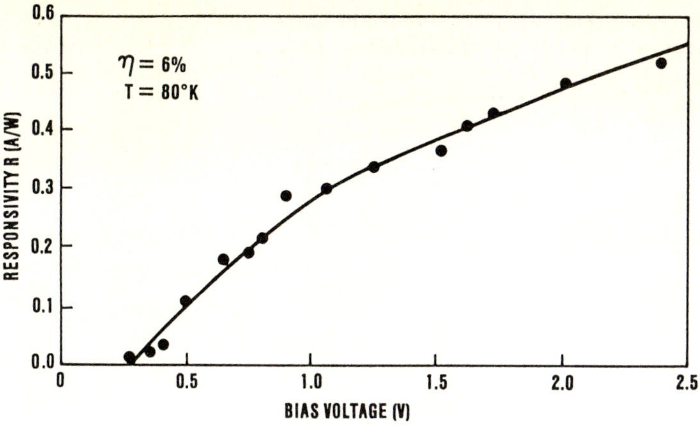

Fig. 1 Responsivity vs. bias voltage. Curve is theory, circles experiment.

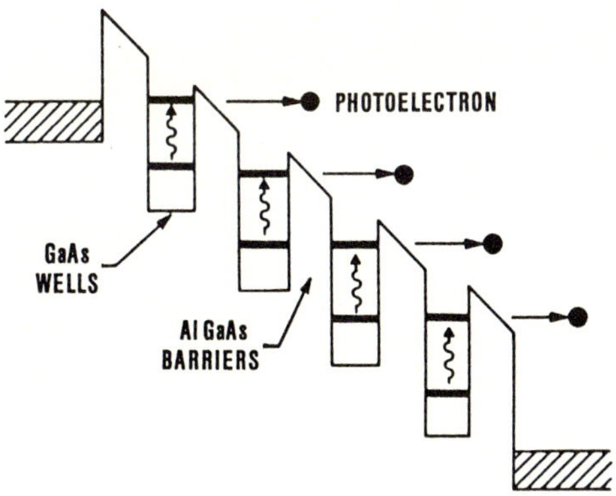

Fig. 2 Energy band diagram of superlattice under bias, showing infrared absorption (wavy arrows) followed by photoexcited electron tunneling.

absorption constant, p is the probability that an electron photoexcited to E_2 will tunnel out of the well, and $L(x) = L(1-e^{-x/L})$ is the transport length where L is the hot electron mean free path. The net photocurrent I_p and thus the responsivity $R = I_p/P$ are obtained by averaging $I_p(x)$ over the length l of the active high field region. This active length l can be directly related to the applied bias voltage V. Under bias, the ground state E_1 of one quantum well and the excited state E_2 of the neighboring well line up (shown in Fig. 2) as the bias is increased, the length l of these resonantly aligned wells increases. From a detailed analysis of this process R can be calculated as a function of V as shown in Fig. 1. The agreement with the data (circles) is excellent, allowing the determination of the mean free path L = 2500 Å. The value R = 0.52 A/W at 2.6V corresponds to a quantum efficiency of $\eta = 6\%$. As expected the detector response was found to be highly polarized.

In conclusion, we have demonstrated the first $10\mu m$ edtector using $GaAs/Al_xGa_{1-x}As$ quantum wells. These STAIR detectors, which are compatible with GaAs IC technology, show promise for long wavelength applications.

1. J. S. Smith, L. C. Chiu, S. Margalit, A. Yariv and A. Y. Cho, J. Vac. Sci. Tech. *B1,* 376 (1983).
2. D. D. Coon and R. P. G. Karunasiri, Appl. Phys. Lett. *45,* 649 (1984).
3. L. C. West and S. J. Eglash, Appl. Phys. Lett. *46,* 1156 (1985).
4. B. F. Levine, R. J. Malik, J. Walker, K. K. Choi, C. G. Bethea, D. A. Kleinman and J. M. Vanderberg, Appl. Phys. Lett. *50,* 273, (1987).

Part VII

Photoconductive Switching Devices

Carrier Lifetime Dependence on Ion Implantation in Silicon

F.E. Doany, D. Grischkowsky, and C.-C. Chi

IBM T.J. Watson Research Center, P.O. Box 218, Yorktown Heights, NY 10598, USA

The generation of subpicosecond electrical pulses has recently been demonstrated using fast photoconductive switches driven by short laser pulses [1]. In these measurements electrical pulses on the order of 0.6 psec were obtained by shorting a charged transmission line fabricated on an ion-implanted silicon-on-sapphire (SOS) wafer. The major factors determining the shape and duration of these electrical pulses were the laser pulsewidths, the circuit characteristics of the photoconductive gaps and the transmission line, and the carrier lifetime of the semiconductor [2]. Since laser pulses shorter than 100 fsec are routinely obtained from colliding-pulse-modelocked dye lasers, the laser pulsewidth can be made negligible compared to the generated electrical pulse. In addition, the limiting factors on the time response due to the circuit reactance may be eliminated if the capacitance of the generation site is reduced to negligible amounts. Under these conditions the duration of the electrical pulse would be mainly determined by the carrier lifetime in the ion-implanted SOS. However, no systematic measurements of this lifetime are available.

In this paper we present measurements of the carrier lifetimes of a systematic series of SOS samples with increasing amounts of ion implantation. The samples consist of 0.5 μm silicon on sapphire wafers that were implanted with O^+ ions of doses ranging from 1 x $10^{12}/cm^2$ to 7.5 x $10^{15}/cm^2$ at 200 keV (and 100 keV in some cases). The carrier lifetimes are measured by time-resolved reflectivity and transmission measurements following excitation with femtosecond optical pulses.

The laser source is a compensated colliding-pulse-modelocked dye laser producing 70 fsec pulses at 625 nm (2.0 eV) at a repetition rate of 100 MHz. These pulses are split into two beams to provide the pump and probe pulses. The average power of the pump beam was 0.1 - 2 mW (1 - 20 pJ/pulse) and is focused to a diameter of about 10 μm. The probe beam is attenuated to less than 5% of the pump beam power and is focused to a diameter of about 5 μm. The polarization of the probe beam is rotated by 90° to suppress interferences between the two beams on the sample surface. The pump is incident on the sample at an angle of 30° relative to the normal while the probe is incident on the opposite side of the normal at an angle of 60°. The pump beam is chopped at 1 kHz. Before reaching the sample, the probe beam is additionally split into two beams: one is reflected off the sample onto a photodiode detector and the second, incident directly onto a matched photodiode, is used as a reference. The reflection signal is derived as the difference of these two photocurrents and is detected, using a lock-in amplifier, as a function of time delay between the pump and probe pulses. The sensitivity of our apparatus to changes in the probe beam reflectance is approximately 1 part in 10^6.

Without pumping, the reflectivity at normal incidence of the samples used in this study at 625 nm is relatively constant at a value of 0.2 and increases only slightly at the highest implantation dose. The absorption, however, increases monotonically from 0.3 at a dose $1 \times 10^{12}/cm^2$ to 0.65 at the highest dose of $7.5 \times 10^{15}/cm^2$. For the unimplanted SOS sample the reflectivity and absorption are 0.2 and 0.25 respectively.

Excitation of the silicon with 625 nm optical pulses generates an electron-hole plasma. The associated index change results in a reflectivity/absorption change. Because the pump intensities used in this study are 4 to 5 orders of magnitude below the melting threshold of silicon (ca. $0.1\ J/cm^2$), lattice heating effects are negligible. The low carrier densities in our experiment (10^{16} - $10^{17}/cm^3$) also minimize possible effects due to carrier-carrier processes (e.g. Auger recombination) and saturation of occupied traps. Thus, the reflectivity and transmission measurements are representative of carrier recombination rates.

The change in reflectivity following photoexcitation of two representative samples is shown in Fig. 1. Figure 1a was obtained for SOS sample ion-implanted at a dose of $2 \times 10^{15}/cm^2$. The smooth curve also shown in Fig. 1a is a convolution of our 0.14 psec pump/probe instrument response function with a 0.65 psec exponential decay. Figure 1b shows similar data obtained at a dose of $1 \times 10^{13}/cm^2$ which is closely simulated by an exponential decay of 4.8 psec. The peak changes in reflectivity ($\Delta R/R$) in the data shown in Fig. 1 is approximately 10^{-5}.

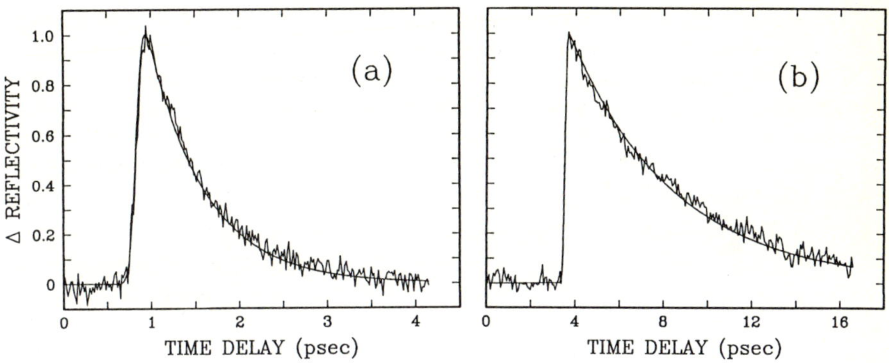

Figure 1. Measured change in reflectivity for SOS samples ion implanted at doses of (a) $2 \times 10^{15}/cm^2$ and (b) $1 \times 10^{13}/cm^2$. The smooth curves are fits obtained using a 0.14 psec instrument response convoluted with (a) 0.65 psec and (b) 4.8 psec exponential decays.

Femtosecond resolved reflectivity data were obtained for a series of SOS samples with a range of implantation doses covering 4 orders of magnitude and also for unimplanted SOS. In attempting to interpret these data, each curve was fit to a single exponential decay time, which varied from >300psec for the unimplanted SOS wafer down to about 600 fsec for the highly implanted samples. These results are summarized in Fig. 2. There are two major regions of interest: the variation of the lifetime at low ion-implantation doses, $< 1 \times 10^{14}/cm^2$, and the lack of dependence at the higher doses, $> 3 \times 10^{14}/cm^2$.

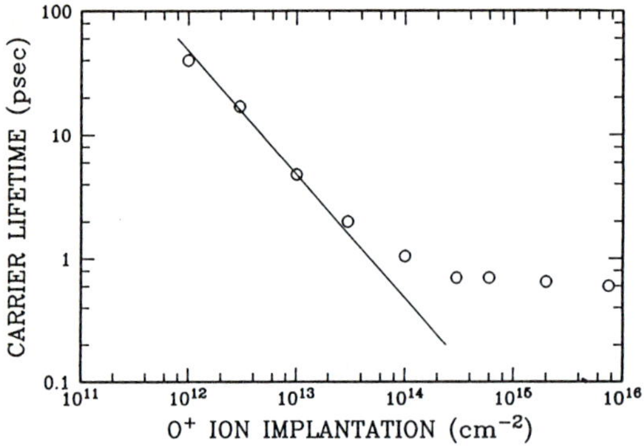

Figure 2. Carrier lifetime vs. ion implantation dose. Data points are derived from individual reflectivity curves at each dose.

At low implant levels, the lifetime decreases linearly with the ion-implantation dose. This is clearly demonstrated by the fit of the low dose data points in Fig. 2 to a slope of unity. The linear dependence in this region is consistent with recombination rates proportional to the density of traps introduced by the ion-implantation. In this case, the trap density and therefore the carrier lifetime will vary linearly with ion-implantation level.

The variation of lifetime with implantation dosage saturates at a level 3 x $10^{14}/cm^2$. Above this dose, the lifetime reaches a limit of 600-700 fsec. This lifetime does not vary even as the implantation dose is increased by more than a factor of 20. These lifetimes are clearly resolvable by our experimental resolution as evidenced by the $<$150 fsec risetimes of the data presented in Figure 1. The approximate 600 fsec limit is therefore due to the response of the semiconductor itself. Possible interpretations of this lack of dependence on ion-implantation level include saturation of effective trap densities and hot carrier relaxation.

One possible explanation for the lack of carrier lifetime dependence on ion implantation dose is hot carrier relaxation since the photoexcitation energy (2.0 eV) is significantly larger than the band gap (1.15 eV for pure Si). The 600 fsec lifetime may represent the thermalization time of the photoexcited carriers. However, no experimental evidence such as non-exponential behavior in the recovery of the reflectivity signal was observed.

The most probable process limiting the carrier lifetime is amorphization of the crystalline silicon. Above a critical dose of 3 x $10^{14}/cm^2$ the density of effective traps introduced by implantation saturates and the sample is essentially amorphous. This interpretation is supported by third harmonic generation [3] and electron spin resonance [4] studies on ion implanted SOS which exhibited similar saturation behavior occurring at critical implantation doses of 10^{14} - $10^{15}/cm^2$.

It is informative to compare the above results with reflectivity measurements on amorphous silicon films deposited on sapphire wafers. Using samples prepared by sputtering under low Argon pressure and by e-beam evaporation under vacuum, we have obtained carrier lifetimes of ~600 fsec and ~400 fsec, respectively. Figure 3 shows the reflectivity signal from a 0.4 μm amorphous silicon on sapphire sample

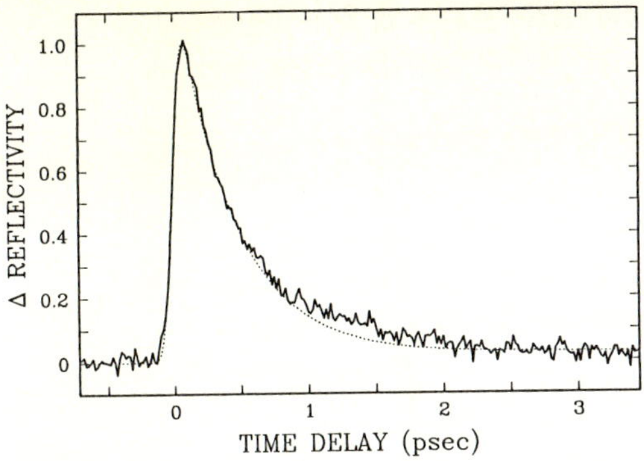

Figure 3. Measured change in reflectivity for an amorphous silicon sample. Smooth curve is a convolution of the experimental response function with a 0.42 psec exponential decay.

prepared by evaporation under vacuum. Previous measurements of the carrier lifetime of amorphous silicon varied from 0.8 psec to several 100 psec [2,5,6] depending on the method of fabrication. We believe the variation in lifetime is due to different contaminants (particularly hydrogen) introduced by each method of preparation.

This research was partially supported by the U. S. Office of Naval Research.

1. M.B.Ketchen, D.Grischkowsky, T.C.Chen, C-C.Chi, I.N.Duling, N.J.Halas, J-M.Halbout, J.A.Kash, and G.P.Li, Appl. Phys. Lett. **48**, 24 (1986).
2. D.H.Auston, in *Picosecond Optoelectronic Devices*, edited by C.H.Lee (Academic, London, 1984), pp. 73-116.
3. C.C.Wang, J.Bomback, W.T.Donlon, C.R.Hou and J.V.James, Phys. Rev. Lett. **57**, 1647 (1986).
4. J.R.Dennis and E.B.Hale, J. Appl. Phys. **49**, 1119 (1978).
5. Z.Vardeny and J.Tauc, *Semiconductors Probed by Ultrafast Laser Spectroscopy*, vol. II, edited by R.R.Alfano (Academic, New York, 1984) pp. 23-53.
6. J.Kuhl, E.O.Gobel, Th. Pfeiffer and A.Jonietz, Appl. Phys. **A34**, 105 (1984).

An Investigation of Photoconductive Picosecond Microstripline Switches on Self-Implanted Silicon on Sapphire (SOS)

P. Polak-Dingels[1], *G. Burdge*[1], *Chi H. Lee*[2], *A.C. Seabaugh*[3*], *R.T. Brundage*[3**], *M.I. Bell*[3], *and J. Albers*[3]

[1]Laboratory for Physical Sciences, College Park, MD 20740, USA
[2]Department of Electrical Engineering, University of Maryland, College Park, MD 20742, USA
[3]National Bureau of Standards, Gaithersburg, MD 20899, USA

Optoelectronic switches fabricated on amorphous and implantation-damaged semiconductor thin films have proven to be among the fastest of the picosecond photoconductors, with switching times less than 10 ps [1-3]. Ion implantation offers a systematic way to vary the disorder in materials from partially damaged to amorphous. However, foreign ions may alter the chemical composition as well as the crystallinity of the silicon, so we have used silicon implantation to study the effect of damage and amorphization on the photoresponse of silicon on sapphire (SOS). Raman scattering and resistivity measurements were made to characterize the material. Picosecond laser excitation was used to study the response of photoconductive switches as a function of bias voltage, gap size, and implantation dose. Comparison of resistivity, Raman, and photoconductive measurements indicates that optimum response and resistivity are achieved at implant doses below that required to fully amorphize the silicon.

The SOS starting material consisted of a 0.6-μm layer of 50 Ω-cm Si on a 0.457-mm (18-mil) thick sapphire substrate. Implantation with 270 keV Si ions was performed at room temperature, after deposition of the metal photoconductive devices, with the beam current kept below 3 μA to minimize sample heating. The implantation doses were 10^{12}, 10^{13}, 10^{14}, 10^{15}, and 2×10^{15} cm^{-2}. Possible contamination by N_2^+ or CO^+ was checked by implanting one sample (10^{13} cm^{-2}) with ^{29}Si.

A Monte Carlo calculation of the energy deposited via displacement damage by the implant was used in choosing the implant energy and doses [4]. The critical energy density for amorphization (E_c) increases with temperature, rising from 8.5×10^{23} eV-cm^{-3} at 300 K to 6×10^{24} eV-cm^{-3} at 380 K [5]. The latter temperature is the maximum expected for the implant conditions used in our studies, as calculated from a steady-state heat flow analysis [6]. Our Monte Carlo calculation indicates E_c is reached for an implant of 1.8×10^{14} cm^{-2} at 300 K and 1.3×10^{15} cm^{-2} at 380 K. The energy deposited as a function of depth for three values of the doses used is shown in Fig. 1, as well as the E_c values mentioned above.

The resistivity of samples was measured by the van der Pauw method, with the results shown in Fig. 2. The resistivity peaks at a dose of 10^{14} cm^{-2} and then decreases with increasing dose. The decrease in resistivity at high implant dose is similar to that reported for amorphous Si with an increase in dangling bond density [7]. The resistivity of the ^{29}Si sample is equal to that measured for the ^{28}Si within the experimental error. We attempted to measure the Hall mobility of the implanted SOS material, but were unable to observe a Hall voltage greater than 5 mV for magnetic fields as large as 1 T and currents of 1 nA. This result together with the measured resistivity puts an upper limit on the mobility of 3 cm^{2}V$^{-1}\cdot$s^{-1}, which is a typical value for amorphous silicon.

* Present address Texas Instruments, Box 655936, M.S. 134, Dallas, TX 75265.
** NBS-NRC Postdoctoral Research Associate

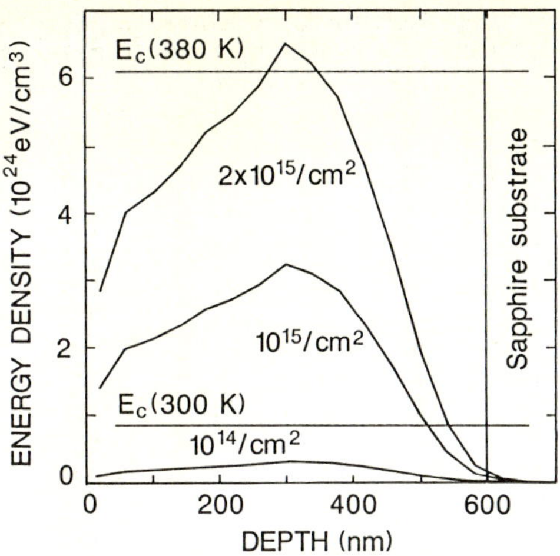

Fig. 1. Calculated energy deposited as displacement damage by 270-keV Si ions , compared with the critical energy density for amorphization, E_c.

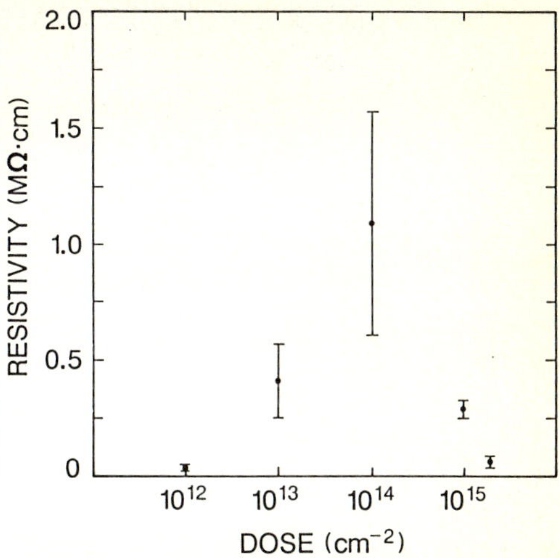

Fig. 2. Resistivity versus implant dose for 270-keV Si into 50 Ω-cm SOS. Error bars are ±1 standard deviation.

In order to determine the extent and distribution of the damage induced at each implant dose, Raman spectra were taken in a backscattering geometry using argon ion and cw dye lasers. The crystalline content of ion implanted silicon [8,9] and partially annealed amorphous silicon [10], can be determined from the ratio of the areas of the sharp crystalline optic phonon peak to the broad feature caused by amorphous silicon. Initial experiments at 514.5 nm indicated no contribution to the Raman spectra from amorphous material up to doses of 10^{15} cm^{-2}. Measurements taken with a cw dye laser operating at 590 nm showed some broadening to low frequency in the 10^{14} and 10^{15} cm^{-2} samples, as found in heavily damaged silicon [9].

The 2×10^{15} cm^{-2} sample showed a large signal from amorphous material with excitation at 514.5 nm, but measurements at 457.9 nm indicated the presence of a crystalline layer on the surface of the silicon film. This result is predicted by the Monte Carlo calculation, as less energy is deposited at the surface of the silicon film than in the bulk. Between 600 and 450 nm, the absorption coefficient of amorphous silicon is more than one order of magnitude larger than that of crystalline silicon [11]. Raman spectra taken at different pump wavelengths can be used to determine the distribution of the amorphous material as a function of depth into the sample [8]. Spectra were taken using the dye laser and several argon ion laser lines, and the measured crystalline-to-amorphous signal ratio was used to calculate the thickness of the crystalline film. Correction was made for the wavelength dependence of the ratio of scattering cross-sections for crystalline and amorphous silicon [12]. The data are best described by a crystalline layer 50 nm thick on top of the amorphous film, as shown in Fig. 3.

The Monte Carlo calculation (Fig. 1) predicts that the implant does not fully penetrate the silicon film. Raman spectra characteristic of crystalline silicon were observed from the sapphire/silicon interface of the 2×10^{15} cm^{-2} sample when illuminated through the sapphire substrate. The thickness of the crystalline layer on the back could not be determined from the wavelength dependence of the Raman spectrum because of negligible contributions to the spectra from amorphous material, even using the dye laser as the pump source. This is surprising, as the calculation indicates this layer is no more than 0.2 μm thick, and the relatively long absorption length in crystalline silicon at 590 nm (1.6 μm) should allow penetration to the amorphous layer. Our result may be

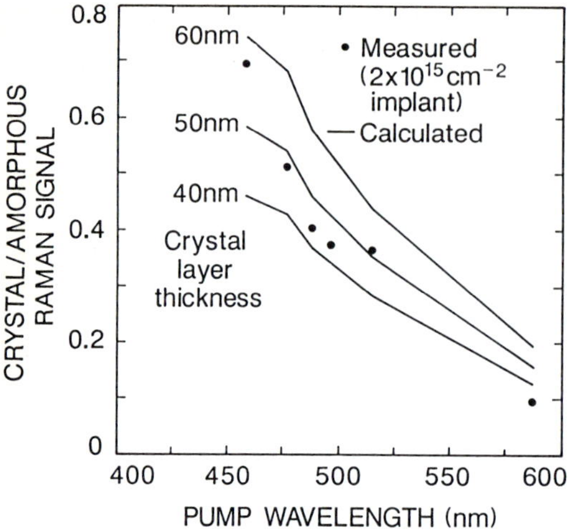

Fig. 3. Ratio of crystal-to-amorphous Raman signal for a crystalline layer of varying thickness on an amorphous film.

caused by an enhancement of the Raman signal from the crystalline layer due to the presence of the sapphire substrate.

Raman scattering has also been used to measure the strain induced in SOS materials by implantation and differential thermal expansion [13,14], as revealed by the shift of the frequency of the optic phonon peak. In all of the samples studied here, the crystalline optic phonon frequency was measured to be 522 ± 1 cm^{-1}, indicating the strain in these samples is less than 4×10^8 N-m^{-2} [13].

The device design used in the picosecond photoconductive sampling measurements was similar to that introduced by Auston [1], consisting of two photoconductive switches. One switch, connected to the bias voltage, functioned as a pulse generator, and the second switch, connected to the lock-in amplifier, acted as a sampling gate. Planar 50-Ω microstriplines, 450 μm wide, were deposited on the SOS wafers. The metallization for the striplines was Ti/Au, formed by an oxide-enhanced lift-off process. Gold was then deposited on the back surface of the wafer and the metallization sintered at 300°C for 30 minutes. Three different device sets were fabricated with gap lengths of 2, 5, and 10 μm. After implantation, the devices were mounted in brass holders and attached to K connectors using silver epoxy.

Picosecond photoconductivity measurements were made by the electronic auto-correlation technique developed by Auston [1]. The laser system consisted of a dye laser synchronously pumped by the frequency doubled output of a cw mode-locked Nd:YAG laser. The pulse train had a repetition rate of 100 MHz, a pulse duration of 3 ps, and an average power of 120 mW. The output of the dye laser was split into two beams, one illuminated the sampling gate while the other passed through a variable delay line before illuminating the pulse generator. A stepper motor varied the length of the delay line in steps of 0.4 μm over a total travel distance of 5 cm. The optical pulse train illuminating the pulse generator was chopped at 74 Hz, and the resultant signal from the sampling gate was fed to the lock-in amplifier.

The results of the picosecond auto-correlation measurements are shown in Fig. 4. There was no difference between the response of the ^{28}Si- and ^{29}Si-damaged samples; both samples with implant dose of 10^{13} cm^{-2} have carrier lifetimes of about 12 ps, as measured from the width of the auto-correlation response. At a higher dose level of 10^{14} cm^{-2}, the carrier lifetime decreases to 9 ps. Samples implanted with higher doses show no significant decrease in carrier lifetime. The estimated gap capacitance of our devices is 0.04 pF [15], which leads to a device-limited HWHM of the auto-correlation

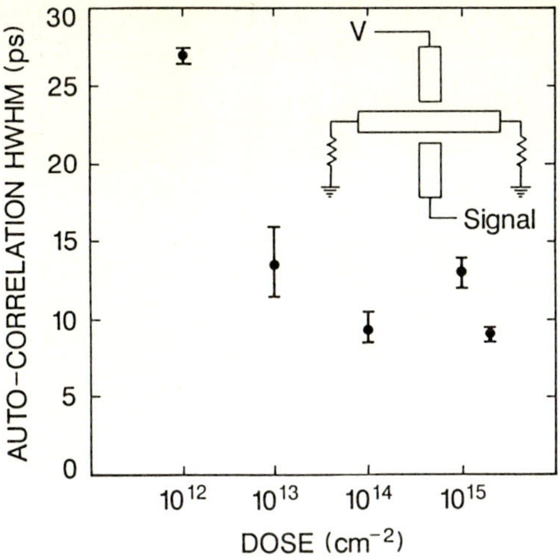

Fig. 4. Time response of photoconductive switches as function of implant dose. Insert shows the circuit for our auto-correlation measurements.

of 3.5 ps [1], compared to 9 ps for the fastest device studied. Although this estimate is a lower bound, the capacitance would have to be more than a factor of 1.7 greater for our measurements to be capacitance limited. There was no change in pulse width with bias voltage or gap width. We conclude that the primary influence on the pulse width is the excess carrier lifetime, with no measurable contribution from the transit time.

The mobility can be estimated from the photoconductive measurements using the output voltage for a given device geometry, bias voltage, and total number of carriers generated by the laser [16]. Our measurements indicate a mobility of 4 $cm^2V^{-1}{\cdot}s^{-1}$ for a dose of 10^{12} cm^{-2}, decreasing to less than 0.5 $cm^2V^{-1}{\cdot}s^{-1}$ at $2{\times}10^{15}$ cm^{-2}, in reasonable agreement with the upper limit set by the dc Hall effect measurement mentioned earlier.

In summary, we have measured picosecond photoconductivity on self-implanted SOS over an implant dose range of 10^{12} to 2×10^{15} cm^{-2}, which produces material ranging from damaged to almost entirely amorphous. The photoconductive response time reaches a minimum for an implant dose of 10^{14} cm^{-2}, where the resistivity reaches a maximum. Raman measurements indicate that the silicon film has not been fully amorphized at this dose, which is consistent with Monte Carlo calculations of the energy deposited in the silicon. The presence of thin crystalline layers in the 2×10^{15} cm^{-1} sample does not affect the time response of the photoconductive devices, both because the layers are thin compared to the carrier diffusion length in the crystalline material, and because the majority of the carriers is produced in the amorphous material. For devices with gaps as small as 2 μm, the response time is still limited by the carrier lifetime and not the transit time across the gap. Our results indicate that superior performance is achieved with heavily damaged, as opposed to fully amorphous, material, as the latter shows no additional decrease in carrier lifetime, and has a reduced on/off ratio arising from both decreased dark resistance and reduced mobility.

1. D.H. Auston: in *Picosecond Optoelectronic Devices*, edited by C.H. Lee (Academic Press, 1984); *IEEE J. Quantum Electron.* **QE-19**, 639(1983).
2. P.R. Smith, D.H. Auston, A.M. Johnson, and W.M. Augustyniak: *Appl. Phys. Lett.* **38**, 47(1981).
3. R.B. Hammond, N.G. Paulter, and R.S. Wagner: *Appl. Phys. Lett.* **45**, 289(1984).
4. J. Albers: *IEEE Trans. on Electron Devices*, **ED-32** 1930(1985).
5. J.R. Dennis and E.B. Hale: *J. Appl. Phys.* **49**, 1119(1978).
6. J. Albers: *NBS Special Publication* 400-76 (U.S. Govt. Printing Office, Washington, D.C., 1984).

7. A.M. Johnson, D.H. Auston, P.R. Smith, J.C. Bean, J.P. Harbison, and D. Kaplan: in *Picosecond Phenomena II*, edited by R. Hochstrasser, W. Kaiser, and C.V. Shank (Springer-Verlag, Berlin, 1980).
8. J.F. Morhange, R. Beserman, and M. Balkanski: *Phys. Stat. Sol.* **23**, 383(1974).
9. K.K Tion, P.M. Amirthraraj, F.H. Pollak, and D.E. Aspnes: *Appl. Phys. Lett.* **44**,
10. J. Gonzales-Hernandez and R. Tsu: *Appl. Phys. Lett.* **42**, 90(1983).
11. *Handbook of Optical Constants of Solids*, edited by E. D. Palik (Academic Press, 1985).
12. D. Bermejo and M. Cardona: J. Non. Cryst. Solids **32**, 405 (1979).
13. K. Yamazaki, M. Yamada, K. Yamamoto, and K. Abe: *Jap. J. Appl. Phys.* **23**, 681(1984).
14. Y. Ohmura, T. Inoue, and T. Yoshi: *J. Appl. Phys.* **54**, 6779(1983).
15. M. Maeda: *IEEE Trans. Microwave Theory Tech.* **MMT-20**, 390(1972).
16. C.S. Chang, V.K. Mathur, M.J. Rhee, and Chi H. Lee, *Appl. Phys. Lett.* **41**, 392(1982).

Picosecond Optoelectronic Switches Using Composite Electronic Materials

E.A. Chauchard[1], *Chi H. Lee*[1], *V. Diadiuk*[2], *and G.W. Turner*[2]

[1]Department of Electrical Engineering, University of Maryland, College Park, Maryland 20742, USA
[2]MIT-Lincoln Laboratory, Lexington, MA 02173, USA

Since the discovery of picosecond photoconductivity, optoelectronic switches made of a variety of materials have been studied [1]. Many new techniques depend on the availability of ultrafast (picosecond) devices. The two types of switches using composite electronic materials studied in this work are Fe doped InGaAs grown on InP substrates and GaAs on silicon-on-sapphire (SOS) substrates. The semi-insulating character of the Fe:InGaAs epilayer makes it a very attractive material for fabrication of integrated devices such as PIN photodetectors and FETs [2]. This material also exhibits a very high mobility (10000 cm^2/Vs) and a light sensitivity extending up to 1.65 μm. Interest in GaAs on SOS is due to its potential for monolithic integration of GaAs and silicon devices [3]. In addition, SOS substrates provide better isolation and are transparent. The silicon-GaAs interface exhibits a high density of defects, allowing faster devices to be fabricated.

Studying the speed of a device often reveals several phenomena which can appear as distinct time constants in the observed time waveform. The phenomena contributing to the device speed can be classified into two main categories: carrier recombination times (bulk recombination, surface recombination, Auger recombination, etc.) and effects related to the applied electric field (carrier sweep-out). In addition, contact fabrication can play a role in the device speed because of the formation of a Schottky barrier at the metal-semiconductor interface.

This paper presents a study of the effect of the applied electric field on the speed of different devices. We point out the distinction between the true carrier recombination time, which is independent of the field, the carrier sweep-out effect due to the field-induced movement of the carriers and the effects due to Schottky barriers at the electrodes. This last effect was investigated by comparing the responses of switches made of the same material but with different electrode fabrication techniques. Two types of contacts were used: non-alloyed gold and alloyed NiGeAu electrodes. Table 1 summarizes the characteristics of the switches. Two different gap geometries-straight gap and interdigitated- were also compared. An estimate of the effective field in the material can be obtained from the measured switch response time.

The light source used in this work was a synchronously pumped CW mode-locked dye laser producing 3 ps pulses at a repetition rate of 100 MHz. The observations were made with a Tektronix sampling scope and/or using a cross-correlation technique for waveform sampling. Fig. 1 shows some examples of waveforms observed for switch #3. For Fe:InGaAs, the switch response time was studied for an applied voltage varying between 0.2 V and 10 V. Fig. 2a and b are plots of the observed decay time constants for the switches with gold and alloyed electrodes respectively. In many instances,

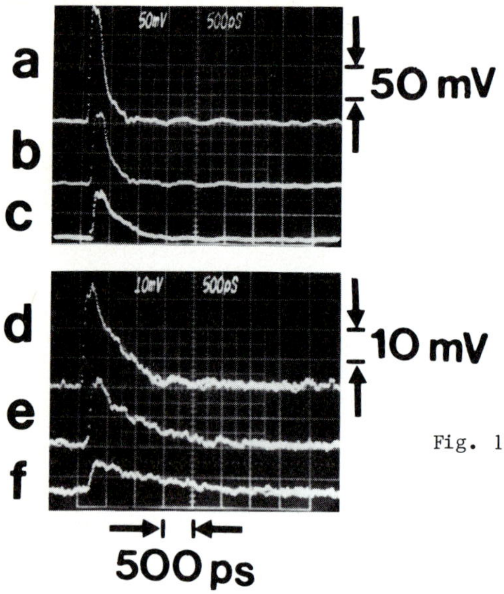

Fig. 1: Waveform observed for switch #3 biased at different voltages. a: 10 V, b: 7 V, c: 4 V, d: 2 V e: 1 V, f: 0.2 V.

Table 1. Description of devices

Switch #	Material	Gap Size (μm)	Gap Geometry	Electrode Type
1	Fe:InGaAs	20	straight	Au
2	Fe:InGaAs	10	straight	Au
3	Fe:InGaAs	3	straight	Au
4	Fe:InGaAs	10	straight	NiGeAu
5	Fe:InGaAs	6	fingers	NiGeAu
6	Fe:InGaAs	5	straight	NiGeAu
7	GaAs on SOS	4	straight	Au

two components of the time constant could be resolved on the oscilloscope trace. The slower component of the decay is represented on Fig. 2, a and b, by a solid line, while the slower component is represented by a dashed line. It can be seen that the slower component of the switches with gold electrodes varied drastically with applied bias voltage. This effect was not observed with any of the alloyed electrode switches. We interpret this fact as follows. The non-alloyed electrodes create a potential barrier at the metal-semiconductor interface where the photogenerated carriers can be trapped. Increasing the electric field lowers the barrier and allows the carriers to flow, increasing the speed of the device. For the switches with alloyed electrodes, the barrier is much lower so there is little effect of the applied field. In order to obtain a fast device even at low bias voltage, alloyed electrodes are required. The faster component of the decay probably reflects the actual carrier recombination time of the material, although this cannot be assured for the following reason. When a voltage is applied across the gap, the electric field sweeps out the carriers, increasing the speed of the device [4]. When determined by sweep-out, the device response time is expected to vary linearly with the applied field. However, the results of Fig. 2 show that the decay time of

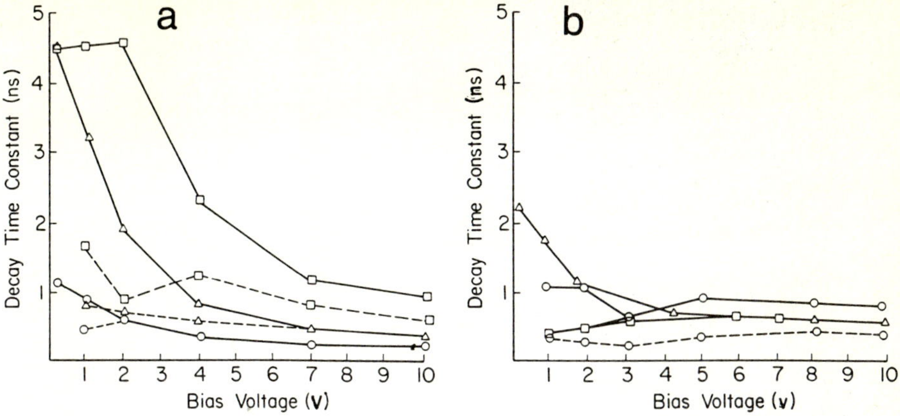

Fig. 2: Decay time constants versus bias voltage for Fe:InGaAs switches with: (a) gold contacts, (b) NiGeAu alloyed contacts. Points joined with solid or dashed lines represent respectively the slowest and fastest components of the decay.

(a): □ : switch #1, Δ : switch #2, 0 : switch #3.

(b): □ : switch #4, Δ : switch #5, 0 : switch #6.

these switches does not decrease further than 150 ps. This could be due to two effects. First, the carrier mobility saturates at high fields, so a saturated value of velocity should be expected. However, calculations using reported mobility data show that the electric fields present in our devices is too low to yield mobility saturation. Second, the photogenerated carriers screen the applied field, decreasing the effective electric field in the gap [5]. Assuming that the response time corresponds to the time needed by the carriers to cross the gap, a crude estimate of the effective field can be made. With 10 V bias in the 3 μm switch, we find the effective field to be 270 times smaller than the field expected from the applied voltage. Thus screening of the field by photogenerated carriers is very important at this relatively high injection level.

In order to use cross-correlation techniques, GaAs-on-SOS switches were fabricated in a cross shape with two 4 μm gaps. The results of cross-correlation measurements indicate a response time of 19 ps. No dependence on the applied voltage was observed, showing that the device response time can be attributed to carrier recombination time only. As the light intensity was increased, the device response became slower. This saturation phenomenon has been explained in [6].

In conclusion, electrode fabrication process can modify drastically the performance of a device. Gold electrodes deposited on the surface were found to increase the response time, particularly at low bias voltages. The screening of the field by the carriers can be estimated by the method presented here and was found to decrease the effective field in the switch by a factor of up to 270. When the carrier recombination time is small (tens of ps or under), the effect of applied electric field was not seen. The fastest devices studied here were GaAs-on-SOS switches; they exhibit a 19 ps response time.

Acknowledgement

The University of Maryland portion of this work was supported in part by the Laboratory For Physical Sciences and the Air Force Office of Scientific Research. The Lincoln Laboratory portion of this work was sponsored by the Department of the Air Force.

References

1. C.H. Lee (Ed.): Picosecond Optoelectronic Devices (Academic, New York 1984)
2. V. Diadiuk, S.H. Groves: Appl. Phys. Lett. 46, 157 (1985)
3. H.K. Choi: In Device Research Conference, (Amherst, MA, June 1986)
4. G.G. Shahidi, E.P. Ippen, J. Melngailis: In Picosecond Electronics and Optoelectronics, ed. by G.A. Mourou, D.M. Bloom, C.-H Lee, Springer Ser. Electrophys. Vol. 21 (Springer, Berlin, Heidelberg 1985) p. 197
5. D.H. Auston: In Picosecond Optoelectronic Devices, ed. by Chi H. Lee (Academic, New York 1984) p. 73
6. Kenneth K. Li, John R. Whinnery, Andrew Dienes, Picosecond Lasers and Applications, Proc. SPIE 322, p. 124 (1982)

High Speed, High Repetition Rate, High Voltage Photoconductive Switching

W.R. Donaldson

Laboratory for Laser Energetics, University of Rochester, 250 East River Road, Rochester, NY 14623, USA

This paper describes a method for constructing reliable high speed photoconductive switches D.H. AUSTON [1], A. ANTONETTI [2]. These switches have been operated at a 1 kHz repetition rate switching 10 kV in 100 ps in excess of 10 million shots without failure or degradation in performance. Once the reliability of these switches had been established, studies were conducted to determine the optimal pumping conditions with regards to spot size and induced carrier density.

The photoconductive switch is a very simple and robust device consisting only of a block of silicon. The primary reason for using intrinsic silicon (30 kΩ cm) is that it has a long carrier recombination time. The high thermal conductivity of silicon insures that the switch can support an on-state current of 13 kA/cm^{-2} W.C. NUNNALLY et al. [3]. The difficulty of constructing a reliable photoconductive switch is primarily associated with the external connections to the circuit. The first problem is to produce an appropriate contact pad on the silicon. This was done by a laser annealing method developed at the Laboratory for Laser Energetics. A coating of gold or aluminium was evaporated on to the surface of the silicon. This coating was then annealed by scanning a tightly focused laser beam over the surface to produce a good contact to the silicon W. DONALDSON et al. [4]. These contacts significantly reduced the dark current leakage of these switches. This is an important consideration because of the relatively low conductivity of intrinsic silicon. The dark current leakage can drive the switch into thermal runaway before it is triggered optically. But a metallic contact on the silicon does not insure a good contact to the external circuit. The external contact must be mechanically sound and it must smoothly blend into the geometry of the external circuit so as not to generate any stray inductance or electrical reflectances. The 100 ps rise time of the electrical pulse requires a very clean geometry.

The switches were fabricated in a microstrip geometry. The lines were laid out on printed circuit board. The gap in the line was bridged by the silicon switch which was pressure contacted to the microstrip line with indium solder. The insulating material directly underneath the silicon was milled out so that when the switch was submerged in silicone oil its surface was uniformly wetted. This was necessary to alleviate the surface electrical stress problem thus reducing the possibility of switch failure due to surface breakdown. The switch and the lines were 2 mm wide and the gap in the silicon was 2 mm. The switch was connected to a charge line, with a length corresponding to a pulse duration of 10 ns which was pulse charged at a 1 kHz repetition rate with a 10 kV pulse. The rise time of the pulse was 140 ns. At the peak of this pulse, the switch was triggered with an optical pulse from a 1 kHz YAG regenerative amplifier. This pulse had an energy of 300 microjoules and a 100 pulse width. This was sufficient to operate the switch at greater than 90% switching efficiency. The switch was left running in the configuration until more than 10 million shots had been recorded. There was no deterioration in switching during these tests or during any subsequent tests.

Next, the operating parameters of these switches were determined. In the lifetime tests, the switches were driven with 300 microjoules of optical energy. This was more than sufficient to drive the switch into saturation. In an application where laser power could be the limiting factor it may be necessary to do the switching with the minimum amount of optical energy. The lowest light level that could be used and still keep the switched voltage within 10% of its saturation level was 70 microjuoles. The laser pulse had a gaussian spatial profile. To get the fastest possible rise time, it was necessary to have the most uniform possible illumination of the switch. But the circular profile did not match the $2x2mm^2$ rectangular switch cross section. The best results in terms of switching were obtained with a spot size chosen to have its $1/e^2$ diameter 20% greater than the separation between the electrodes. These results were independent of voltage. The normalized switched waveform remained unchanged over the range of 100V to 10000V. This established the lower bound on the switch performance parameters.

The upper bound of the optical energy needed was established by the carrier concentration. If too many carriers were generated in the semiconductor the carrier lifetime became dominated by Auger recombination. The carrier density began to fall off very rapidly with increasing carrier density. This resulted in the switch shutting off prematurely. The normal carrier lifetime was measured by activating the switches with an optical pulse without having a voltage applied. Then the electrical conductivity was probed with a short electrical pulse at a later time. This experiment demonstrated that the normal carrier lifetime in these samples was about 70 microseconds. However, there also appeared to be a fast component associated with this decay. To study this component a 100ns charge line was attached to a photoconductive switch and various optical intensities were applied to the switch under bias. Normally, the switched waveform would be a square pulse with a duration equal to twice the electrical transit time of the charge line. A fast component of the carrier recombination time will manifest itself as a drooping of the top of this square pulse as illustrated in the first figure. This

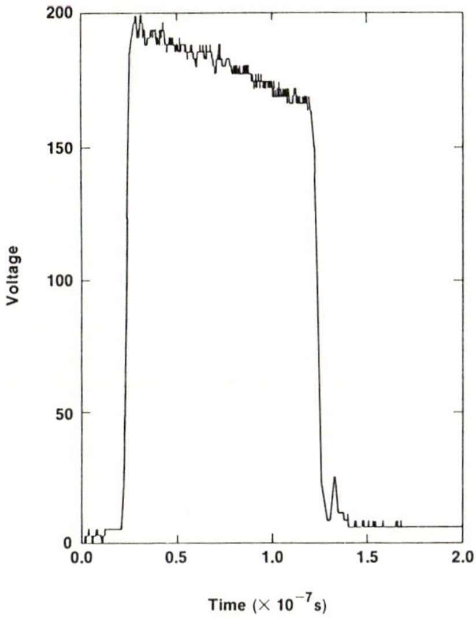

Fig. 1 *Photoconductive switch waveform*

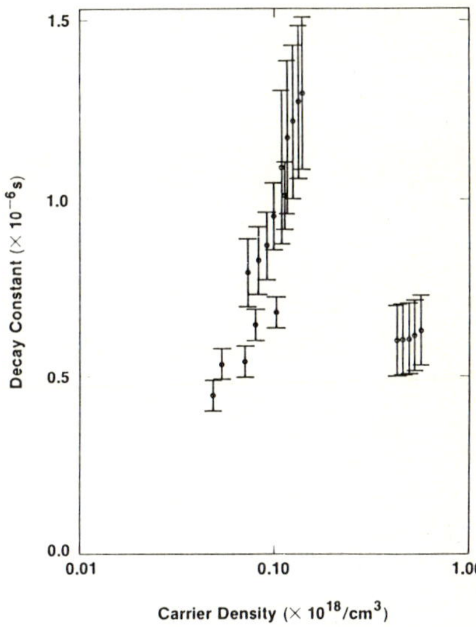

Fig. 2 *Fast component of photoconductivity decay vs. carrier density*

drooping was modeled with a decaying exponential. The lifetime associated with this exponential is plotted versus the initial concentration of carriers in Fig. 2. Below 10^{-17} cm^{-3} the apparent lifetime increases with increasing carrier concentration. This may be associated with some form of carrier sweepout. Inducing more carriers compensates for the lost carriers. As the concentration increases even further the lifetime starts to decrease again. This is probably due to Auger recombination.

Thus we have shown that very reliable silicon photoconductive switches can be constructed. These switches can be operated with a minimal optical energy of 70 mV in a 100 ps pulse. When the switches are pumped with very high optical densities, the lifetime of the conductivity first increases with increasing carrier concentration and then above 10^{18} cm^{-3} the lifetime tends to decrease. This is probably due to Auger recombination.

Acknowledgment

This work was supported by the Laser Fusion Feasibility Project at the Laboratory for Laser Energetics which has the following sponsors: Empire State Electric Energy Research Corporation, General Electric Company, New York State Energy Research and Development Authority, Ontario Hydro, the University of Rochester, and AFOSR. Such support does not imply endorsement of the content by any of the above parties.

References

1. D.H. Auston, "Picosecond Optoelectronic Switching and Gating in Silicon," Appl. Phys. Lett. 26, 101 (1975).

2. A. Antonetti, M.M. Mally, G. Mourou, and A. Orszag, "High Power Switching with Picosecond Precision," Opt. Commun. 23, 453 (1977).

3. W.C. Nunnally and R.B. Hammond, "Optoelectronic Switch for Pulsed Power," Picosecond Optoelectronic Devices, ed. Chi H. Lee (Academic Press, New York, 1984), p. 509.

4. W. Donaldson and G. Mourou, "Improved Contacts on Intrinsic Silicon for High Voltage Photoconductive Switching," Fifth IEEE Pulsed Power Conference, ed. P. Turchi and M. Rose.

Generation and Forming of Ultrashort High Voltage Pulses

H.A. Sayadian, S.T. Feng, J. Goldhar, and Chi H. Lee

Department of Electrical Engineering, University of Maryland, College Park, MD 20742, USA

High voltage pulse generation using thyrotrons and spark gaps, e.g., Blumlein stacked lines and Marx banks, is a mature field [1]. The replacement of spark gaps by photoconductive switches allows picosecond synchronization [2]. In addition, photoconductive switches can be readily incorporated into various types of pulse generating and forming structures. Two desired properties in photoconductive switches are: (i) very high resistivity so as to hold HV dc bias, and (ii) long lifetime for the photo-generated carriers so as to produce clean pulses devoid of trailing structures associated with reopening of the switches. However, these two properties seem to be mutually exclusive. We report here on a scheme that uses two materials in a composite structure that has both desired features. Our device generates pulses of over 4 kV, with controllable width, starting from dc bias of only 2.5 kV.

Schematic of the HV generation and shaping structure is shown in Fig. 1. This voltage multiplication structure uses the principle of parallel charging and series discharging of pulse-forming lines. The first stage is a parallel plane capacitor, $Z_o \simeq 1\Omega$, that is dc biased to high voltages. The capacitor is connected to the second stage through a bulk GaAs switch. The second stage is composed of three semi-rigid coaxial lines, $Z_o = 50\Omega$, that are connected in parallel to the capacitor and in series to the third stage. The third stage is comprised of a coplanar line, on bulk Si, with impedance equal to the series impedance of the coaxial lines, $Z_o = 150\Omega$. Optically triggering the GaAs switch sends HV pulses, 5 nanoseconds long, to the three parallel coaxial lines. The HV pulse is nearly doubled in the coaxial lines due to the first/second stage impedance mismatch. The series connection at the second/third stage interface further multiplies the voltage of the pulse. The width of the voltage-multiplied 5 nanosecond pulses is then decreased to the desired value by illuminating the gap between the coplanar lines by another laser pulse that is properly delayed with respect to the first laser pulse.

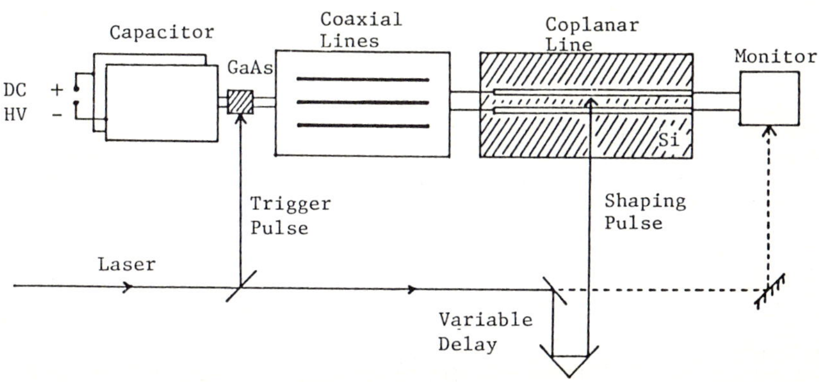

Figure 1 Schematic of HV pulse generating, multiplying, and shaping apparatus.

We note that the voltage multiplication scheme presented here has the advantage of using only one HV photoconductive switch as compared to N switches for N fold multiplication Marx banks. Also, this scheme uses half as many coaxial cables as a stacked line based scheme.

An active-passive mode-locked Nd:YAG laser was used to generate $\sim 1mJ$ of $1.06\mu m$ radiation of 100 picosecond width, using second harmonic generation in KD*P, $0.53\mu m$ radiation was generated from the $1.06\mu m$ pulses. A single laser pulse is divided into two, the first (trigger pulse) is used to trigger the GaAs switch initiating the pulse forming and voltage multiplication process. The second laser pulse (shaping pulse) switches the Si between the coplanar lines, thus shaping the final HV pulse. The lower limit on the width of the generated electric pulses is set by the width of the laser pulse used, 100 picoseconds in our case.

The performance of GaAs switches was tested under dc bias of up to 4 kV, with $1.06\mu m$ pulses for several thousand consecutive shots without any observable degradation in performance. Similar tests with $0.53\mu m$ radiation yielded low efficiencies after the first few shots due to damage to the surface. Figures 2a,b show HV pulses generated with and without the shaping laser pulse. Over 4 kV pulses were generated with a dc bias of only 2.5 kV.

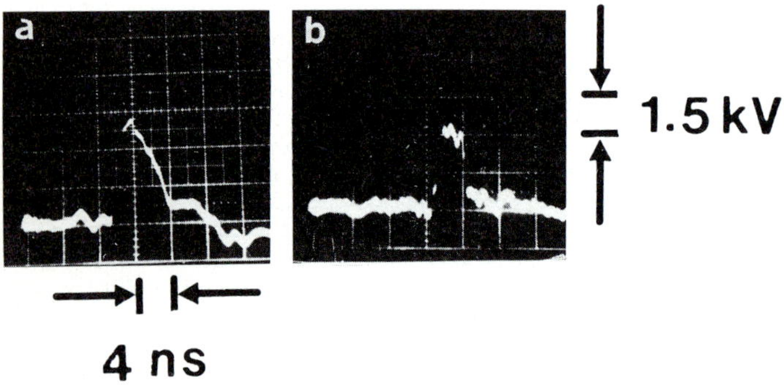

Figure 2 (a) HV pulse generated without the shaping laser pulse; (b) HV pulse generated with the shaping laser pulse. The timing of the shaping pulse controls the width of the HV pulse.

This set-up, with minor modifications, can also generate more complicated pulse shapes. Indeed, a structure based on a minor variation of the above theme generates picosecond monocycles. In Fig. 3, we show a schematic of the monocycle generator and the miniature longitudinal Pockels' cell monitor. The electro-optic material used was KD*P. A single optical pulse was split into two, one triggering the GaAs switches thus generating the picosecond electrical pulses; the other optical pulse was made to go through the pockels' cell so as to probe the electrical pulses. Triggering only switch 1 generates a rectangular electrical pulse, whereas simultaneously triggering switches 1 and 2 generates a monocycle pulse with a period equal to the width of the rectangular pulse. In Fig. 4, we show the generated pulse forms.

A microstrip-based structure was also studied. The pulse generator consisted of a 5 mm microstrip, with composite GaAs-glass dielectric with a single GaAs switch, whose effective round-trip length is $\sim$ 100 picoseconds. In Fig. 5, we show the generated electrical pulses, monitored by a miniature Pockels' cell as in above, when the structure was biased to 2 kV dc.

In summary, we have demonstrated the generation and shaping of multikilovolt pulses of variable width. This is achieved through a composite structure, utilizing GaAs and Si photoconductive switches along with voltage multiplication and pulse-forming lines.

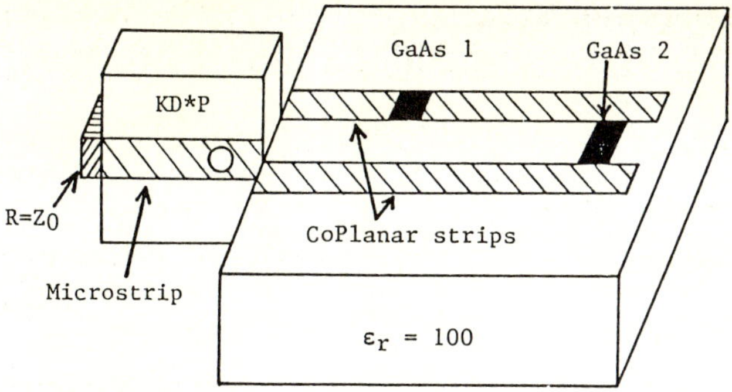

Figure 3 Schematic of the monocycle pulse generator and the miniature longitudinal Pockels' cell monitor. The character impedances of the coplanar and microstrip lines are equal.

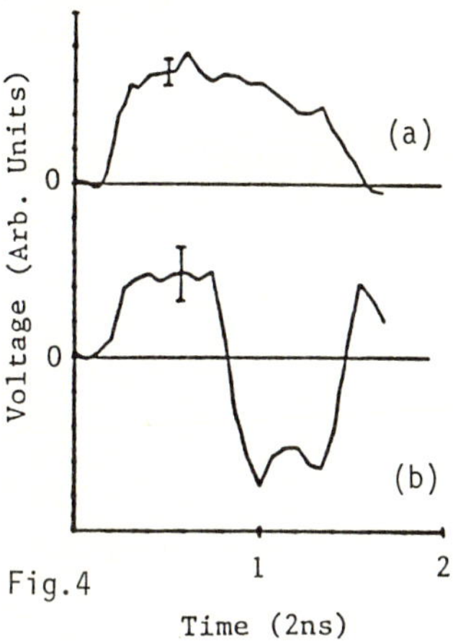

Fig.4

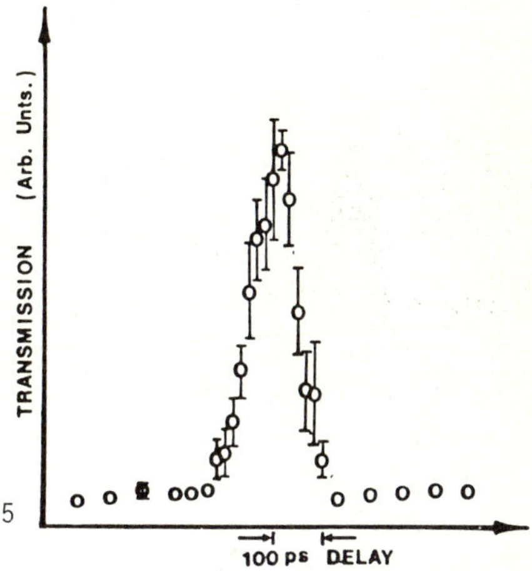

Fig.5

Figure 4 (a) HV pulse generated by triggering only GaAs switch 1; (b) HV pulse generated by triggering switches 1 and 2 simultaneously.

Figure 5 Transmission of $0.53 \mu m$ radiation through KD*P, corresponding to a 1 kV pulse of ~ 100 ps width.

This work is supported by AFOSR.

1. R. A. Fitch and V. T. S. Howell, Proc. IEE, 111, 849 (1964), and Pulse Generators, Edited by G.N. Glasoe and J.V. Lebacqz (Boston Technical Publications, 1964).
2. G. Mourou, W. H. Knox, and S. Williamson, Ch. 7 of Picosecond Opto-Electronic Devices, edited by Chi H. Lee (Academic Press, 1984).

Silicon Pulse Sharpening Diodes – Switching Kilovolts in Tens of Picoseconds *

M.D. Pocha[1], *J.D. Wiedwald*[1], *C.G. Dease*[1], *and M.M. Contreras*[2]

[1]Lawrence Livermore National Laboratory, P.O. Box 5504, L-156, Livermore, CA 94550, USA

[2]Department of Elect. and Comp. Engineering, University of New Mexico, Albuquerque, NM 87131, USA

Silicon pulse sharpening diodes were first described by Grekhov et al. [1] in the USSR. Grekhov's work described generation of multi-kilovolt pulses with rise times in the order of 200 ps. We have been studying these diodes for applications requiring fast, high voltage, low jitter, electrically triggered pulses. The fastest rise times we have observed to date are approximately 70 ps for 1-2 KV pulses. Theoretical calculations using a numerical simulation of the operation of these devices indicates that rise times shorter than 30 ps should be achieved with optimally designed structures.

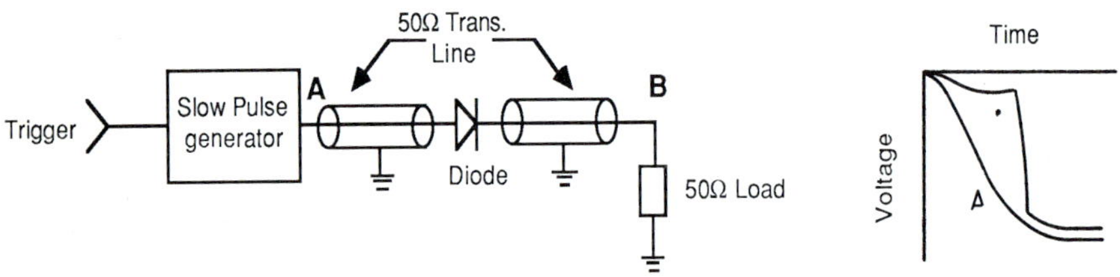

Figure 1. Schematic diagram of pulse-sharpening diode circuit and typical voltage waveforms at points A and B in the circuit.

Figure 1 shows the basic concept of pulse sharpening using these diodes. A silicon diode with reverse breakdown of greater than 1000 volts is connected in series with an electrically triggered slow pulse generator and a load. The slow pulse generator produces a pulse rising in 1-3 ns (trace A in Fig. 1), which substantially exceeds the breakdown voltage of the diode. The current through the diode follows trace B in Fig. 1. Initially a small amount of displacement current flows due to the capacitance of the diode. Once the diode's breakdown voltage is exceeded, avalanche second breakdown occurs and the current through the diode increases rapidly. Figure 2 shows one of the fastest rise times we have generated to date. It also is an example of a measurement to help verify the behavior described above. Note the characteristic initial displacement current followed by the fast transition at the leading edge of the left-hand waveform. By using passive circuit elements, the capacitive pre-cursor can be greatly diminished. Pulse-forming networks have also been used to generate short duration, flat-topped pulses.

We have used a general one-dimensional numerical analysis to model the diode's behavior and to attempt to determine the physical mechanisms involved. The numerical model includes a solution of Poisson's equation and the transport equations including a generation term using the standard avalanche formulation [2]. Our computer code allows

* This work performed under the auspices of the U.S. Department of Energy by Lawrence Livermore National Laboratory under contract No. W-7405-Eng-48.

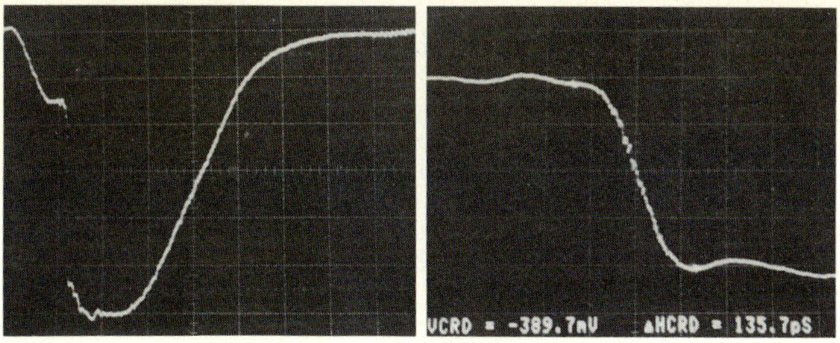

500 V/Div Vertical
1.0 ns/Div Horizontal

500 V/Div Vertical
50 ps/Div Horizontal

Figure 2. Fastest rise (fall) time waveform generated (63 ps - fast transition). This is also an example of a measurement to verify the theoretical behavior being modeled.

display of plots of the internal values for free carrier densities, current density, electric field, etc. at any time step during the calculation. This simulation has proved to be very useful in visualizing how these parameters are changing as the diode switches into second breakdown and has led to a qualitative understanding of the operation of these devices in this unusual mode. We are making measurements and refinements to the theoretical calculations to attempt to gain good quantitative agreement between theory and experiment.

Qualitatively, it appears that due to the high electric field, avalanche generation floods the interior of the device with an electron-hole plasma. This causes the electric field to collapse resulting in a large current flow in the external circuit. Often this electron-hole plasma first forms at one end of the depletion layer (where the electric field is the highest) and propagates to the other end as the electric field collapses. Figure 3 is a plot of the calculated electric field in the depletion layer for several different times showing the propagation of the plasma wave. Our theoretical calculation shows that the rise time of the current pulse is coincident with this propagation of the plasma wave across the depletion layer. Intuitively one could expect that the velocity of this plasma wave across the device would determine the rise time of the electrical pulse. We have found, however, that this is not the case. The electron hole plasma grows not only spatially, but also in density. It is the increase in the density of the plasma that determines the electrical rise time, not the growth spatially. Under different conditions

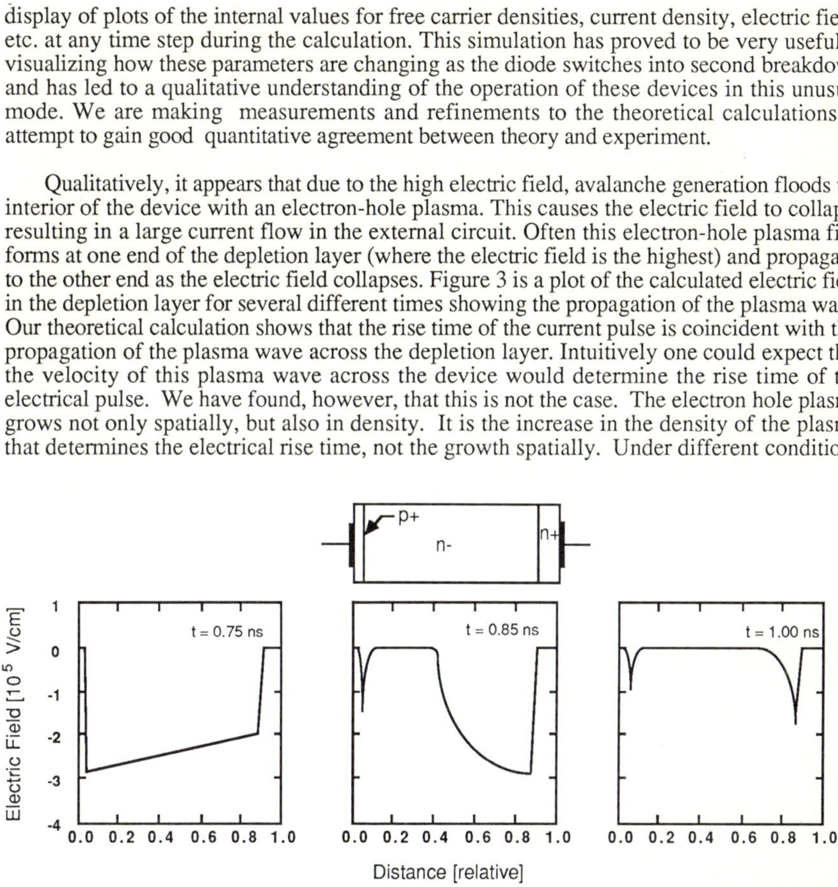

Figure 3. One-dimensional diode structure and calculated electric field profiles at three different times during the breakdown process. See Fig. 4 for resulting current and voltage pulses.

of doping and current density, the simulation shows that the plasma begins initially at the left side and spreads to the right of the device, or begins at the right side and spreads to the left, or grows everywhere simultaneously.

In general, the growth of this electron hole plasma is a complex function of the dynamic electric field and carrier density profiles. We are performing a number of calculations to separate out the effects of various parameters such as the current density, doping concentrations, and physical width of the depletion layer on this plasma growth to get a better understanding of the key contributing factors. Figure 4 shows an example of calculated current and voltage waveforms. Since current densities are very high, there is a possibility that the current flow is filamentary. We feel that an accurate model will require two dimensional calculation. We are therefore in the process of extending our model to two dimensions.

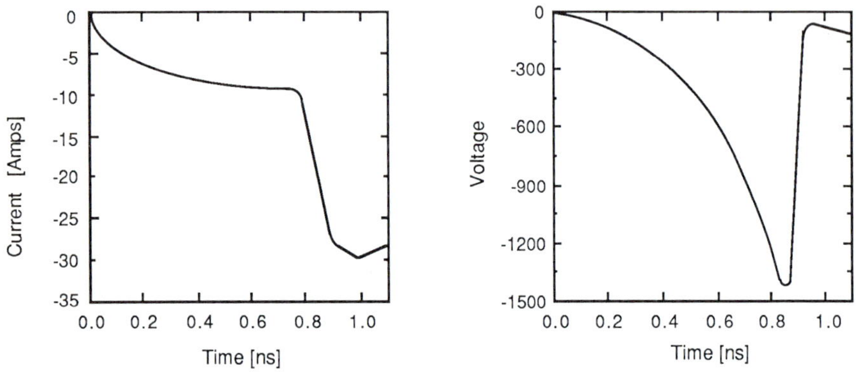

Figure 4. Calculation of current through diode (left) and voltage across diode (right) corresponding to Fig. 3. Input drive for this calculation was a linear ramp from 0 to 3000 volts rising from 0 to 1.0 ns, then remaining constant at 3000 volts after 1.0 ns.

While we have been making measurements specifically directed at verifying and improving the avalanche diode model (see for example Fig. 2), there have been several applications of these diodes implemented as well. The result of one implementation is shown in Fig. 5.

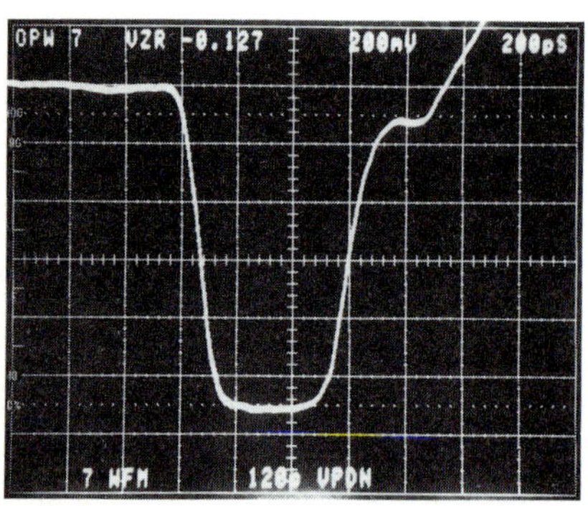

100 V/Div Vertical
200 ps/Div Horizontal

Figure 5. Flat-topped pulse having an amplitude of 550 volts into 6.25 ohms and a width of 520 ps.

This application [3] required the generation of an electro-optic gating pulse having a flat top and capable of delivering several hundred volts into a 6.25-ohm load. The pulse was developed within a 25-ohm structure, shaped by a charge line and small (≈20 nh) inductors, and transformed by a transmission-line transformer to drive the 6.25-ohm load. The driving pulse (input to the diode circuit) had a rise time of about 1.5 ns. The system jitter was quite small, nearly indistinguishable from that of the sampling system (≈30 ps peak-to-peak).

Other implementations have resulted in pulses as narrow as 87 ps and amplitudes exceeding 1300 volts into 25-ohm loads. Diode shapers have also been combined in parallel to generate higher power pulses.

References:

1. I. V. Grekhov et al: Electronics Letters, 17, 7, pp. 422-423, 11 June, 1981.

2. W. J. Orvis, J. H. Yee: Lawrence Livermore National Laboratory, UCID 20602, Aug. 1985, available from National Technical Information Service, U.S. Department of Commerce, 5285 Port Royal Road, Springfield, VA., 22161.

3. Pulse shaper built by R.L. Hanks, Lawrence Livermore National Laboratory.

Graphite as a Picosecond Laser Activated Opening Switch

E.A. Chauchard[1], *Chi H. Lee*[1], *C.Y. Huang*[2], *and A.M. Malvezzi*[3]

[1]Department of Electrical Engineering, University of Maryland, College Park, MD 20742, USA

[2]Los Alamos National Laboratory, Los Alamos, NM 87545, USA

[3]Gordon McKay Laboratory, Harvard University, Cambridge, MA 02138, USA

In a recent experiment, HUANG et al. [1] have observed a large decrease in the reflectivity of a semimetallic sample of highly oriented pyrolytic graphite (HOPG) as it is illuminated by an intense picosecond laser pulse when the fluence is above a well-defined threshold of 140 mJ/cm^2 at 566 nm. They have found that the imaginary part of the refractive index reaches lower than 0.5 at 566 nm. This unexpected result indicates that a phase transformation occurs and that the high-temperature phase, which they believe to be liquid, is non-metallic. This experimental observation is consistent with the result of the recent pseudo-potential calculation that there is an energy gap in "isotropic" carbon [2]. The time-resolved experiments [1] also show that the phase transformation is completed in time scales as short as ~ 10 ps and that the new phase lasts for approximately 3 ns. In the work presented here, we directly assess the change of resistivity of a HOPG sample used as a laser-activated switch. We confirm the results reported in [1] by observing a large increase in the sample's resistivity under intense picosecond laser illumination. This unique property could allow one to use graphite as an opening switch since its conductivity in the c-plane, in the dark state ($\sigma \cong 10^5$ (ohm.cm)$^{-1}$) is almost as good as that of pure copper. It could then be compared with semiconductor optoelectronic switches, noting that its operation is exactly inverse: it is conductive in the dark and its resistance increases when illuminated above the threshold. The main application of this type of switch is in the pulse power technology where the recent interest for inductive storage systems has created a need for an efficient, repetitive, jitter-free opening switch [3]. In [4,5], we proposed to use a semiconductor material as opening switch. In this paper, we demonstrate that graphite is a suitable alternate material for opening switch applications with some advantages.

To demonstrate the operation of HOPG as opening switch, a circuit configuration of current-charged transmission line as described in [4] was used (Fig. 1). In this type of circuit, the energy is stored in the magnetic form in a transmission line of impedance 50 ohms. As the swtich opens and remains open, the current flows towards the 50 ohm load of the oscilloscope and a square pulse is observed. If the switch is ideal (infinite resistance in the open state), the current flowing towards the load is one half of the charging current, and the totality of the energy stored in the transmission line is transferred to the load. This ideal case defines 100 percent switching efficiency. The duration of the square pulse (30 ns) is determined by the length of the transmission line (3 m). The rise and fall time of the square pulse are equal to the switch opening time. A Q-switch mode-locked Nd:YAG laser, providing 90 ps pulses at 1.06 µm was used to illuminate the HOPG sample.

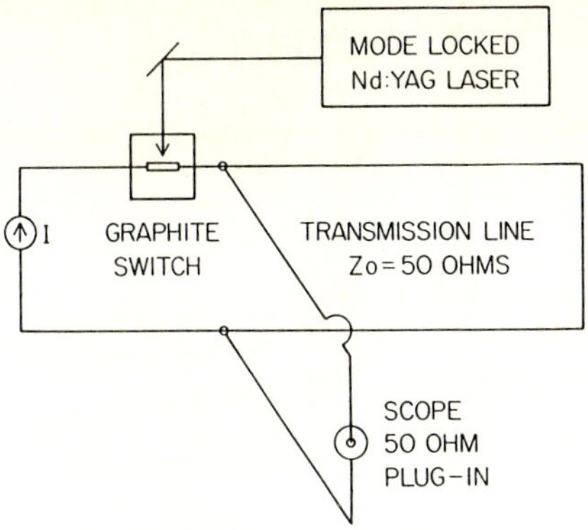

Fig. 1: Experimental set-up for the generation of square pulses using a laser-activated opening switch and a current charged transmission line

The switch was fabricated in a thin (4-6 µm) sheet of material. This sheet is parallel to the c-plane of the material. A narrow slab of ~100 µm is then cut, mounted on a piece of sapphire and connected to microwave SMA connectors. The typical switch resistance, including the contact resistances, was approximately 10 ohms. The small dimensions of the sample are required for the following reason. In order to observe a significant change of resistivity by the method described here, the total sample cross-section needs to undergo the phase transformation. The sample thickness should thus be no larger than a few times the linear light penetration depth which is only 0.04 µm for the 1.06 µm light used. We found that this difficulty can be overcome by using the high-energy laser pulses to etch the sample surface until a suitable thickness is reached and the signal starts being visible. Because of the high laser fluences used, it is likely that the actual penetration depth is larger than 0.04 µm: the first molten layer formed is more transparent and allows the light to penetrate deeper.

The waveforms shown in Fig. 2 indicate that the opening switch action was obtained. It was found as in [1] that a high laser fluence is

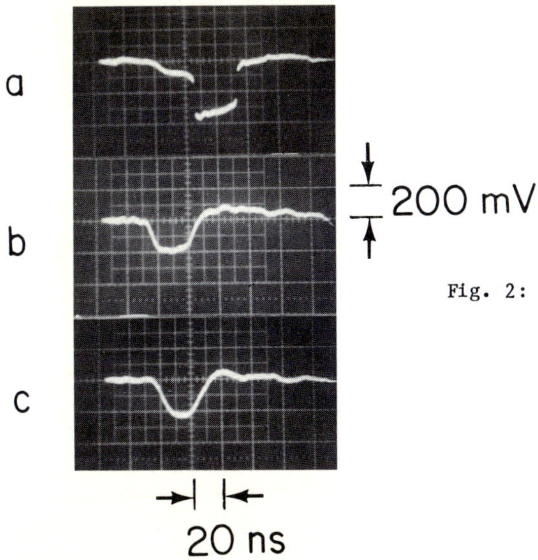

Fig. 2: Three examples of waveforms generated as the graphite switch opens. Each trace is obtained with a different laser pulse but the sample of graphite is the same. The laser pulse energy is: a: 0.16 mJ, b: 0.14 mJ, c: 0.16 mJ. The charging current is 16 mA.

necessary to obtain this result (> 140 mJ/cm^2). The shape of the electrical pulse obtained was not reproducible from shot to shot or from sample to sample. The fastest speed observed on Fig. 2a was less than 1 ns (oscilloscope limited). The switching efficiency was quite high: 90%, corresponding to a resistance of the sample of 450 ohms after illumination.

When the switching efficiency was lower, typically 5-10%, the output waveform had a "staircase" shape due to the successive reflections of the wave in the transmission line. In this case, the waveforms observed were quite reproducible from shot to shot. The rate of decay of the successive pulses can give a measurement of the closing speed of the switch which is directly related to the disappearance of the new phase. This measurement can also be performed without the transmission line configuration. In both cases, a decay time of the order of 100 ns was found, in disagreement with the value of ≈ 3 ns given in [1]. This may be due to the fact that the measurements of [1] were only probing the sample's surface which is likely to recover faster than the bulk.

The maximum DC current in the c-plane that the sample can sustain was estimated to be 160 kA/cm^2. The lack of reproducibility observed in the high efficiency switching case shows that it is difficult to control the formation of the liquid phase and its decay. Although it is reported in [1] that the new phase is liquid, these first experiments do not allow to accurately identify its nature. A better understanding of the new phase would certainly allow us to better determine the switching parameters. Further research is necessary in this direction.

The subnanosecond opening speed, the fact that the switching efficiency of this material can be very good, and the fact that it can sustain a high current density make it a very promising candidate for opening switch applications. From a more fundamental point of view, it is of particular interest to note that HOPG is, to our knowledge, the only material exhibiting this type of behavior.

Acknowledgements

We are grateful to Professor N. Bloembergen for many stimulating discussions. We would also like to thank Professor M.L. Cohen for sending his paper prior to publication. One of us (CYH) is supported by the US Department of Energy. This work was supported by the Air Force Office for Scientific Research.

References

1. C.Y. Huang, A.M. Malvezzi, N. Bloembergen and J.M. Lin: In Mat. Res. Symp. Proc. 51, 245 (1986); A.M. Malvezzi, N. Bloembergen and C.Y. Huang: Phys. Rev. Lett. 57,146-149 (1986).
2. S. Fahy, S.G. Louie and M.L. Cohen: preprint.
3. K.H. Schoenback, M. Kristiansen and G. Shoefer: Proc. IEEE 72, 8 (1984).
4. E.A. Chauchard, M.J. Rhee and Chi H. Lee: Appl. Phys. Lett. 47, 12 (1985).
5. E.A. Chauchard, C.C. Kung, M.J. Rhee, Chi H. Lee and V. Diadiuk: Proc. of the Conference On Lasers and Electrooptics, (San Francisco, CA 1986).

Part VIII

Optical Microwave Techniques

Radar and Electronic Warfare Applications of Multigigahertz Optical Components and Systems

H.F. Taylor

Department of Electrical Engineering, Texas A&M University, College Station, TX 077843, USA

Introduction

Recent advances in wideband components for use in optical fiber systems have been very impressive. Laser diodes have shown 3dB modulation bandwidths near 20 GHz at room temperature /1,2/ and external modulators with similar high-frequency performance have been reported /3/. Bandwidths as high as 100 GHz have been achieved in semiconductor photodiodes /4/. Performance of these components at such high frequencies opens a new realm of possibilities for the use of optics in military systems. These include not only signal transmission over optical fibers, but also signal generation, signal processing, and optical control of microwave devices. Coverage spanning the 0-18 GHz range, where most radar and electronic warfare (EW) systems operate, is now possible. Some of these techniques can also be applied in the millimeter wave frequency range (> 30 GHz), which is receiving increased emphasis for military applications. This paper discusses some generic radar and EW applications for multigigahertz technology, and gives several examples of how optical techniques might be used to advantage in such systems.

Radar and Electronic Warfare System Needs

Most radar systems operate at modest bandwidths (e.g., 10 MHz), so that the signal processing does not require gigahertz technology. However, there is a strong interest in using fiber optics for transmitting baseband microwave signals and local oscillator waveforms in radar systems. With a trend in recent years towards the use of phased array radars for their superior detection and tracking performance, it is hoped that fiber signal transmission can reduce the cost of controlling the frequency and phase of the transmitting elements and of processing the return signals. Interest in phased-array systems extends throughout the 0 - 18 GHz frequency range and into the millimeter-wave regime.

Some radar systems which do employ high bandwidths utilize chirps, pseudorandom digital sequences, or frequency-hopping pulses as the signal waveforms. Broadband radars provide high spatial resolution and immunity from jamming. Present signal generation and processing technology limits the bandwidths of such systems to a few hundred megahertz.

Electronic warfare systems are designed to make use of the electromagnetic spectrum to gain advantage against an enemy. Passive EW is concerned with making use of the enemy's emissions to determine the nature of the threats and their locations and velocities. Active EW seeks to defeat the enemy's radars by jamming or by generating false radar returns.

Bandwidth is at a premium in signal processing for electronic warfare. A typical scenario for defense of an aircraft carrier task force might have a large number of threats coming from different directions and broadcasting in different frequency regimes. Spectrum analysis has been a mainstay of EW systems for signal sorting, but this is less useful against some of the more modern broadband radars. There is currently interest in techniques for storing wideband radar signals and broadcasting them back to the transmitter after a delay to generate false targets. Both analog and digital storage techniques are explored for this application.

Optical Fiber Applications

Optical fibers provide an attractive alternative to metallic waveguides for signal transmission in many microwave systems. The fibers are small, flexible, relatively inexpensive, and have enormous information capacity. For example, a single-mode fiber with a source wavelength near the dispersion minimum can carry a signal with a 20 GHz bandwidth for distances up to 50 km. The fibers can also be used in signal processing systems for the delay and storage of wideband signals, providing time-bandwidth products in excess of 10^6 vs $10^3 - 10^4$ for conventional acoustic-wave delay lines, as illustrated in Fig. 1.

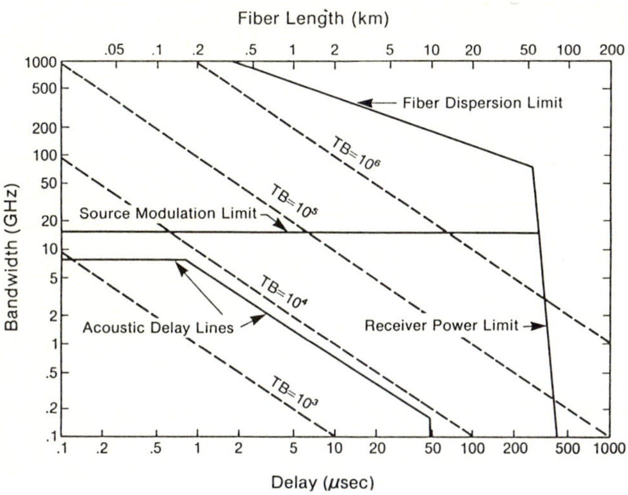

Fig. 1. Limitations on bandwidth and time delay for optical and acoustic delay lines.

The signal processor for a noncoherent radar moving target indicator is an application in which a wideband signal is stored for a long time period. The return signal from a pulsed radar is stored for a time of the order of 1 msec and subtracted from the next return /5/. In an experiment directed towards realization of such a processor, a 2 nsec pulse was stored for 0.5 msec by recirculating 5 times through an analog repeater and 20 km of single-mode fiber /6/.

The tapped delay line signal processor, or transversal filter, can be designed for use as a bandpass or matched filter, or as a waveform generator. Several fiber optic versions of the transversal filter have been demonstrated /7/. To date, however, no suitable method has been devised for providing the large numbers (e.g. hundreds) of taps needed in many applications.

A different type of fiber optic filter can be used for direction finding in radio-frequency or microwave systems /8/. In that case signals from an array of antennas are transmitted optically over fibers to a single photodetector. With appropriate choice of fiber lengths, the photodetector signal amplitude is a function of the angle of arrival of the incoming wavefront.

Generation and Control of Microwave Signals

Several optical techniques for the generation of microwave signals have been investigated. One method involves the excitation of a photoconductive stripline gap with picosecond optical pulses /9,10/. A second utilizes illumination of a microwave p-i-n diode in a resonant circuit with the light from a directly modulated semiconductor laser /11/. In that case, a signal at the third harmonic of the laser modulation was obtained at 10.2 GHz. A third technique utilizes injection locking to sidebands of a modulated master laser. Two different slave lasers /12/, or two different longitudinal modes of a single slave laser /13/, are locked to different sidebands of the master laser. As illustrated in Fig. 2, a signal at 35 GHz with a spectral width of less than 10 Hz was obtained /13/.

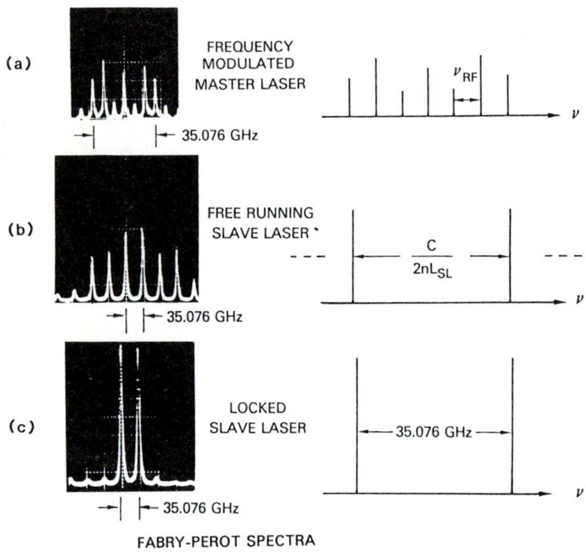

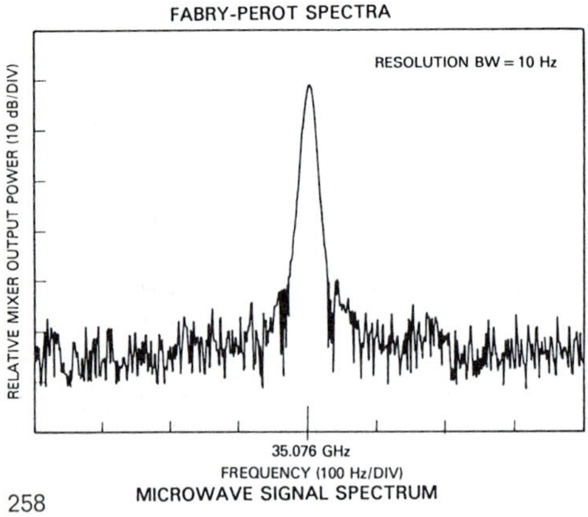

Fig. 2 Millimeter–wave signal generation with injection–locked laser diode.

The frequency and phase of a microwave oscillator can be controlled by injection locking with a modulated optical carrier. Numerous demonstrations of this have been carried out at frequencies as high as 10 GHz /11,14,15/. For FET oscillators the light can be directly coupled into the gate region of the active device, but for IMPATT oscillators obtaining optical access to the active region generally requires a redesign of the package. One of the attractive features of direct optical locking is that it makes possible the use of fibers to transmit local oscillator signals to remotely located elements in phased-array radars and direction-finding systems without the need for separate optical receivers. Monolithic antennas with optically injected control signals can be envisioned.

Other Applications

Some other uses of optics in microwave systems should also be mentioned. Bulk acoustooptic spectrum analyzers based on Bragg diffraction of a light beam are capable of processing analog signals with gigahertz bandwidths /16,17/. Analog-to-digital conversion of wideband signals using the electrooptic effect in an optical waveguide device has been achieved /18/. Phase-shifters for microwave devices can be made by illuminating a photosensitive dielectric waveguide with a CW light beam /9/.

Conclusions

In conclusion, optical techniques are reaching the point of utility in military radar and electronic warfare systems operating in the multi-gigahertz regime. Transmission, generation, control, and processing of signals in 0 - 18 GHz and 35 GHz frequency ranges are possible using present technology. Expansion of research in this field appears certain, with emphasis on establishing the systems utility of some of these techniques and extending them to other spectral regions of interest (e.g. 60 and 94 GHz).

References

1. C.B. Su, V. Lanzisera, W. Powazinik, E. Meland, R. Olshansky, and R.B. Lauer: Appl. Phys. Lett 46, 344 (1985).
2. J.E. Bowers, B.R. Hemenway, D.R. Witt, T.J.Bridges, and E.G.Burkhardt: Electron. Lett. 21, 1090(1985).
3. C.M. Gee, G.D.Thurmond, and H.W. Yen: Appl. Phys. Lett. 43, 998 (1983).
4. B.H. Kolner, D.M. Bloom, and P.S. Cross: Electron. Lett. 19, 574 (1983).
5. C.T. Chang, D.E. Altmann, D.R. Wehner, and D.J. Albares: IEEE Trans. Circuits and Systems CAS – 26, 1132 (1979).
6. C.C.Wang, R.P. Moeller, W.K. Burns, and I.P. Kaminow: Electron Lett. 20, 486 (1984)
7. K.P. Jackson, S.A. Newton, B. Moslehi, M. Tur, C.C. Cutler, J.W. Goodman, and H.J. Shaw: IEEE Trans. Microwave Theory Tech. MTT – 33, 193 (1985).
8. S.A. Pappert, M.N. McLandrich, and C.T.Chang: J. Lightwave Technol. LT – 3, 273 (1985).
9. C.H. Lee and V.K. Mathur: IEEE J. Quant. Electron QE – 17, 2098 (1981).
10. P. Paulus, W. Brinker, and D. Jager: IEEE J. Quant. Electron. QE – 22, 108 (1986).
11. A.S. Daryoush and P.R. Herczfeld: Electron. Lett. 22, 133 (1986).
12. L. Goldberg, H.F. Taylor, J.F. Weller, and D.M. Bloom: Electron. Lett. 19, 491 (1983).

13. L. Goldberg, A.M. Yurek, J.F. Weller, and H.F. Taylor: Electron. Lett 21, 814 (1985).
14. H.W. Yen, Appl. Phys. Lett. 36, 680 (1980).
15. L. Goldberg, C.Rauscher, J.F. Weller, and H.F. Taylor: Electron. Lett. 19, 848 (1983).
16. T.M. Turpin, Proc. IEEE 69, 79 (1981).
17. A.E. Spezio, J.N. Lee, and G.W. Anderson: Microwave J., Feb. 1985, p. 157.
18. R.A. Becker, C.E. Woodward, F.J. Leonberger, and R.C.Williamson: Proc. IEEE 72, 802 (1984).

Characteristics and Applications of Wideband Guided-Wave Devices

C.M. Gee, R.J. Morrison, G.D. Thurmond, and H.W. Yen

Hughes Research Laboratories, 3011 Malibu Canyon Road, Malibu, CA 90265, USA

Design and fabrication of wideband and high-speed guided-wave electro-optic and electroabsorption devices will be reviewed. Their applications in rf fiber-optic links will be presented.

High-speed guided-wave devices are important for fiber-optic systems and high-speed signal processing applications. These devices are considerably more compact and require lower drive power than bulk electro-optic devices. Over the past several years, a number of high-speed guided-wave modulators and switches have been demonstrated [1-13]. Table 1 lists a few

TABLE 1. Various high-speed modulators

WAVELENGTH	ELECTRODE STRUCTURE	SWITCHING VOLTAGE	SPEED	REF.
LiNbO3:				
0.63 μm	lumped coplanar		110 ps	Ref. 1
0.63 μm	asym coplanar TW		12 GHz	Ref. 2
0.83 μm	asym coplanar TW	7V	17 GHz	Ref. 3
0.84 μm	coplanar TW	2V	13 GHz	Ref. 4
0.85 μm	coplanar TW	4.5V	16 GHz	Ref. 5
1.3 μm	coplanar TW	3.5 V	8 GHz	Ref. 6
1.3 μm	asym coplanar TW	4.5 V	7.2 GHz	Ref. 7
	EO switch		20 GHz	Ref. 8
GaAs:				
0.851 μm	lumped microstrip	8 V	100 ps	Ref. 9
0.865 μm	lumped microstrip	1.4V	900Mb/s	Ref. 10
0.789 μm	lumped microstrip	4 V	3 GHz	Ref. 11
1.3 μm	TW microstrip	8 V	4 GHz	Ref. 12
1.3 μm	coplanar TW		20 GHz	Ref. 13

of these devices which operate at multi-gigahertz frequencies. The widest bandwidth modulator is 20 GHz [8,13] and the lowest switching voltage is 1.4V [10]. In this paper, we will review design considerations for fabricating high-speed modulators.

Most of the high-speed guided-wave modulators are based upon the electro-optic effect in $LiNbO_3$. The electro-optic effect in GaAs material systems is much weaker, but still has exhibited excellent performance [12,13]. The $LiNbO_3$ device is fabricated by first photolithographically defining the waveguide pattern in Ti metal on top of the $LiNbO_3$ substrate. The Ti metal is then diffused into the $LiNbO_3$ to create a guiding channel with a refractive index larger than that of the surrounding substrate and top clad. The dimensions of the guide and the value of the refractive indices determine the number of guiding modes. Single-mode waveguides in $LiNbO_3$ are typically designed to be a few microns in dimension. Coplanar stripline electrodes of length L are then deposited over the waveguides by photolithographic techniques. A voltage V applied across the electrodes separated by a gap d would induce an index change given by

$$\Delta n = n_o{}^3 rV/2d\eta,$$

where n_o is the optical effective index of the guiding mode, η is a factor which accounts for incomplete overlap of the optical and electrical fields, and r is the relevant electro-optic coefficient which is determined by the crystal orientation and the direction of the applied electric field.

The bandwidth of the modulator depends upon the electrode and packaging design. Two popular electrode configurations are the traveling-wave (TW) and lumped-element designs, sketched in Fig. 1(a,b). In the traveling-wave design, the electrodes form a transmission line which is ideally driven by a matching transmission line and terminated in a matched load. The device bandwidth is ultimately limited by the difference in velocity between the microwaves and the light waves. The 3-dB bandwidth is given by

$$\Delta f = 1.4c/[\pi(n_o - n_m)L],$$

where n_m is the effective microwave index. For a $LiNbO_3$/air interface, n_o= 2.2 and n_m= 4.2, giving a velocity mismatch bandwidth of 6.6 GHz-cm. The effective microwave velocity can be increased by introducing lower index dielectric layers around the guiding layer [14]. In this way, the bandwidth is broadened at the expense of higher drive power since only the microwave field in the $LiNbO_3$ guiding region contributes to optical modulation. Besides limitation from velocity mismatch, stripline attenuation and impedance mismatch can reduce the bandwidth. With a lossy transmission line, any impedance mismatch causes resonances and limits the bandwidth.

In the case of a lumped-element modulator, an electrode structure with a capacitance C is driven by a transmission line terminated in its characteristic impedance. The speed of the device is limited by the capacitor charging time:

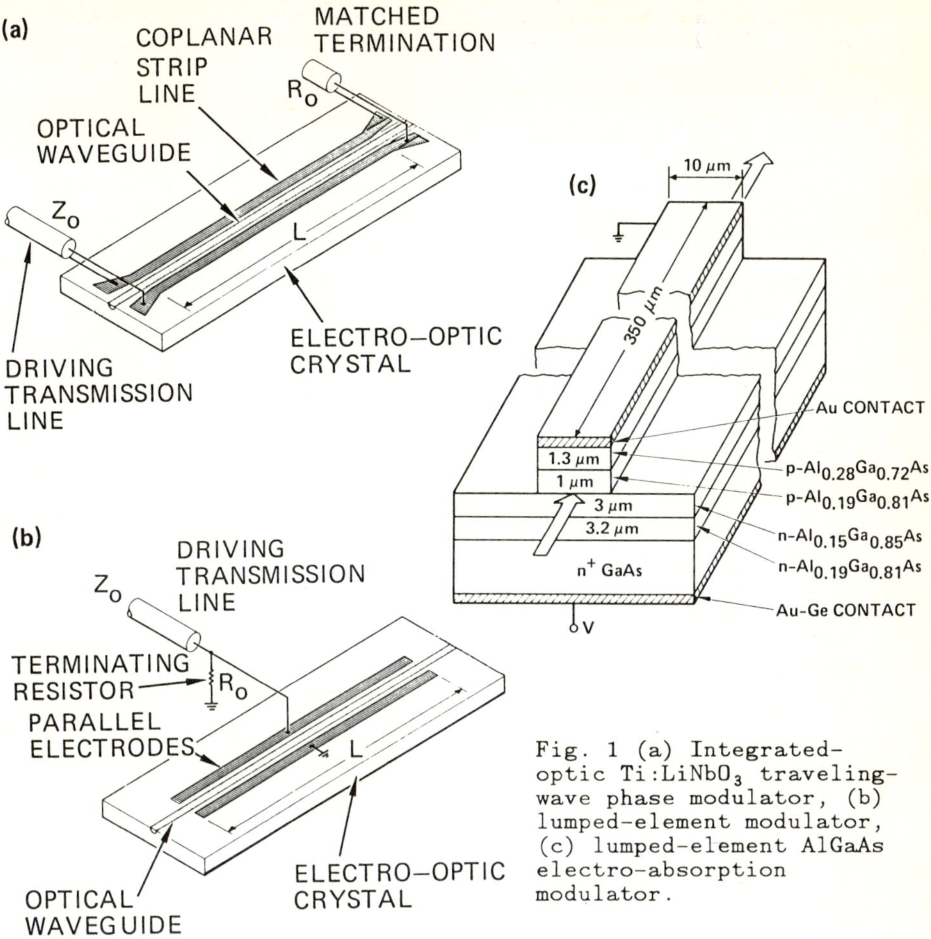

Fig. 1 (a) Integrated-optic Ti:$LiNbO_3$ traveling-wave phase modulator, (b) lumped-element modulator, (c) lumped-element AlGaAs electro-absorption modulator.

$$\Delta f = 1/\pi RC.$$

For dimensions identical to those of the traveling-wave structure, the bandwidth of a lumped-element $LiNbO_3$ modulator is 2.3 GHz-cm.

Alternative to $LiNbO_3$ devices, GaAs multi-quantum well (MQW) and bulk electroabsorption modulators have recently demonstrated performance up to a few gigahertz (Table 1). The principle of operation is based on the Franz-Keldysh effect in which the absorption edge of semiconductors shifts to longer wavelengths with an applied electric field. In MQW modulators, a Stark shift to lower energies of the exciton resonance is observed at room temperature. Hence, light passing through a piece of GaAs can be modulated by an applied electric field.

The bulk electroabsorption modulator structure consists of strip-loaded channel waveguides formed by etching ridges in epitaxial layers grown on n^+GaAs [9-11]. An electric field is applied vertically across the guiding layer using a microstrip electrode structure (Fig. 1(c)). The MQW double-heterostructure waveguide modulator consists of periodic thin QW layers sandwiched by thick cladding layers. The bandwidth of these devices is limited by their capacitance to a few gigahertz. Higher bandwidth can be achieved with shorter devices.

Extending the operation of integrated-optic devices out to the millimeter frequency regime requires innovative package designs. One technique for coupling millimeter wave energy into a lumped element modulator is by using a millimeter waveguide taper as illustrated in Fig. 2(a). Optical transit time effects become appreciable when the signal frequency is sufficiently high to have a period comparable with the time required for an optical wavefront to traverse the electro-optic medium. When the optical transit time is one quarter of a microwave signal period, the phase shift is reduced by 10% compared to a constant signal strength. The frequency response limited by transit time must exceed the carrier frequency, but the bandwidth limited by the RC time constant need only exceed the system bandwidth.

Limitations on the modulating frequency imposed by load capacitance and transit time effects may be circumvented by a traveling wave structure. If the signal phase velocity matches the light beam, then the optical wavefront sees a constant field strength throughout the modulator length. A conceptual configuration for a millimeter wave traveling-wave integrated-optic modulator is illustrated in Fig. 2(b). The millimeter waveguide contains a fin line which has been characterized by HOEFER AND BURTON [15]. The modulator is placed in the waveguide over the fins. The electrodes of the modulator extend to the edge of the substrate and the gap over the optical waveguide forms an iris loaded by the dielectric substrate. Input and output waveguide matching network couple energy into and out of the circuit. The output port can be terminated in a matched load for wideband operation or short circuited to form a fin line resonator with limited bandwidth but reduced drive power requirement. The optical input and output can be fiber pigtails.

Access to the transmission lines of an integrated-optic device by millimeter waveguide can also be provided by an adapter. NEIDERT [16] has demonstrated such an adapter between coaxial cable and W-Band waveguide (75-110 GHz). One end of the center conductor of a coaxial cable extends as a probe into the waveguide while the other attaches to the integrated-optic device striplines using pressure contact and silver paint. The coaxial cable should be small to minimize discontinuities at the device transmission line, but should be large enough to minimize rf loss. NEIDERT has measured less than a 1 dB loss for these adapters.

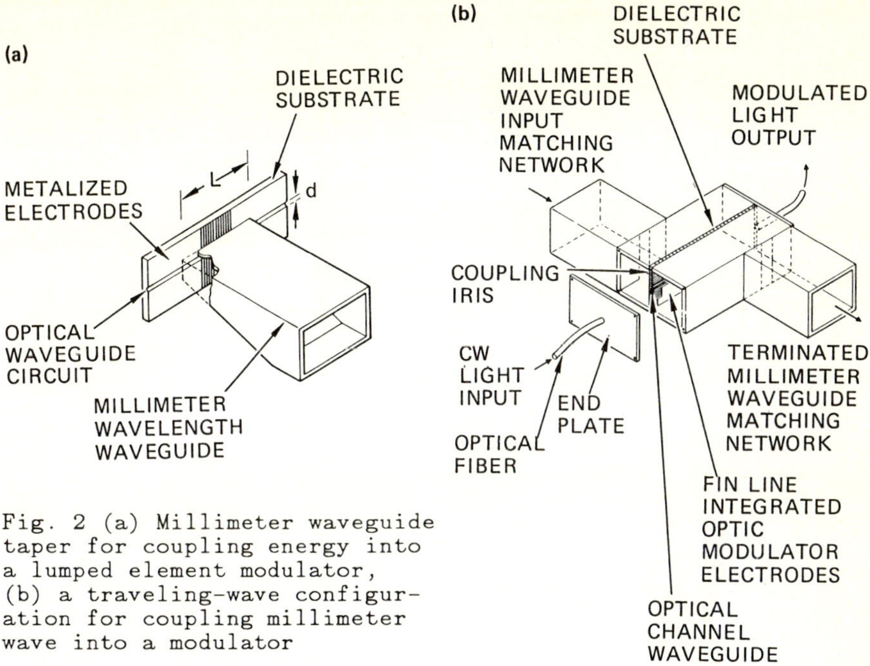

Fig. 2 (a) Millimeter waveguide taper for coupling energy into a lumped element modulator, (b) a traveling-wave configuration for coupling millimeter wave into a modulator

Due to the advancement of high-speed modulators, we have implemented external modulators in microwave fiber optic links. Although direct modulation offers simplicity and low drive power, certain high-speed applications might prevent their use. For example, we have assembled and characterized an external modulation link at 10 GHz for radar applications. The performance of the link was primarily limited by laser noise. Noise in the frequency range close to the rf carrier is particularly important for doppler radar applications. The use of external modulators relaxes the demand for high-frequency lasers and provides flexibility in selecting a low-noise laser. Some systems sensitive to laser wavelength chirping under high-speed modulation might also require external modulators.

Figure 3 illustrates the external modulator link and summarizes its performance. The rf insertion loss is the square of the efficiencies of the optical losses. The total rf insertion loss includes 10 dB loss from the detector efficiency, 10-15 dB loss from optical losses and 25 dB loss from the modulator efficiency. A higher power laser or a more efficient modulator would reduce the overall rf insertion loss below 50 dB.

The signal-to-noise (S/N) of the link far from the carrier (5-25 MHz away) was measured directly with a high-speed detector and a spectrum analyzer. The S/N of 130 dB/Hz is quite adequate for many radar applications. The AM and FM

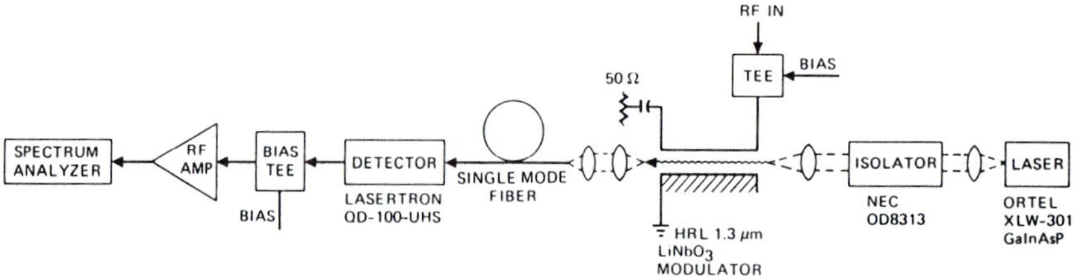

CARRIER FREQUENCY	RF DRIVE	RF INSERTION LOSS	SIGNAL–TO–NOISE		
			5–25 MHz	5–100 kHz	
				AM	FM
9.6 GHz	20 dBm	50 dB	130 dBc/Hz	125 dBc/Hz	130 dBc/Hz

Fig. 3 X-band external modulation link and its performance.

noise at 5-100 kHz from the carrier was measured with an X-band noise test set. The phase noise measurement was made by applying the output from the rf signal generator which drives the link and the output of the link to a mixer used as a phase detector. The AM noise was measured with a microwave crystal detector. The output of the phase detector or AM detector was measured with a spectrum analyzer and compared to a calibration signal. The AM and FM noise close to the carrier is approximately equal to that measure far from the carrier (5-25 MHz away).

Slightly better S/N performance could be achieved by driving the external modulator harder. Distortion does become apparent though. If two signals are present within the passband of the link, then the third order intermodulation product distortion can limit the link before the laser noise. For this link, the maximum dynamic range for the link is 39 dB in a 1 GHz bandwidth. This occurs at an input rf level of 19 dBm into the modulator.

The application of high-speed guided wave devices for long haul optical fiber transmission at 4 Gbit/s has been demonstrated by KOROTKY and coworkers [17]. They reported repeaterless transmission over 117 km at 1.53 μm wavelength using a Ti:$LiNbO_3$ traveling-wave directional coupler switch as an external modulator. They found good opening for the optical eye pattern at the output of the modulator with an on/off ratio of 12:1. The received power for 10^{-9} BER was -28.2 dBm for the link.

Guided-wave devices also promise unique applications in high-speed signal processing. Figure 4 illustrates the use of an array of GaAs modulators for interconnecting integrated circuit chips with fibers. The array of electro-absorption modulators are reflection coated on their back sides.

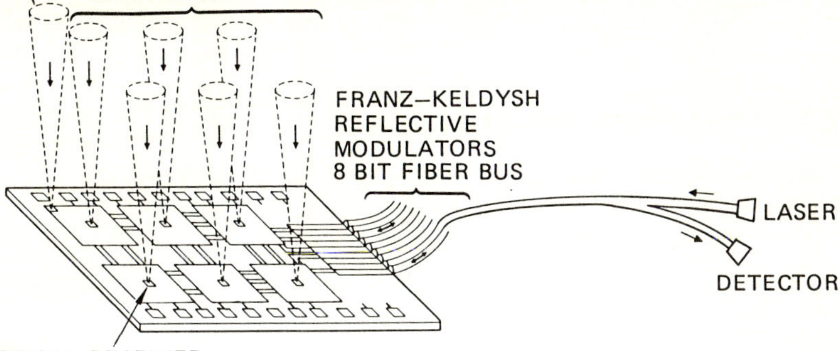

Fig. 4 Signal processing and optical interconnects using an array of reflective electro-absorption modulators.

Depending upon the electrical state of the modulator (whether or not voltage is applied to it), incoming light from a fiber will be either absorbed by the modulator or reflected back out of the chip.

In conclusion, several advantages and potentials for high-speed guided wave devices have been demonstrated. We believe improved packaging, by making the devices more rugged and operable in military environments, will enable their implementation in many systems outside of the laboratory.

References

1. R.C. Alferness N.P. Economou and L.L. Buhl, Appl. Phys. Lett. 38, 214 (1981).

2. T. Sueta and M. Izutsu, J. Opt. Comm. 3, 52 (1982).

3. C.M. Gee, G.D. Thurmond and H.W. Yen, Appl. Phys. Lett. 43, 998 (1983).

4. P.S. Cross, R.A. Baumgartner and B.H. Kolner, Appl. Phys. Lett. 44, 486 (1984).

5. R.A. Becker, Appl. Phys. Lett. 45, 1168 (1984).

6. D.W. Dolfi, App. Opt. 25, 2479 (1986).

7. R.C. Alferness, C.H. Joyner, L.L. Buhl and S.K. Korotky, IEEE J. Quantum Electron. QE-19, 1339 (1983).

8. S.K. Korotky, G. Eisenstein, R.S. Tucker, J.J. Veselka and G. Raybon, paper FB4, Picosecond Electronics and Optoelectronics Conf., 1987.

9. T.H. Wood, C.A. Burrus, R.S. Tucker, J.S. Weiner, D.A.B. Miller, D.S. Chemla, J.C. Damen, A.C. Gossard, W. Wiegmann, Elec. Lett. 21, 693 (1985).

10. S. Tarucha, H. Iwamura, T. Saku, H. Okamoto, Jap. J. of Appl. Phys. 24, L442 (1985).

11. C.M. Gee, G.D. Thurmond, H.W. Yen and H. Blauvelt, paper ThAA3, Topical Meeting on Integrated and Guided-Wave Optics, Atlanta, Feb 1986.

12. S.H. Lin, S.Y. Wang and Y.M. Houng, Electron. Lett. 22, 934 (1986).

13. S.Y. Wang, S.H. Lin and Y.M. Houng, paper WK3, OFC/IOOC'87

14. K. Atsuki and E. Yamashita, to be published.

15. W.J.R. Hoefer and M.N. Burton, IEEE Trans. Micro. Theory MTT-30, 2190 (1982).

16. R.E. Neidert, Microwave Journal, June 1983.

17. S.K. Korotky, G. Eisenstein, A.H. Gnauck, B.L. Kasper, J.J. Veselka, R.C. Alferness, L.L. Buhl, C.A. Burrus, T.C.D. Huo, L.W. Stulz, K. Ciemiecki Nelson, L.G. Cohen, R.W. Dawson and J.C. Campbell, J. of Lightwave Tech LT-3, 1027 (1985).

A New Optoelectronic CW Microwave Source

C.J. Clark[1], *E.A. Chauchard*[1], *K.J. Webb*[1], *K. Zaki*[1], *C.H. Lee*[1], *P. Polak-Dingels*[2], *H.A. Hung*[3], *and H.C. Huang*[3]

[1]Electrical Engineering Department, University of Maryland, College Park, MD 20742
[2]Laboratory for Physical Sciences, 4928 College Avenue, College Park, MD 20740, USA
[3]COMSAT Laboratories, Clarksburg, MD 20871, USA

Ultrafast switching utilizing picosecond photoconductors finds many applications, including optoelectronic sampling and characterization of high-speed electronic devices [1]. A major area of application that has not been explored is in microwave and millimeter-wave generation and control. The low-noise switching characteristic of these devices is very promising, since the microwave or millimeter wave signals can be generated in complete time synchronization with the exciting optical pulses. The time synchronization opens the door to unique applications. For example, it becomes possible to use optoelectronic techniques to modulate the CW microwave signal, allowing high data-rate signaling. Another potentially powerful application is in phased array antennas, where the central issue is to provide the correct phase and amplitude at each element of the array in a manner which allows rapid scanning.

A picosecond photoconductor can respond to optical pulses almost instantaneously, and with an appropriately engineered photoconductor, the photoconductive signal will decay rapidly after the termination of each optical pulse. As a result, electrical transients with extremely fast rise and fall times and with very small jitter can be generated. The spectral content of such electrical transients extending from DC to terahertz has already been observed [2]. Fast transitions in the electrical pulse from the photoconductor are necessary to generate sufficiently large spectral components at the higher frequencies. For rapid fall times it is necessary to use photoconductors with carrier lifetimes in the picosecond range. In this paper, we report the first optoelectronic realization of a CW microwave source at 10 GHz. This represents an electrical signal frequency multiplication of 100. The technique can be extended to higher frequencies and the microwave source (including a diode laser) can be integrated on a single chip.

A biased picosecond photoconductor is illuminated by high repetition rate picosecond laser pulses, producing a train of electrical pulses. A single RF component is selected out of the spectrum of the electrical pulses and amplified. A CW microwave source can thus be realized. The experimental configuration is shown in Fig. 1. A CW mode-locked dye laser operating at 590 nm generates 3 ps FWHM pulses at a repetition rate of 100 MHz. The laser beam is focused on a biased semiconductor switch, producing electrical transients with picosecond rise times. The photoconductive switch consists of a microstrip line with a $100\mu m$ gap deposited on Cr doped GaAs. Its photoconductive lifetime is 200 ps. The spectral components are spaced 100 MHz apart, corresponding to the repetition rate of the generated electrical pulses. The output of the switch is matched to 50 Ohm and fed to a cascaded high-Q 10 GHz bandpass filter and a lowpass filter. The lowpass filter was necessary to remove spurious signals from the output of the bandpass filter, which is shown in Fig. 2 together with the switch output spectrum. A three stage X-band GaAs MESFET low-noise amplifier designed at COMSAT provides the signal with adequate gain for display on the sampling scope. The sampling scope signal is shown in Fig. 3. At

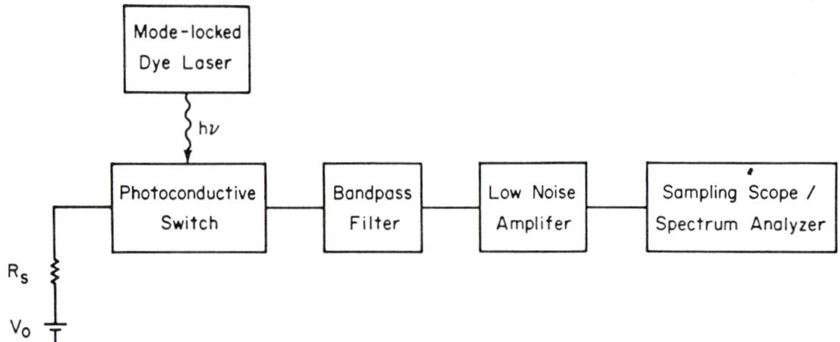

Fig. 1 Experimental set-up for the generation of CW microwave signals

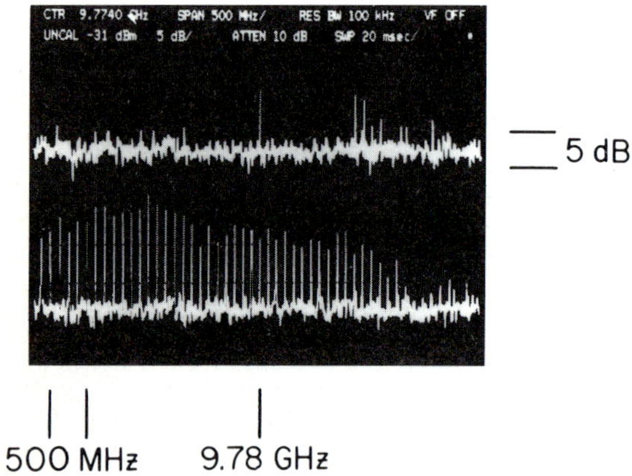

Fig. 2 Instantaneous power spectrum of the switch output (bottom trace) and bandpass filter output prior to lowpass filtering (top trace)

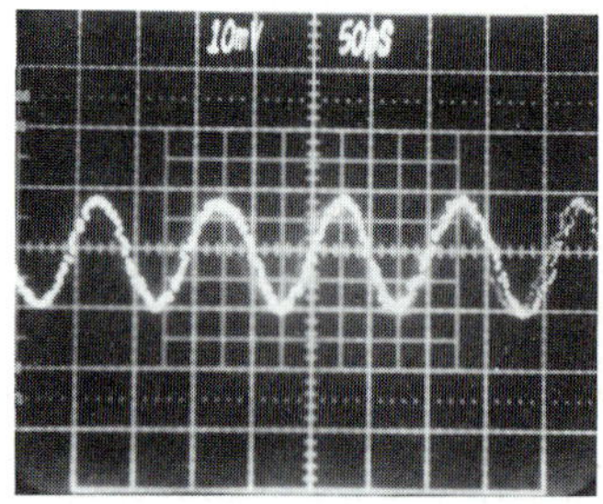

Fig. 3 Output of the GaAs MESFET amplifier when using both the bandpass and lowpass filters. Vertical scale: 10 mV/larger div, horizontal scale: 50 ps/-larger div

the selected frequency, the amplifier has a noise figure of 2.2 dB and a power gain of 30 dB. A power of -18 dBm was obtained with a 15 V DC switch and a light power of 150 mW at a wavelength of 590 nm. With the particular switch used, the spectrum extended to at least 25 GHz.

The 10 GHz signal must satisfy certain stability requirements so that it may be used in the applications mentioned above. It is thus of great interest to measure its total rms time jitter, which may be done by measuring the FM noise in the frequency band [3]. The 10 GHz spectral component shown in Fig. 4, as measured at the output of the amplifier with a HP8566B spectrum analyzer, exhibits a continuous noise pedestal as well as discrete sidebands. The total system time jitter, including discrete noise and continuous noise from the laser, was found to be quite small: 3.53 ps (30 Hz to 5 kHz deviation). The discrete sidebands could be detected at the output of the RF source driving the laser mode-locker and could be attenuated by using a more suitable driver. The continuous noise was mostly attributed to the laser fluctuations and could be greatly improved by stabilizing the laser [4].

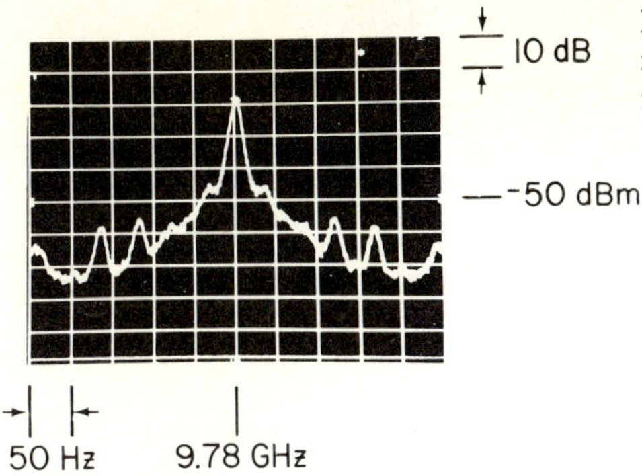

Fig. 4 Power spectrum showing phase noise of the 9.78 GHz spectral component

Our results can be compared with other recent work. In [5], the generation of a CW 3 GHz microwave signal is reported. The authors have used a mode-locked Krypton-ion laser and an avalanche photodiode or an optoelectronic switch. No measurement of microwave power or source noise is given. In [6], a technique of FM sideband injection locking of laser diodes has been used to generate a CW 35 GHz microwave signal. This technique requires careful temperature stabilization of the lasers.

In conclusion, we have demonstrated a new method of generating microwave signals in the multigigahertz range by a picosecond optoelectronic technique. The source phase noise is 3.53 ps, primarily due to noise in the lasers. The microwave signal generated is in time synchronization with the laser, allowing further optoelectronic processing. Higher power can be obtained by driving arrays of such devices and/or using higher gain amplifiers. With a tunable bandpass filter, the source frequency can be varied in 100 MHz increments. Although 10 GHz generation is reported here, the extension of the method to generate signals in the 100 GHz range is in principle straight forward, since femtosecond lasers and photoconductors with picosecond lifetimes are currently available.

Acknowledgements

The authors would like to thank Dr. Geoffrey Burdge and Dave Rush for technical discussions and measurement assistance. This work was supported in part by the Laboratory for Physical Sciences and the Air Force Office of Scientific Research.

References

[1] For a review, the reader is referred to the book Picosecond Optoelectronic Devices, Chi H. Lee, Ed., Academic Press (1984).

[2] D.H. Auston, K.P. Cheung, J.A. Valdmanis, and P.R. Smith, in Picosecond Electronics and Optoelectronics, G.A. Mourou, D.M. Bloom, and Chi H. Lee (Eds.), Springer Series in Electrophysics, 21, pp. 2-8, (1985).

[3] V.P. Kroupa (Ed.), Frequency Stability: Fundamentals and Measurements, IEEE Press, New York (1983).

[4] M.J. Rodwel, K.J. Weingarten, D.M. Bloom, T. Baer, and B.H. Kolner, submitted to Optics Letters (May 1986).

[5] A. Mooradian, Appl. Phys. Letter 45, 5, pp. 494-496, (1984).

[6] L. Goldberg, A.M. Yurek, H.F. Taylor, and J.F. Weller, Elect. Lett., 21-18, pp. 814-815 (1985).

Optical Intensity Modulation to 40 GHz Using a Waveguide Electrooptic Switch

S.K. Korotky, G. Eisenstein, R.S. Tucker, J.J. Veselka, and G. Raybon

AT&T Bell Laboratories, Crawfords Corner Road, Holmdel, NY 07733, USA

We report the intensity modulation of an optical carrier at frequencies as high as 40 GHz using a Ti:$LiNbO_3$ optical waveguide switch. A self-electrooptic sampling technique is used to observe the modulation.

1. Introduction

Optical intensity modulators operating from below 1 GHz to the millimeter wave regime are of interest for many applications such as analog microwave fiber communication [1], general purpose electrooptic sampling [2], and spectroscopy using shaped optical pulses [3]. Employing modulators external to the laser source offers the flexibility of optimizing the optical transmitter for high modulation bandwidths and high optical power output, while maintaining excellent spectral purity, as has been demonstrated in multi-gigabit/sec digital fiber communication [4]. Electrooptic modulators, based on single-mode titanium indiffused lithium niobate (Ti:$LiNbO_3$) optical waveguides, with 3 dB electrical bandwidths of approximately 16 GHz have been demonstrated and directly evaluated to about 18 GHz [5]. The extension of these results to higher frequencies has been hindered by stringent device design requirements and a lack of sufficiently high-speed detection systems.

Here we report a Ti:$LiNbO_3$ foreshortened electrode directional coupler modulator [6] with a small signal 3 dB electrical bandwidth of ~20 GHz and the operation of this modulator to 40 GHz. To evaluate the modulator at these frequencies the self-sampling method [7] was used to observe the optical modulation in the time domain.

2. Device Design and Measurements

The Ti:$LiNbO_3$ modulator (Fig. 1) was designed for 1.5 μm wavelength and consists of a 1 cm optical waveguide directional coupler of single coupling length and a 2.5 mm traveling-wave electrode located midway along the length of the coupler. Such a structure has a switching characteristic and frequency response similar to a Mach-Zehnder interferometric intensity modulator. The expected small-signal 3 dB electrical bandwidth of the modulator, taking into account optical/electrical walkoff (velocity mismatch) [8] and electrode transmission loss, is approximately 22 GHz.

The modulator electrode consists of a 35 Ω asymmetric coplanar transmission line with a strip width of 15 μm and gap of 5 μm in the active portion of the device and over an additional passive length of 7.5 mm. Connections to 50 Ω coaxial cable at the input and output ends of the microscopic electrode were made using 2.5 mm coplanar tapers and Wiltron K-connectors. The electrode gap at the coaxial to coplanar transition is ~250 μm . The measured switching voltage, V_π , of the modulator is 26 V . The electrical characteristics of the entire structure were measured from 45 MHz to 22 GHz using an HP-8510 vector network analyzer and to 40 GHz using a Wiltron 5669 scalar network analyzer. Little structure is found in the RF transmission and group delay below 22 GHz indicating fundamental mode operation and negligible dispersion into the 20 GHz range. At frequencies above 22 GHz oscillations in the transmission indicate that smaller connection pad dimensions are required to optimize the performance to 40 GHz.

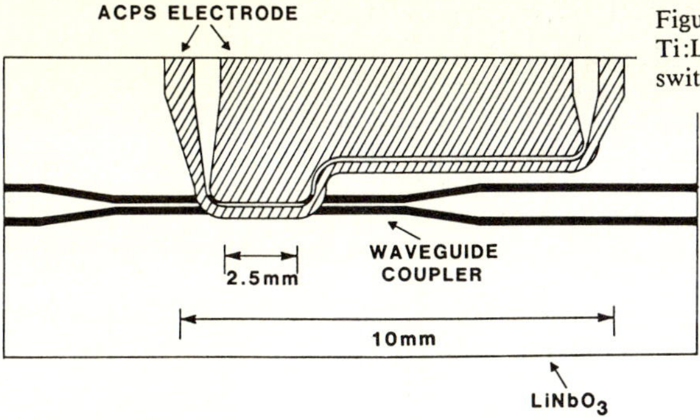

Figure 1 Schematic diagram of the $Ti:LiNbO_3$ waveguide electrooptic switch

The *electrooptic* frequency response of the modulator between 2 and 20 GHz was evaluated using a cw injection laser, a very high-speed InGaAs photodiode, and the HP-8510 network analyzer. For this power transfer-function measurement, two HP-8349 2-20 GHz amplifiers were used to provide a drive power of ~20 dBm to the modulator and a gain of ~15 dB following the detector. The network analyzer compared the transmitted and received electrical power as a function of frequency in relation to a test path to determine the relative small signal frequency response. This method provides an accurate assessment of the response of the device. We show the measured electrooptic modulator response in Fig. 2 . The rolloff observed from 2 to 20 GHz is 1-1.5 dB, indicating the usable bandwidth extends to higher frequencies. However, it is clear from the figure that the signal-to-noise ratio of the setup is degrading as a function of frequency, thereby precluding measurements above 20 GHz. The source of this degradation is the rapidly decreasing gain of the amplifiers near 20 GHz.

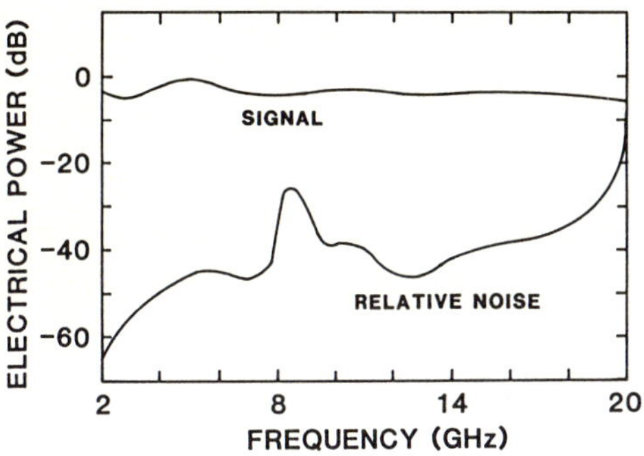

Figure 2 Electrooptic power transfer function

Because of the lack of suitable broadband amplifiers above 20 GHz, it was necessary to employ an alternate approach to evaluate the device above this frequency. The method used here is an adaptation of a previously demonstrated optical sampling technique for characterizing electrooptic modulators [7] . We illustrate the experimental arrangement in Fig. 3 . The cw laser light source of the previous measurement is replaced by a pulsed laser. In the experiments described here we used an actively mode-locked extended fiber cavity semiconductor laser [9] operating at 1.3 μm wavelength. The laser was driven from a microwave synthesizer to generate optical pulses at a repetition rate of ~2 GHz. The pulse width was determined by measuring the pulse auto-correlation using second harmonic generation in $LiIO_3$ and was 15 ps for a laser drive frequency, f_L, of 2.047 GHz . A slow detector was used to measure the average optical power arriving at the output of the crossover waveguide of the coupler. The modulator was driven at the n'th harmonic of f_L ($n \sim$ 1-19) with a power of ~10 dBm using a second synthesizer phase-locked to the first. In this manner, the average optical power level at the detector contains a component that is proportional to the modulator response at a point of the RF cycle determined by the relative phase of the modulator drive signal and the optical sampling pulses.

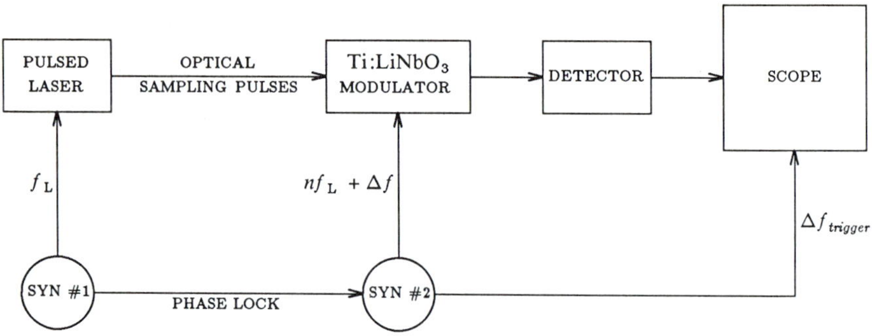

Figure 3 Arrangement for self-sampling evaluation of modulator

The time-domain response of the modulator is mapped out by measuring the optical power as a function of the relative phase of the laser and modulator drives. For convenience, this relative phase is changed by slightly offsetting the frequency of the modulator drive signal with respect to a harmonic of 2.047 GHz. [10] In Fig. 4 we display the detected waveforms for modulation frequencies of 26.6 and 38.9 GHz. Using this measurement technique, the frequency response was evaluated by driving the modulator at different harmonics of the laser pulse rate. The observed modulation in the frequency range of 2-40 GHz was corrected for the frequency content of the laser pulse (~ -4 dB at 40 GHz) and is plotted in Fig. 5 . We observe a rolloff of ~3 dB at 20 GHz and a further decrease to ~-10 dB at 40 GHz. To our knowledge this is the first report of the operation of an optical waveguide modulator to 40 GHz. Factors affecting the extension of this measurement arrangement to still higher frequencies are the laser pulse width, synthesizer frequency range, and the frequency response of coaxial components such as connectors and the bias "T" circuit.

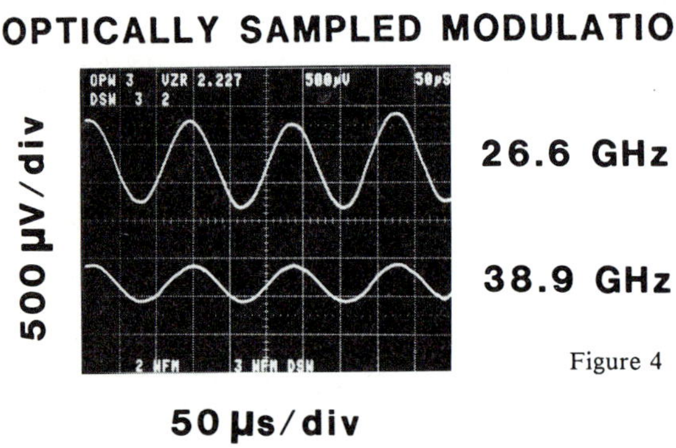

Figure 4 Optically sampled waveforms

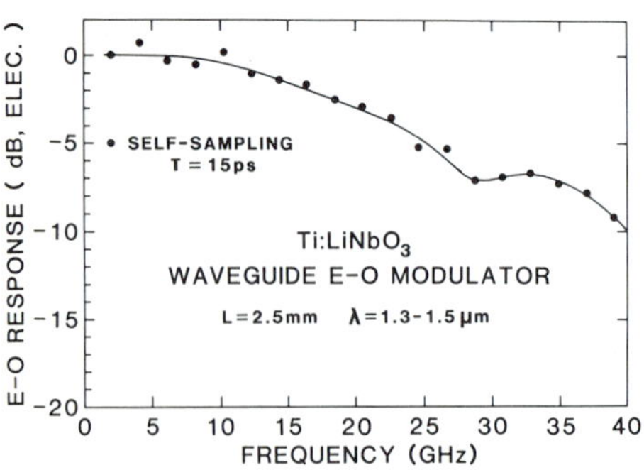

Figure 5 Electrooptic response to 40 GHz

3. Summary

In conclusion, we have demonstrated a Ti:$LiNbO_3$ waveguide switch with a small signal electrical bandwidth of ~20 GHz and the evaluation of this device in the time domain to 40 GHz using a self-sampling arrangement. Modulators operating to higher frequencies appear to be possible with further work, but corresponding improvements in the supporting component technologies will also be required.

Acknowledgement

We are grateful to C.A. Burrus for providing the fast InGaAs photodiode.

REFERENCES

1. H. Blauvelt, J. Parsons, D. Lewis, and H. Yen, in Proc. Conf. on Opt. Fiber Comm., New Orleans, 1984, paper TuH3.
2. J.A. Valdmanis, G.A. Mourou, and C.W. Gabel, IEEE J.. Quantum Electron. **QE-19**, 664 (1983).
3. W.S. Warren, J. Chem. Phys. **81**, 5437 (1984).
4. S.K. Korotky, G. Eisenstein, A.H. Gnauck, B.L. Kasper, J.J. Veselka, R.C. Alferness, L.L. Buhl, C.A. Burrus, T.C.D. Huo, L.W. Stulz, K.C. Nelson, L.G. Cohen, R.W. Dawson, and J.C. Campbell, J. Lightwave Technol. **LT-3**, 1027 (1985).
5. R.A. Becker, Appl. Phys. Lett. **45**, 1168 (1984).
6. S.K. Korotky, IEEE J. Quantum Electron. **QE-22**, 953 (1986).
7. R.C. Alferness, N.P. Economou, and L.L. Buhl, Appl. Phys. Lett. **37**, 597 (1980).
8. I.P. Kaminow and J. Liu, Proc. IEEE **54**, 1374 (1963).
9. G. Eisenstein, R.S. Tucker, G. Eisenstein, U. Koren, and S.K. Korotky, IEEE J. Quantum Electron. **QE-22**, 142 (1986).
10. M.J.W. Rodwell, K.J. Weingarten, J.L. Freeman, and D.M. Bloom, Electron. Lett. **22**, 499 (1986).

Picosecond Response of an Optically Controlled Millimeter Wave Phase Shifter

C.-K.C. Tzuang[1], *D. Miller*[1], *T.-H. Wang*[1], *P. Cheung*[1], *T. Itoh*[1], *D.P. Neikirk*[1], *P. Williams*[2], *and M. Downer*[2]

[1]Department of Electrical and Computer Engineering, University of Texas at Austin, Austin, TX 78712, USA

[2]Department of Physics, University of Texas at Austin, Austin, TX 78712, USA

1 Abstract

The dispersion of a picosecond pulse propagating along a coplanar waveguide on a semiconductor substrate is studied. Numerical calculations are used to predict the pulse shape for varying carrier concentrations in a multilayer AlGaAs/GaAs/AlGaAs structure.

2 Introduction

The introduction of electro-optic sampling techniques now allows the study of very high speed pulse propagation on planar transmission lines. In the frequency domain, recent work on microstrip and coplanar waveguides (CPW) on semiconductor substrates has shown the existence of slow-wave phenomena [1, 2]. Since these phenomena are frequency dependent, strong dispersive effects on picosecond pulses would be expected. For a lossless substrate, a spectral domain technique has been used to calculate pulse dispersion for a coplanar waveguide (CPW) and coplanar strips [3], and for a particular geometry and lossy layer both mode matching and finite element techniques have been used to predict pulse behavior in microstrip and CPW [4]. In a typical structure exhibiting strong slow wave effects there are three distinct layers in the substrate: first, a thin, lossless spacer layer immediately below the CPW; second, a somewhat thicker lossy (i.e. doped) layer; and third, a very thick lossless layer. The slow-wave phenomenon (and corresponding variation in effective dielectric constant) occurs as a result of the different interactions of the electric and magnetic fields of the propagating wave with the lossy layer below the CPW. The complex effective dielectric constant of the guide is a function of the conductivity of the lossy layer, the separation of the CPW from this layer, the transmission line dimensions, and the frequency. For frequency domain applications, we have recently proposed a new method to control the slow wave factor in these structures by replacing the doped lossy layer with an (CW) optically generating electron-hole plasma layer in the semiconductor substrate [5]. The device could then serve as a phase-shifter, controlled by varying the optical excitation level, and thus the conductivity of the lossy layer. In this paper we discuss the the effects of the lossy layer on picosecond pulse dispersion.

3 Device Structure and Model

In this device we would generate the lossy layer within the substrate and confine the induced electron-hole plasma through the use of a multilayer AlGaAs/GaAs structure. A cross section of the device is shown in Figure 1. The wide band gap AlGaAs layer immediately below the CPW is undoped, and serves as the lossless spacer layer. The second layer, made of narrower band gap, unintentionally doped GaAs, would be optically excited by continuous wave illumination entering from above (through the gaps between the CPW conductors).The excess carriers generated then cause the formation of the lossy layer. By choosing light of an energy that is below gap for the AlGaAs but well above gap for the GaAs (about 600-700nm for the Al mole fraction shown) the plasma is formed only in the GaAs layer. The band gap discontinuity between the AlGaAs and GaAs presents a barrier to carrier diffusion out of the GaAs layer, so another buffer layer of AlGaAs is used below the GaAs to confine the excess carriers. Based on high-level injection and carrier diffusion calculations, the optically induced electron-hole

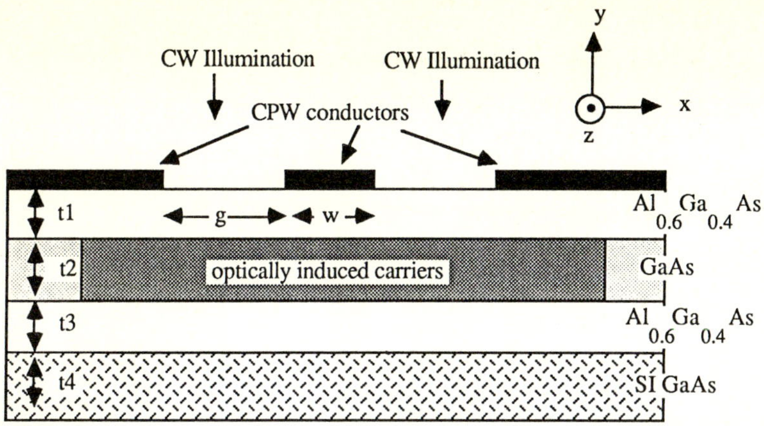

Figure 1: Cross-section used to model pulse propagation (in the z direction) on a coplanar waveguide transmission line; t1 = 0.1μm; t2 = 3.2μm; t3 = 3.4μm; t4 = 250μm; d = 12μm; g = 7μm; w = 10μm

plasma density in the GaAs layer should be fairly uniform within 10μm of the waveguide. The interactions between millimeter waves and such electron-hole plasmas have been studied extensively in dielectric (high resistivity semiconductor) slab waveguides [6], and we follow a similar model. The frequency-dependent complex conductivity of the lossy layer is calculated based on a Drude-Lorentz model and an assumed carrier density.

Once the complex dielectric constant of the lossy layer is found and all guide dimensions are specified, a mode-matching technique [2] is used to calculate the ac fields in the CPW as a function of frequency. The program used is quite general, allowing the calculation of complex propagation constant and impedance for a structure with five different dielectric layers below the waveguide and any dimensions of the guide. At present the calculations do assume perfectly conducting metal layers, although finite thickness of the metal layer is considered. Once the propagation constants and impedances are found a fast fourier transform technique is used to predict picosecond pulse behavior. The pulse dispersion and attenuation can be found for each assumed value of electron-hole plasma density in the lossy layer.

4 Results

Figures 2 and 3 show the results for a gaussian input pulse with a full-width-half-maximum of 20psec for two different carrier concentrations in the lossy layer. For the lower density, $1\times10^{15}cm^{-3}$, the pulse is very rapidly attenuated, and broadens dramatically after propagating only a few millimeters. In contrast, for the higher density case, $1\times10^{17}cm^{-3}$, the pulse does not attenuate as quickly, and only broadens slowly as the pulse propagates. Although this seems somewhat counter-intuitive, due to the presence of the lossy layer, the characteristics of this type of transmission line cannot be obtained by use of quasi-static (i.e. dc) concepts. This lossy layer also presents theoretical and numerical complications in the analysis, so these results should be viewed as illustrative, rather than as quantitatively correct predictions. One experimental advantage of the structure studied here is that by varying the CW optical illumination intensity the lossy layer carrier concentrations can be easily varied. In the experimental set-up currently being tested this should allow these theoretical results to be tested with only one device structure.

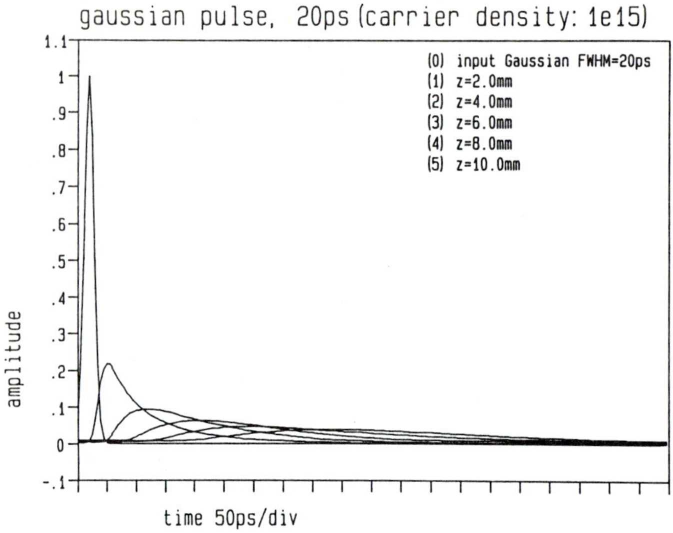

Figure 2: Calculated pulse evolution as the pulse propagates along the coplanar waveguide shown in Fig. 1 with a carrier concentration in layer 2 of $10^{15}cm^{-3}$

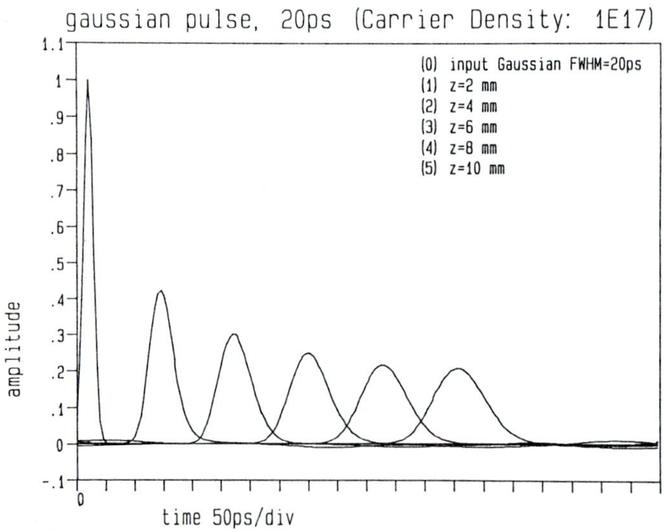

Figure 3: Pulse evolution with a carrier concentration in layer 2 of $10^{17}cm^{-3}$

5 Conclusions

In this paper we have calculated the effects of a lossy layer on the propagation characteristics of picosecond pulses on coplanar waveguides. Dramatic changes in pulse dispersion and attenuation are seen as the conductivity of the layer is varied. Experimental verification of these results is being pursued. By using the optically controlled structure discussed here it should be possible to measure pulse dispersion as a function of lossy layer conductivity with the same device. Although there are severe theoretical difficulties in analyzing wave propagation in the presence of highly lossy layers, it is clear that the presence of such layers in the substrate supporting a planar transmission line must be considered when studying the behavior of picosecond pulses.Such studies must be performed if high speed pulse waveforms are to be understood for transmission lines constructed on semiconductor substrates.

This work was supported by the Texas Advanced Technology Research Program and by the US Air Force under grant number AFOSR 86 0036.

1. D. Jager, IEEE Trans. Microwave Theory Tech., MTT-24, 566 (1976).
2. Y. Fukuoka, T. Itoh, IEEE Trans. Microwave Theory Tech., MTT-31, 567 (1983).
3. G. Hasnain, A. Dienes, and J.R. Whinnery, IEEE Trans. Microwave Theory Tech., MTT-34, 738 (1986).
4. C.-K. Tzuang, and T. Itoh, IEEE MIC Conf., June 25-26, 1985.
5. P. Cheung, D. Fun, D. Miller, C.K. Tzuang, D.P. Neikirk, and T. Itoh, Tenth International Conference on Infrared and Millimeter Waves, Lake Buena Vista, Florida, Dec. 9-13, 1986.
6. C. H. Lee, P. S. Mak, and A. P. DeFonzo, IEEE J. Quantum Electron., QE-16, 277 (1980).

Picosecond Optoelectronic Transceivers

A.P. DeFonzo, C.R. Lutz, and M. Jarwala

Department of Electrical and Computer Engineering,
The University of Massachusetts, Amherst, MA 01003, USA

1. Abstract

Picosecond photoconductivity is used to generate and sense transient millimeter wave radiation from monolithic broadband tapered slotline antennas integrated on ion-damaged silicon on sapphire substrates. The far field is observed to consist of a traveling wave component and a standing wave component. The time-dependent electromagnetic impulse response of these optoelectronic transceivers is also modeled.

2. Introduction

Recent studies have demonstrated that picosecond photoconductivity can be used to generate electromagnetic transients of picosecond duration [1]. The use of photoconductivity to generate millimeter wave radiation is a relatively new concept in picosecond electronics. It has many practical and spectroscopic applications, including developing a new class of devices capable of generating, controlling, and detecting millimeter waves. Investigations of such transients from discontinuities in microstrip lines [2] deposited on semiconductor substrates indicate the potential for producing intense coherent pulses with durations less than one picosecond, however, the structures used in these previous experiments were not optimized for efficient radiation of the resulting transient. To provide some control of the spatial and temporal distributions of the photoconductively generated radiation, more sophisticated antenna elements must be employed.

We present the results of a study of the picosecond impulse response of photoconductively driven integrated microelectronic transmitter and receiver antennas fabricated on a common semiconductor substrate. In addition, further experiments were conducted in which an intervening air gap separated the transmitting and receiving antennas. The advantages of this method can be summarized as follows: Unlike the transmission line gaps of the past [1,3], our antennas are designed for radiation of picosecond transients. We have incorporated a pump/probe sampling technique in the design of the experiment, hence our measurements will be more accurate and we expect to observe high-frequency components. The photoconducting gap driving the antenna is an integral part of the antenna structure and there is no aperture between transmitter and receiver. Thus, these structures are completely planar and can be easily fabricated on semiconductor substrates using conventional photolithographic techniques. Also, the antennas are small in size and are consequently compatible with other integrated millimeter wave circuits.

3. Experimental Design

The principle of our experiments is illustrated in Fig. 1. Identical transmitting and receiving antennas are fabricated on silicon on sapphire (SOS) substrates. The dimensions of the antenna were obtained by scaling down the design of the Vivaldi aerial first proposed by Gibson [4]. The length l, height h, and aperture sizes a_1 and a_2 were as follows: l= 2.96mm, h=3.8mm, a_1=25um, a_2=1.9mm. The distance between generator and detector was 9mm and the separation between transmitting and receiving antennas when fabricated on the same substrate was 3mm. Two antenna shapes were investigated. One was an Exponentially Tapered Slot Antenna, ETSA. The other was a Linearly Tapered Slot Antenna, LTSA.

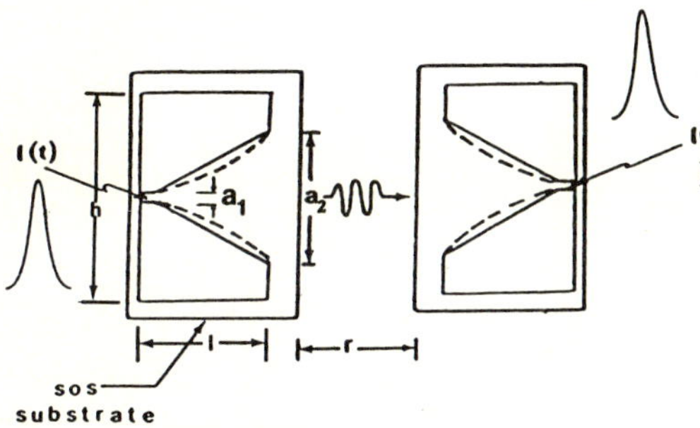

Fig. 1. Schematic of experimental configuration using matched optoelectronic antenna pairs. Dotted line indicates Exponentially Tapered Slot Antenna (ETSA). Solid line indicates Linearly Tapered Slot Antenna (LTSA).

The duration of the photoconductive transients is controlled by bombarding the silicon epilayer with energetic ions. In our experiments we bombarded the silicon with 100Kev and 200Kev O+ ions to a dosage of 10^{15} cm^{-2}. The antennas were fabricated on the radiation damaged substrates from aluminum films using standard photolithographic techniques. The photoconductive generator for the transmitting antenna is comprised of a short segment of slot line. This slot is DC biased and discharged photoconductively by illuminating the gap with a picosecond optical pulse. The discharge current pulse propagates into the antenna region where it dissipates radiatively. The radiated field which is emitted along the endfire direction of the antenna propagates to the opposing antenna. The received field results in transient bias voltage across the receiver slot line at a time equal to the propagation delay. This voltage is sampled photoconductively by illuminating the semiconductor material within the slot line with a second picosecond optical pulse derived from the same source as the exciting pulse and delayed by a variable time τ. The time dependence of the received signal is obtained by varying the time delay over the duration of the received transient.

4. Results: Transceivers on Common Substrate

In Fig. 2 we show a trace of the correlation for the ETSA pair. Superimposed is a direct correlation of the photoconductive transient measured on the same material using a conventional microstrip correlator [5]. The data clearly demonstrates that the signal received is the time derivative of the photoconducting driving pulse. Earlier work at much longer time scales [6] indicates that the ETSA takes the time derivative of the generator signal on transmission.

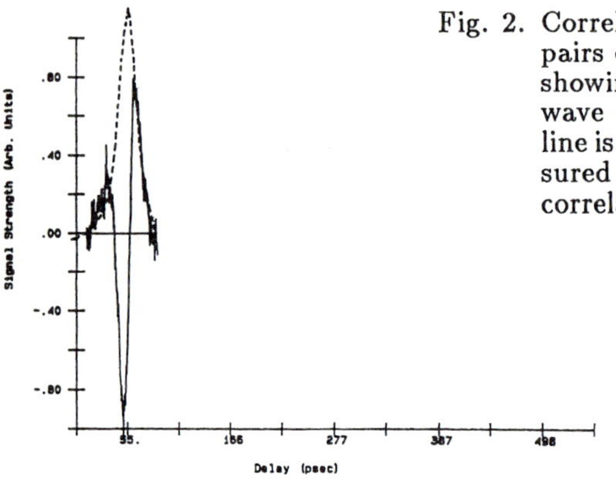

Fig. 2. Correlation trace of ETSA pairs on common substrate showing isolated traveling wave component. Dotted line is the independently measured photoconductive pulse correlation.

5. Transceiver Pairs Separated by Air Gap

The measured correlation traces shown in Figs. 3(a) and 3(b) are the results obtained when the transmitting and receiving antennas were on different substrates separated by air gaps of approximately 1 cm and 0.7cm respectively. Figure 3(a) is the signal obtained for the LTSA and Fig.3(b) is the result for an ETSA. The dashed curve in each figure corresponds to the correlation of the photoconductive driving pulse.

The results for the air-spaced ETSA and LTSA are clearly different from those obtained for the same antennas fabricated on a common substrate. Each correlation trace indicates the presence of a fast transient followed by a decaying oscillation of much lower frequency. When the antennas are on the same substrate the oscillatory component is largely suppressed due to the guiding effects of the closely spaced ends of the antenna structure. The data can be accurately predicted using a model response function derived from antenna theory [7].

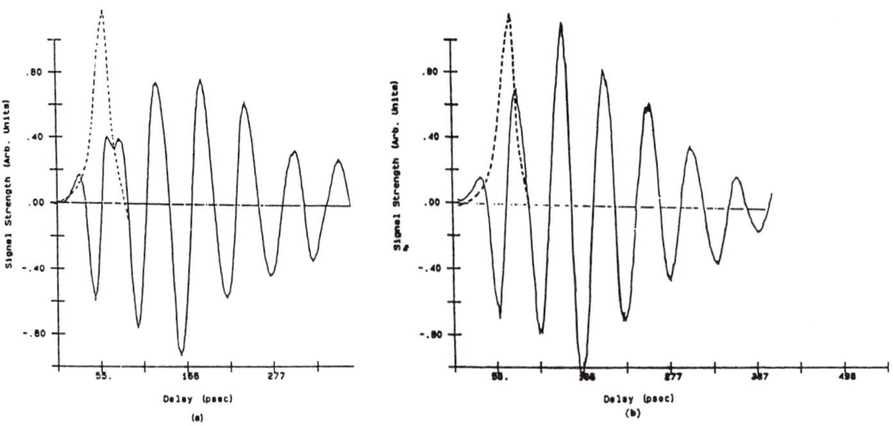

Fig. 3. Correlation traces with antennas on seperate substrates (a) LTSA (b) ETSA

6. Modeling of Transient Behavior

A model based on concepts found in the antenna literature has been developed which accurately predicts the picosecond photoconductive response of these structures. Two distinct radiation mechanisms operate in these devices - a travelling wave mechanism which produces the initial transient and a standing wave mechanism which is associated with the longer time oscillatory radiation. The total measured response is the sum of the traveling wave response and the standing wave response delayed by the difference in propagation times in the antenna. The calculated response based on this model is shown in Fig. 4. Recent experiments have measured responses in which the single transient has been isolated from the standing wave component and is in fact the only signal radiated [8].

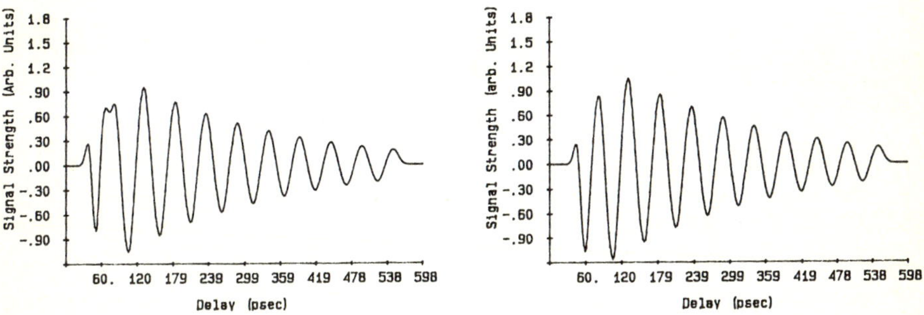

Fig. 4. Calculated response based on theoretical model for (a) LTSA (b) ETSA

7. Conclusions

We have shown that a new class of picosecond optoelectronic devices, fabricated using planar antenna technology, can be used to generate and sense millimeter wave radiation. Our results indicate that these structures can be readily scaled down to the dimensions required for operation with femtosecond transients. In addition, a time domain model of the broadside transient radiation from these antennas has been proposed. The development of these devices will provide insights into a variety of transient electromagnetic problems of practical and fundamental interest. This work was directly supported by the Air Force Office of Scientific Research, (Grant AFOSR-84-0377). Additional support for our general efforts was provided by the National Science Foundation (Grant# NSFECS-8404495).

References

1. D.H. Auston, K.P. Cheung, and P.R. Smith, Appl. Phys. Lett., 45, 284, (1984).
2. J.R. Karin, P.M. Downey, and R.J. Martin, IEEE J. of Quantm. Electrns., QE-22, 677, (1986).
3. D.H. Auston, and P.R. Smith, Appl. Phys. Lett., 43, 631, (1983).
4. P.J. Gibson, Proc. 9th European Microwave Conference, Brighton, 101, (1979).

5. D.H. Auston, A.M. Johnson, P.R. Smith, and J.C. Bean, Appl. Phys. Lett., 37, 371, (1980).
6. R. Heidemann, Th. Pfeiffer, and D. Jager, Electrncs. Lett., 19, 316, (1983).
7. A.P. Defonzo, M. Jarwala, and C.R. Lutz, submitted IEEE/OSA Conference on Lasers and Electro-Optics, 1987.
8. A.P. Defonzo and C.R. Lutz to be published.

High Speed Opto-isolator for Radar Applications

J.P. Anthes[1], *P. Garcia*[1], *K.Y. Lau*[2], *and I. Ury*[2]

[1]Sandia National Laboratories, P.O. Box 5800,
Albuquerque, NM 87185, USA

[2]Ortel Corporation, 2015 W. Chestnut St., Alhambra, CA 91803, USA

1. Introduction

We have designed, fabricated and tested a high-speed optical isolator utilizing state-of-the-art optoelectronic components operating at frequencies up to 6 GHz. This opto-isolator may become an indispensible component in modern radar systems. It is used for preventing frequency-pulling of a low-power radar oscillator due to reflected signals.

1.1 Design Approach

The present prototype design incorporated a directly modulated buried heterostructure GaAlAs laser diode [1] coupled to a GaAs photodiode detector using a very short length (1 cm) of optical fiber. The key to the design is to maximize the forward transmission (from laser to photodiode) and minimize reverse rf leakage. The forward transmission can be maximized by choosing a highly efficient laser and detector, a large NA and large core size fiber, and by appropriately lensing the tips of the optical fiber. The reverse rf leakage can be minimized by placing the laser and photodiode and their associated circuits into two separate compartments of a metallic enclosure. Ideally, the electromagnetic leakage between the two compartments can be made to approach zero if they were placed far apart. But for reasons of size compatibility with modern radars, the compartments must be placed as close as possible. A compromise was made such that the physical separation between the laser and the photodiode is 11 mm, separated by a metallic wall 3-mm-thick. An 11-mm-fiber of NA 0.3, core size 100 μm and outside diameter of 140 μm guides the laser light onto the photodetector. A 250 μm hole in the partition wall allows the fiber to link the components and the hole is sealed by conductive epoxy after the fiber is in place.

1.2 Design Prototype

The completed assembly is shown in Fig. 1. The laser diode compartment is equipped with a thermistor and thermoelectric cooler for temperature stabilization of the laser. The photodiode sensitive area is 60 μm in diameter. Since the fiber core size is 100 μm, a microlens was formed on the tip of the fiber to focus the light onto the detector surface.

2. DC Characteristics

The laser diode dc injection current versus photodiode output current is shown in Fig. 2. The corresponding photocurrent output signal, when biased

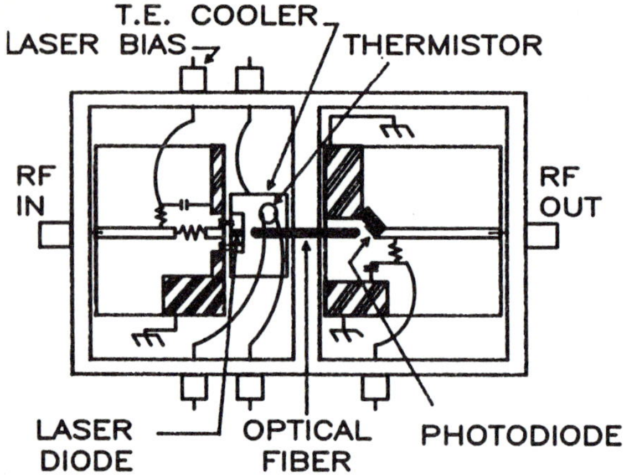

Fig. 1 Schematic top view of the high-speed opto-isolator

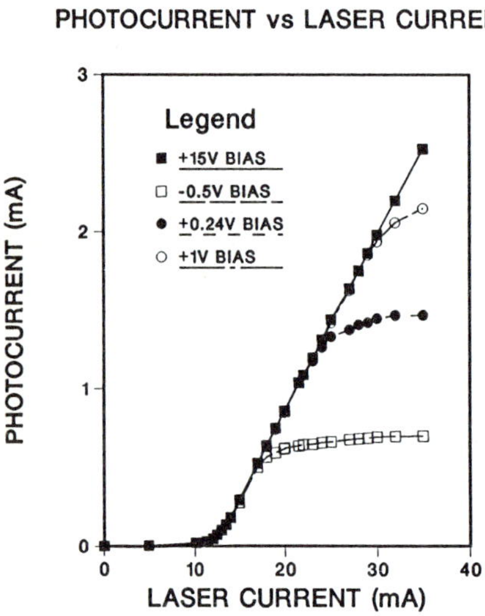

Fig. 2 DC transfer characteristics of the opto-isolator at various photodiode bias levels

at 15 volts, varied between 0.3 - 2.5 mA. The forward dc signal transmission was -19.0 dB. Typical values for quantum efficiency of the laser diode and the photodiode indicated that the overall optical fiber coupling efficiency was $\simeq$ 62%. Also shown is the effect of changing the photodiode dc bias level. Switching action was produced based upon the detector-bias dependence of the photodiode sensitivity. When biased at a forward voltage greater than 1.16 V, the microwave photocurrent generated by the laser light was switched [2] off.

3. High-Frequency Characteristics

Shown in Fig. 3 are the typical microwave scattering (S-parameter) data when the optical isolator was operated at 0-dBm input modulation power. Forward transmission losses, in the present prototype, were predominantly due to wide bandwidth (resistive) impedance-matching circuits and intrinsic

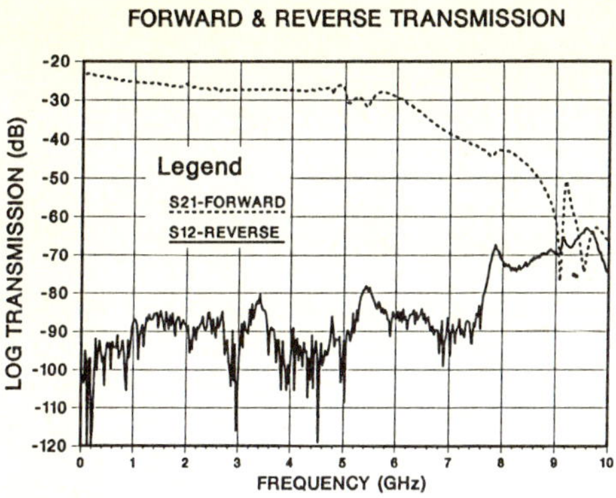

Fig. 3 Microwave forward and backward transfer characteristics

quantum efficiencies of the optoelectronic components. These losses can be further minimized by impedance matching both the laser and the detector at a specific frequency. With the photodiode switched off in the manner described above, the forward transmission drops to within 5 dB of the reverse leakage over the active frequency band.

3.1 Effects of Optical Resonance

A sharp feature is evident in Fig. 3 forward transmission data near 9.3 GHz. This characteristic frequency corresponds to the time required for laser light to travel to the fiber end near the photodiode receiver and then back again to the laser diode. A dielectric antireflection coating on the photodiode surface would reduce the retroreflected light intensity by $\simeq$15 dB and would therefore reduce the effects of optical resonance. Evidence of a harmonic signature, near 5 GHz, was also visible. It is therefore very important that, for operation at a frequency of X GHz, integer multiples of the fiber length L must be chosen so that $L \neq C/nX$ where C is the velocity of light and n is the refractive index of the fiber core.

4. Effects of rf Leakage

Unwanted rf signals can leak through improperly sealed cracks or holes that connect the compartments of the optical isolator. The rf power W radiated into the laser diode compartment through the hole which passes the fiber can be computed using standard electromagnetic analysis. The analytical results illustrate three relevant trends: attenuation for present optical fiber aperture dimensions was very high (attenuation >> 100 dB), small changes in aperture dimensions strongly influence microwave transmission, and higher microwave frequencies produce higher reverse transmission. The third point confirms the trend of S_{12} data for increasing frequency shown in Fig. 3.

Reproducible structure on the Fig. 3 reverse transmission data are also present. Wavelengths associated with increased microwave leakage at 3.5, 5.3, and 7.8 GHz are correlated with the width, length, and depth of the compartment that contains the photodiode. Microwave leakage between compartments takes place at an imperfect seal joining the vertical wall that separates the laser-diode/photodiode pair and the top cover plate of the opto-isolator.

5. Conclusions

Overall, the potential for significant device improvements seem likely. Techniques to improve the value of the forward rf transmission power will focus on improved microwave circuits designed to decrease opto-electronic losses. An opto-isolator can be theoretically configured (in an optimized narrow bandwidth version) to exhibit modest rf power gain by carefully matching electrical impedances with signal transformers, biasing the receiver photodiode, and minimizing optical losses at all interfaces. An extended response opto-isolator could conceivably operate in the 20-40 GHz frequency regime.

6. References

1. Kam Y. Lau and Amnon Yariv, IEEE J. Quantum Electron. QE-21, No. 2, 121, 1985

2. R. A. Kiehl and D. M. Drury, IEEE Trans. on Microwave Theory Tech. MTT-29, No. 10, 1004, 1981

Index of Contributors